IN SITU EVALUATION OF BIOLOGICAL HAZARDS OF ENVIRONMENTAL POLLUTANTS

ENVIRONMENTAL SCIENCE RESEARCH

Series Editor:
Herbert S. Rosenkranz
Department of Environmental Health Sciences
Case Western Reserve University
School of Medicine
Cleveland, Ohio

Founding Editor:
Alexander Hollaender

Recent Volumes in this Series

A Continuation Order Plan is available for this series. A continuation order will bring delivery of each new volume immediately upon publication. Volumes are billed only upon actual shipment. For further information please contact the publisher.

IN SITU EVALUATION OF BIOLOGICAL HAZARDS OF ENVIRONMENTAL POLLUTANTS

Edited by

Shahbeg S. Sandhu
US Environmental Protection Agency
Research Triangle Park, North Carolina

William R. Lower
University of Missouri
Columbia, Missouri

Frederick J. de Serres
Research Triangle Institute
Research Triangle Park, North Carolina

William A. Suk
National Institute of Environmental Health Sciences
Research Triangle Park, North Carolina

and

Raymond R. Tice
Integrated Laboratory Systems
Research Triangle Park, North Carolina

PLENUM PRESS ● NEW YORK AND LONDON

Library of Congress Cataloging-in-Publication Data

Symposium on In Situ Evaluation of Biological Hazards of Environmental
  Pollutants (1st : 1988 : Chapel Hill, N.C.)
    In situ evaluation of biological hazards of environmental
  pollutants / edited by Shahbeg S. Sandhu ... [et al.].
      p.   cm. -- (Environmental science research ; v. 38)
    "Proceedings of the First Symposium on In Situ Evaluation of
  Biological Hazards of Environmental Pollutants, held December 5-7,
  1988, in Chapel Hill, North Carolina"--T.p. verso.
    Includes bibliographical references ·index.
    ISBN-13: 978-1-4684-5810-7   e-ISBN-13: 978-1-4684-5808-4
    DOI: 10.1007/978-1-4684-5808-4
    1. Environmental monitoring--Congresses.   I. Sandhu, Shahbeg S.
  II. Title. III. Series.
  QH541.15.M64S96 1988
  628.5--dc20   90-7466
                                                                    CIP

The research described in this volume has been reviewed by the Health Effects Research
Laboratory, U.S. Environmental Protection Agency, and approved for publication.
Approval does not signify that the contents necessarily reflect the views and policies
of the U.S. Environmental Protection Agency, nor does mention of trade names or
commercial products constitute endorsement or recommendation for use

Proceedings of the First Symposium on *In Situ* Evaluation of Biological Hazards of
Environmental Pollutants, held December 5–7, 1988, in Chapel Hill, North Carolina

© 1990 Plenum Press, New York
Softcover reprint of the hardcover 1st edition 1990
A Division of Plenum Publishing Corporation
233 Spring Street, New York, N.Y. 10013

PREFACE

The study of the relationship between environmental pollution and human
health is in its infancy. The number of substances and mixtures that have
been identified in uncontrolled hazardous waste sites or that have been in-
advertently released into the environment is large and data on how these
substances are modified as they interact with one another as they migrate
through soil, air, and water are limited. There are also limits on our understanding of how these substances may be ingested, inhaled, or absorbed
by people. The complexity of possible interactions between biological,
chemical, and physical components in a given environment makes it virtually
impossible to evaluate the potential for adverse biological effects adequately in the laboratory. Other, more comprehensive methods which provide
realistic and interpretable results must be used. Many scientists believe
that humans represent the ultimate sentinel species of a toxic exposure resulting from environmental pollution, however such exposures may also severely impact environmental health. There exists a wide variety of organisms in the natural environment that could be used to provide an early
warning for potential human health effects as well as to indicate adverse
ecological effects.

The issue of effective utilization of sentinel species for environmental monitoring is a rapidly developing area of research which has grown in
importance during the last decade. In addition to acting as an indicator
of potential human health and environmental health effects, the in situ biomonitoring with sentinel species experiencing acute or chronic multimedia
exposure in polluted areas can be used to prioritize the contaminated area
for further in-depth studies. Also, in situ batteries of tests conducted
before and after the cleanup of polluted sites can be useful in determining
the degree of success achieved. Complementary to sentinel species is sentinel bioassay, a measurable biological function which is particularly suited to measure particular effects of environmental pollution. Thus, the combined use of sentinel species and sentinel bioassay can be a valuable tool
in the integration of ecologic and hydrogeologic data and chemical analyses
with human health related research.

The purpose of this Symposium was to address the application of in situ
bioassay and chemical analyses for assessing hazards to the ecological
health in marine, freshwater, and terrestrial environments, to discuss the
application of in situ monitoring to human health in the workplace and living space, and to review the regulatory aspects related to hazardous waste
programs. The ultimate goal of this Symposium was to encourage the active
integration of biomedical research and ecological effects research, an integration which would allow for a more rapid advancement in new techniques,
and the development of a broader data base on both biomedical and ecotoxicological effects of environmental pollution. The level of scientific and
regulatory interest in in situ monitoring was indicated by the variety and
quality of the research efforts presented at this meeting and by the formal

involvement in the program of the U.S. Environmental Protection Agency, the National Institute of Environmental Health Sciences, and the National Institute of Occupational Safety and Health.

Shahbeg S. Sandhu
William R. Lower
Frederick J. de Serres
William A. Suk
Raymond R. Tice

CONTENTS

RELEVANCY AND FUTURE

INTRODUCTORY REMARKS

IN SITU EVALUATION OF BIOLOGICAL HAZARDS OF ENVIRONMENTAL POLLUTANTS

INTRODUCTORY REMARKS

Anne P. Sassaman

Division of Extramural Research and Training
National Institute of Environmental Health Sciences
Research Triangle Park, North Carolina 27709

The National Institute of Environmental Health Sciences, although a relative newcomer to some of the areas to be discussed, is looking forward to new opportunities for the Institute as a result of authorities contained in the last Superfund reauthorization. Although this introduction is not meant to focus solely on the NIEHS Superfund Basic Research Program, or to be limited to human health consequences of environmental pollutants, these two particular areas will be emphasized because of the NIEHS's particular responsibilities and mission. The Institute appreciates the opportunity to be involved in this meeting, for we believe that the inclusion of human health aspects is an important and essential part of the topics to be discussed.

The purpose of this Symposium is to discuss the application of the currently available bioassays for in situ environmental assessment and to evaluate the utility of the integrated chemical and biological data observed under real world conditions for the assessment of human health effects from exposure to a given environment. The concept of assessment at the origin of pollution is intriguing and important since, as stated in the announcement of this meeting, the complex interaction of chemical and physical components from which most adverse biological effects result cannot be reproduced--at least not easily or in all cases--in the laboratory. Thus, the concept of using plants, terrestrial and aquatic animals for in situ environmental assessment is sound and offers potential for insight into both risks and mechanisms for damage to ecosystems and human health.

Although not explicitly stated in the program announcement, the contents suggest that the focus of the Symposium will be biologic hazards of hazardous waste, specifically, the explosive, reactive, and toxic substances resulting from the manufacture and use of chemicals. Thus, I believe it is reasonable to consider the relationship of the problems being addressed here to the Superfund program, both the original CERCLA and later the SARA legislation.

In 1979, the first report on the potential health effects of toxic chemical dumps was released by the then-Department of Health, Education, and Welfare. This report was written at the request of the Secretary of HEW by the Committee to Coordinate Environmental and Related Programs. An ad hoc subcommittee of Public Health Service scientists and policy experts, joined by two scientists and an engineer from the Environmental Protection Agency,

In Situ Evaluations of Biological Hazards of Environmental Pollutants
Edited by S. S. Sandhu *et al.*
Plenum Press, New York, 1990

3

analyzed the problem of abandoned and uncontrolled hazardous chemical waste in terms of the potential risk to human health. If its conclusions and recommendations are reviewed against the health- related authorities of the original Superfund law, it is clear that the HEW report had considerable influence on the authors of the Act. Three examples follow.

First, the HEW Report concludes that "a registry of potentially affected people should be established." Superfund says "establish and maintain a national registry of serious diseases and illnesses and a national registry of persons exposed to toxic substances." Second, the Report states that "[T]oxicologic data gathered from waste dumps should be integrated with information from the literature in readily retrievable form so that it will be available to public and private sectors." The Superfund response: "Establish and maintain inventory of literature, research and studies on the health effects of toxic substances." And finally, the Report concluded that "[A] set of decision criteria needs to be mutually developed by EPA and HEW to establish the degree of health hazard...include additional monitoring to quantify the chemicals and determine degree of exposure...include health examinations to assess if an exposed population is experiencing disease or adverse health effects." The legislation responded by requiring the HEW "conduct periodic survey and screening programs to determine relationships between exposure to toxic substances and illness."

Thus, there is no doubt that the HEW report was influential in framing the health authorities of the Superfund law, which in its original form set up the Agency for Toxic Substances and Diseases Registry as the focus for the activities cited. Both documents, the Report and CERCLA, cite the importance of genetic toxicology as a marker of human exposure, the role of epidemiologic studies in the assessment of human health effects, the need for clinical evaluation and medical care for persons exposed to hazardous substances, and perhaps most important, a mandate that the public health expertise available in the Public Health Service be added to the vast technical, scientific, and programmatic resources in the Environmental Protection Agency in the program to respond to the adverse health consequences and environmental damage from uncontrolled hazardous substances.

In the Superfund Amendments and Reauthorization Act of 1986, the health-related components were strengthened and expanded. Furthermore, the complexity of the problems being encountered in the clean-up effort and the need for a new approach were recognized and were addressed in the creation of the NIEHS Superfund Basic Research and Training Program. The legislation specifically called for inclusion of research in ecology, engineering, and hydrogeology in a basic biomedical research program. One can interpret this as a realization of the importance of an "in situ" approach, if you will, looking at what goes on in a particular area and how various perspectives can be combined to measure what is there, where and how it is moving, and what might be the toxicity and to which biological systems, including humans.

Figure 1 is a representation of one way of looking at the biologic targets of hazardous substance pollution and may be a useful framework for considering the papers to follow. Various routes of transport are depicted, and each has its own "biological target." For example, volatile emissions (one of the routes most commonly measured at hazardous waste sites) may expose not only on-site workers or other individuals, but may have an effect on vegetation, populations at distant locations (shown here as both humans and animals), and may enter another phase, such as through dissolution in surface water. Contamination through both surface water and ground water is of much concern, and we are all aware of the increasing instances of contamination of drinking water from dump sites or from nonpoint sources. A later session will focus on use of aquatic systems in field

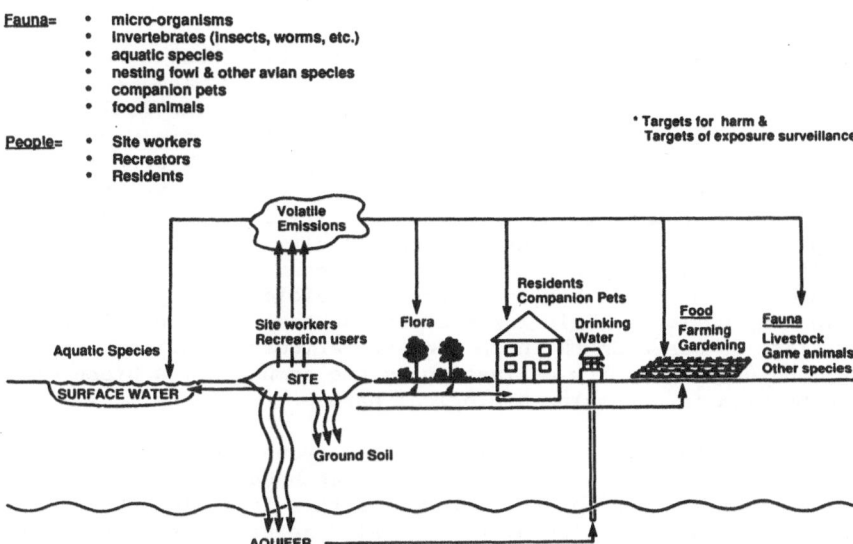

<u>Flora</u>= food, vegetation

<u>Fauna</u>=
- micro-organisms
- invertebrates (insects, worms, etc.)
- aquatic species
- nesting fowl & other avian species
- companion pets
- food animals

<u>People</u>=
- Site workers
- Recreators
- Residents

* Targets for harm &
  Targets of exposure surveillance

Figure 1. Biologic targets* of hazardous substance pollution.

studies, and use of aquatic species will be discussed in the session on sentinel surveillance systems. Transport through the soil may also lead to exposure of humans and animal species, and this will be covered in the session on field studies in terrestrial systems.

From the figure, two types of biological targets can be identified: targets for harm and targets for exposure surveillance. From the NIEHS perspective, the important idea is that environmental agents may produce harm to a particular target population--humans. It is recognized, however, that harm to the ecosystem itself is of concern, and we know that measurement of this harm or exposure is useful in human risk assessment. This concept will be covered in the session on integration of data for effective problem solving and assessment.

Finally, we should focus briefly on some of the unique aspects of the NIEHS Superfund Research Program, for it is because of this program that the NIEHS is participating in the Symposium. The Program allows one to step back from a specific site and look at it from a number of different perspectives. It takes an integrative, multidisciplinary approach to the problem, with representatives from each scientific discipline in a given institution's program bringing their own expertise and viewpoints. The program itself is not discipline specific nor is it site specific. It is the only one of this kind within the various research and development activities related to Superfund or other hazardous substance programs. While it is small, the uniqueness of the approach is expected to contribute significantly to the body of knowledge required to deal with the problem of hazardous substances. Although the NIEHS Superfund program is targeted toward humans and the potential harm environmental agents may have on them, it is made up of studies of a number of biological targets and these targets may provide pathways to humans. Six of these studies will be discussed during the course of the Symposium.

The Environmental Protection Agency and the Public Health Service have worked productively together in the past decade, but problems remain. Among these are the limits of the science base itself. The goal of this meeting is to describe the scientific tools available to evaluate the bio- logic hazards of hazardous chemical waste, with particular emphasis on the hazard to the health of people who have been or are being exposed to these wastes. One of the goals of the interdisciplinary research being conducted through Superfund is to provide data which can ultimately be used in a specific site to permit evaluation of human exposure and ecological damage prior to remediation and to assess exposure to biological systems for various clean-up options. This information can then be used to prioritize sites for clean-up based on the harm done to various biological targets. Therefore, the use of in situ evaluation is one component, a piece of the puzzle, to be used in assessing biological harm.

It would seem that a key objective of this symposium is to provide a framework to help define the field of in situ methods and evaluation. The organizers must have plans for the future, as this is called the FIRST Symposium on In Situ Evaluation of Biological Hazards of Environmental Pollutants. Therefore, it will be interesting to see what the SECOND symposium will address, how far the science base will have advanced, what more we will know about the usefulness of sentinel species in human risk assessment, and whether there will be extrapolations and links between what is learned from in situ evaluation and methods for in situ treatments.

INDUSTRIAL AND REGULATORY NEEDS

TOXICOLOGICAL RESEARCH AND RELATED PROGRAMS UNDER SUPERFUND

Barry L. Johnson

Agency for Toxic Substances and Disease Registry
Atlanta, Georgia   30333

Anne P. Sassaman

Division of Extramural Research
National Institute of Environmental Sciences
Research Triangle Park, North Carolina   27709

INTRODUCTION

We must be one with our environment. With industrialization and modern agriculture have come great benefits, but with these advances also have come burdens on the fragile environment that sustains us. We derive great economic gain from the industrial and agricultural products that we in industrialized countries have come to expect. Products that make our lives easier, safer, and longer: automobiles, convenience foods, packaged products, pharmaceuticals and such. But these commercial successes and products of industrialization have an environmental consequence--the dimensions of which are still being assessed. One of the environmental consequences has been the dumping of unwanted, sometimes hazardous, waste into the environment. In many instances this has resulted in contamination of soil, ground water, air, and the human food chain. What gets into these media has the potential for getting into people. And that has implications for public health.

Numerous hazardous chemical substances have entered the environment through improper waste disposal and accidental spills and remain there today. Many of these substances are toxic under experimental conditions to laboratory animals, and some have caused adverse health effects in humans, primarily in workers, but in some instances, in community residents. Effective programs to prevent human exposure and resulting adverse health effects are necessary to ensure that the chemical legacy left in the environment does not harm current and future generations. Given this public health concern, federal and state legislation has been enacted into laws that are intended to provide the programs and resources needed.

One of the major federal statutes that addresses the health impact of toxicants in the environment is the Comprehensive Environmental Response, Compensation, and Liability Act of 1980 (CERCLA). It is one of the most sweeping laws concerned with hazardous substances in the environment. This law directs the Environmental Protection Agency (EPA) to identify waste sites that present environmental and public health hazards, to remediate the sites, to recover the costs of remediation from those parties principally responsible, and to conduct emergency removal of hazardous substances

In Situ Evaluations of Biological Hazards of Environmental Pollutants
Edited by S. S. Sandhu *et al.*
Plenum Press, New York, 1990

when necessary. The CERCLA, (also referred to as Superfund) also created the Agency for Toxic Substances and Disease Registry (ATSDR). In establishing the ATSDR, the Congress intended a nonregulatory, public health organization that would work closely with a regulatory agency, the EPA, to deal with hazardous substances in the environment. This arrangement is akin to the pairing of the Occupational Safety and Health Administration, a regulatory agency, with the National Institute for Occupational Safety and Health (NIOSH), a nonregulatory public health organization.

The ATSDR has stated that its mission is to prevent or mitigate adverse human health effects and diminished quality of life resulting from environmental exposure to hazardous substances in the environment. The Superfund Amendments and Reauthorization Act of 1986 (SARA) expanded greatly the ATSDR's directives and responsibilities toward a greater fulfillment of its mission. Some of the Agency's expanded responsibilities under SARA include toxicological research and related programs, health assessments, information development, and health education. Other ATSDR responsibilities that carry over from CERCLA of 1980 include health surveillance, exposure and disease registries, emergency response activities, and dissemination of information about sites closed to the public because of toxic contamination. The Agency's programs that most directly relate to toxicological interests are described here. A more detailed description of ATSDR's programs and activities is available in Johnson (1988).

In addition to ATSDR's responsibilities that are relevant to toxicology, SARA authorizes a program of basic research to be undertaken by the National Institute of Environmental Health Sciences (NIEHS). This program, which is being administered as research grants, is an important component of toxicological and health support under Superfund. The NIEHS grants program is summarized in this article.

HEALTH ASSESSMENTS

The SARA requires the ATSDR to conduct a public health assessment (health assessment) of every site on or proposed for inclusion in EPA's National Priorities List (NPL). (The NPL is a ranking of those uncontrolled waste sites in greatest need of remediation from the approximately 30,000 sites identified by EPA as being of concern). An ATSDR health assessment is a site-specific evaluation of data and information on the release of hazardous substances into the environment in order to: assess any current or future impact on public health, develop health advisories or other health recommendations, and identify studies or actions needed to evaluate and mitigate or prevent human health effects.

Health assessments are prepared by a multidisciplinary team of toxicologists, physicians, epidemiologists, engineers, and public health advisers. The toxicology of substances associated with individual waste sites is at the center of the ATSDR health assessment, since the exposure of humans to known or putative toxicants constitutes a potential cause of adverse health effects. Our current knowledge about the toxic properties of many hazardous substances and their human health effects is limited. For this reason, SARA provides a mechanism to add to our collective knowledge about the toxic properties of those substances and human exposure limits for which significant data gaps exist. How these gaps are identified is discussed later in the article.

In addition to those sites listed on or proposed for the NPL, ATSDR also considers those sites for which an individual petitions the ATSDR to conduct a health assessment of a site of concern. Furthermore, Section 3019 of the Resource Conservation and Recovery Act (RCRA) authorizes the

ATSDR to conduct a health assessment of a RCRA facility if requested by
EPA, a state, or an individual. (RCRA facilities are sites permitted by
EPA to store, transport, and dispose of hazardous waste).

Health assessments are normally based on Remedial Investigation/Feasi-
bility Study (RI/FS) data furnished by EPA. The RI/FS report can include
environmental data (what hazardous substances are present and in what concentrations), modelling forecasts (where and by what means the substances
are likely to migrate from the site), numerical estimates of risk associated with specific hazardous substances identified for the site, pathways
of possible human exposure, and occasional estimates of human populations
who may be at special risk of adverse health effects (e.g., children). In
addition, for some sites, ATSDR may receive data from state and local
agencies. When RI/FS data are not available, the ATSDR collects any site
summary data obtainable from EPA and state and local health departments and
conducts a preliminary health assessment. In the case of a petition for a
health assessment, some supplementary data may come directly from the peti-
tioner (e.g., health complaints or medical records).

After evaluating the relevant data, the Agency prepares a health assessment report that expresses its opinion on the public health implications
posed by the site and sets forth recommendations for public health actions
to reduce human exposure to the hazardous substances of concern. These
recommendations are used by EPA and States in designing remedial actions to
clean up sites. In instances of imminent health threat, the ATSDR, EPA,
and State/local authorities work to reduce or prevent further human contact
with hazardous substances of concern. For example, people might be relocated away from contaminated sites or given an alternate supply of water
for drinking. In one instance, the ATSDR worked with EPA and states in
their relocation of persons from two small communities built over mine
tailings. Some children in the communities had been found to have high
levels of arsenic in their urine (Binder et al., 1987). Follow-up health
surveys showed that the arsenic levels decreased to nonhazardous amounts
after the children were relocated to a different geographic area.

In addition to advising EPA and States on health implications of individual sites, ATSDR, through the health assessment process, reviews the
site's data to ascertain when to mobilize traditional public health follow-
ups. Each site is evaluated to identify those for which certain further
activities are appropriate: human exposure assessments (collection of
biological samples for measurement of chemical substances associated with
exposure), health surveillance of citizens around a site, an epidemiological study of health outcomes, or development of an exposure registry (i.e.,
a list of the names of persons known or thought to have been exposed to
hazardous substances of concern).

The ATSDR's health assessments often lead to meetings with concerned
citizens. The agency views this as an important form of health communication and public education. Whenever possible, the Agency actively involves citizen groups in the planning of health surveys, exposure assessments, exposure registries, and other such activities. The result is
better communication and understanding (though not always agreement)
between citizens and government. The problem of presenting often complex,
sometimes incomplete, toxicological material to concerned community groups
in a public meeting tests the best health communicators (Johnson, 1987).
Better environmental science will lead to better health communication.

SARA required that the ATSDR complete, before December 10, 1988, a
health assessment of each site included in or proposed for the NPL as of
October 17, 1986, a total of 887 sites. Each new site proposed to be added
to the NPL must receive a health assessment within one year of its proposed

inclusion. All told, ATSDR by statute was to have completed health assessments of 951 NPL sites on or before December 10, 1988. The Agency met this
mandate. The data base that has been established in ATSDR for the NPL
sites will allow for examinations of interest to toxicologists. For example, what are the principal contaminants across these sites? What patterns
of human exposure predominate? These kinds of questions will be answerable
in the near future.

The ATSDR also conducts annually about 2500 health consultations (less
detailed responses to requests for hazardous substances information) and a
small number of health assessments of non-NPL sites. The health consultations often are in response to specific questions from citizens regarding
toxic properties of hazardous substances of concern to them.

IN SITU MONITORING AND EVALUATION

In situ monitoring data are of considerable importance to ATSDR in
conducting health assessments. Monitoring data specific to individual
waste sites that portray levels of contamination of toxicants in air, soil,
food, and ground water are irreplaceable in terms of forecasting public
health impact. The EPA and some States conduct important in situ monitoring of environmental conditions around waste sites, but resource restrict-
ions sometimes limit the extent of data collection. For example, data concerning the levels of hazardous substances in the food chain of populations
around waste sites are very seldom available. Moreover, in situ evaluations, as characterized by toxic effects in biological systems (domestic
animals, wildlife, fish, microbes) also represent a potentially valuable
data base for the conducting of health assessments. However, for Superfund
purposes, very few in situ, biological evaluations are currently underway.

NATIONAL EXPOSURE REGISTRY

The ATSDR is required by CERCLA to establish a national exposure registry of persons exposed to toxic substances. The Agency has stated the principal purpose for its national exposure registry is to assist in conducting
research on the long-term health effects of chronic exposure to low levels
of select hazardous substances (ATSDR, 1988). For example, ATSDR, working
with affected states, will use its health assessments to identify sites
that are appropriate for the establishment of chemical-specific subregis-
tries. The specific hazardous substances selected for exposure registries
will be chosen from ATSDR's list of Superfund priority substances. The
Agency currently maintains, or is developing, registries of persons exposed
to trichloroethylene, dioxins, and beta- naphthylamine. The use of exposure registries for research pursuit has great potential for adding to the
body of toxicological science. For example, an exposure registry of
persons exposed in Missouri to dioxins in soil has been used extensively by
epidemiologists to study health effects (Hoffman et al., 1986).

PRIORITY HAZARDOUS SUBSTANCES

Some substances at Superfund sites present a greater threat to the public health and the environment than do others. The CERCLA's Section 104
(i)(2) requires the ATSDR and EPA jointly to rank in order of priority
those substances that appear most often at NPL sites and pose the greatest
threat to public health. The statute specifies three criteria to be used
in the ranking: overall toxicity of the substance (acute, subacute, and
chronic toxicities), frequency of occurrence of the substance at NPL sites,

and potential for human exposure to the substance. The EPA and ATSDR derived an algorithm that includes all three criteria. Using the algorithm, the agencies produced a list in April 1987 of the 100 most hazardous substances (ATSDR and EPA, 1987). In October 1988, an additional 100 substances were added to this list (ATSDR and EPA, 1988). The SARA requires that 25 substances will be added each year through 1991. The Appendix lists the 200 most hazardous substances found at NPL sites.

The list of Superfund priority hazardous substances represents a systematic identification and evaluation of those toxicants of greatest environmental concern in a Superfund context. There are three principal consequences of the listing of a chemical as a Superfund priority substance. First, the ATSDR must develop a toxicological profile for the substance. Second, the ATSDR (in consultation with the National Toxicology Program and the EPA) must assess whether adequate information on the health effects of each is available. Third, if adequate information is not available, a plan of research must be developed and initiated. The ATSDR toxicological profiles are therefore the starting point for important efforts in health education and environmental research.

TOXICOLOGICAL PROFILES

The CERCLA, as amended, directs the ATSDR to develop and maintain up-to-date toxicological profiles on the Superfund priority hazardous substances. The profiles must describe levels of exposure that are significant to human health, associated health effects, and the adequacy of the data available for determining the exposure levels of significance (ATSDR and EPA, 1987). The profiles contain other information that is intended to be useful to state agencies, health professionals, and the general public. A list of the contents of an ATSDR toxicological profile is given in Table 1.

The ATSDR is required by SARA to prepare no fewer than 25 profiles each year for substances in Groups 1 through 4 in the Appendix. Profiles are to be completed within 3 years for the substances added to the priorities list in October 1988 (Groups 5 through 8, Appendix). Moreover, the statute directs that ATSDR revise and republish each profile no less often than every 3 years as a means of keeping them current. (However, the Congress, through appropriations legislation, has currently capped the number of toxicological profiles at 40 documents per year.)

Table 1. Contents of ATSDR's Toxicological Profiles on Superfund Priority Hazardous Substances

---

Public Health Statement
Health Effects Summary
Chemical and Physical Information
Manufacture, Import, Use, and Disposal
Environmental Fate
Potential for Human Exposure
Analytical Methods
Regulatory and Advisory Status
References
Glossary

The profiles will be used by the ATSDR, EPA, and the National Toxicological Program (NTP) of the Public Health Service to identify significant data needs for profiled hazardous substances. The ATSDR must ensure a program of research to fill the gaps. Programs of research are to be coordinated with the NTP and with programs of toxicological testing established under the Toxic Substances Control Act and the Federal Insecticide, Fungicide, and Rodenticide Act (FIFRA). The CERCLA's Section 104 (i)(5)(D) states, "It is the sense of the Congress that the costs of research programs under this paragraph be borne by the manufacturers and processors of the hazardous substance in question, as required in programs of toxicological testing under the Toxic Substances Control Act. Within one year after the enactment of the Superfund Amendments and Reauthorization Act of 1986, the Administrator of EPA shall promulgate regulations which provide, where appropriate, for payment of such costs by manufacturers and processors under the Toxic Substances Control Act, and registrants under the Federal Insecticide, Fungicide, and Rodenticide Act, and recovery of such costs from responsible parties under the Act." The administrative procedures to identify significant data needs for each priority hazardous substance, together with mechanisms to fund the research to fill these needs, are currently being developed by ATSDR, EPA, and the NTP.

DECISION GUIDE FOR DATA NEEDS

Although ATSDR has not at this writing announced how it will identify significant data needs for each hazardous substance profiled and has not articulated the criteria for ranking data needs across chemicals, it is nonetheless possible to offer some preliminary observations. The ATSDR plans to use a decision guide for the purpose of identifying data needs for each substance profiled. The elements of the decision guide are likely to build upon the ATSDR process for conducting health assessments. That is to say, the decision guide will ask "What data would be needed for this substance in order to conduct a health assessment?" From such an approach comes the observation that health assessors usually lack data bearing on levels of human exposure to the Superfund priority hazardous substances. This represents one clear and significant data need for many Superfund priority substances.

The decision guide for identifying significant data needs for profiled substances will use a tiered structure. Although the details of the tiers are still being developed, some directions are already apparent. Tier I will focus on what information needs to be available to constitute a foundation for health assessments. For example: What is known about the chemical and physical properties of a substance? What basic toxicological data are available? Tier II will be directed toward confirming scientific knowledge. Given what is known from Tier I, what needs to be confirmed regarding toxic properties and human health effects of a toxicant? Tier III will address application of Tier I and Tier II knowledge. If a substance has confirmed toxic properties, what needs to be done in order to prevent adverse human health effects? The ATSDR will conclude its decision guide in 1989, after the document has been peer reviewed and comments from the public have been evaluated.

TOXICOLOGICAL RESEARCH UNDER SUPERFUND

In addition to toxicological research that is to be initiated subsequent to the release of ATSDR's toxicological profiles, Superfund has already sponsored much research that is linked to the needs of ATSDR's health assessment program.

Exposure Assessment and Epidemiology

The ATSDR, through a cooperative agreement with the Centers for Disease Control (CDC), has funded a considerable amount of work on the toxic properties of dioxin. The sponsorship has led to analytical methods to measure dioxin in adipose tissue (Patterson et al., 1986, 1987a) and, more recently, in blood serum (Patterson et al., 1987b). These methods have been used by epidemiologists to evaluate the health status of Missouri residents and of workers occupationally exposed to dioxin (Hoffman et al., 1986). The current body of knowledge about dioxin has been developed in large measure by funds from Superfund. Other CDC epidemiology research has focused on patterns of morbidity in human populations who live around specific waste sites. Approximately 20 studies have been sponsored via the Superfund (ATSDR, 1987). In addition, ATSDR, through CDC, is sponsoring state-based health surveillance projects in 10 states. The health surveillance data are addressing chronic diseases and adverse reproductive outcomes. The plan is to link the health outcome data to environmental data and then to examine for associations between the two data sets. In 1989 ATSDR will conclude a retrospective analysis of all epidemiological and toxicological studies that it has sponsored via Superfund. The analysis will be provided to the Congress and the public as part of a biennial report that is required by SARA.

The ability to quantify exposure to specific hazardous substances is vital to successful health studies. The ATSDR is working with the CDC's Center for Environmental Health and Injury Control (CEHIC) to develop analytical chemistry methods to measure exposure to substances on its list of priority chemicals. In addition, CEHIC and CDC's National Center for Health Statistics have undertaken, via funds from the Superfund, a reference range study for 52 environmental hazardous substances. The study will analyze blood samples from a national sample of 1,000 persons. Findings from the methods development and the reference range studies will be enormously useful to health officials and toxicologists in their work on investigating health effects in human populations exposed to hazardous substances.

Toxicological Testing by the NTP

The ATSDR is also using Superfund monies to support testing at the NTP of specific hazardous substances. Initially, approximately 50 chemicals, including a number of complex mixtures, were nominated by EPA, ATSDR, and NIEHS for testing by the NTP. Included among the nominations were a wide range of hazardous substances that, in commerce, were used as solvents, pesticides, chemical intermediates, heat transfer fluids, metal salts, gasoline antiknock compounds, and acids. The substances selected for study were either known or thought to be present at waste sites. To date, 23 compounds and a chemical mixture representative of what is found at some Superfund sites are undergoing testing. A list of the chemicals under test is given in Table 2. Results from individual toxicity tests are available from the authors. Future nominations to the NTP testing program will be guided by data needs of the Superfund priority hazardous substances.

Findings from the NTP studies and the CDC research on exposure assessment methods will be highly useful to ATSDR and others conducting health assessments and health studies of persons exposed to environmental toxicants. Improved scientific data will reduce uncertainty in risk assessments. Less uncertainty translates into more informed risk management.

NIEHS Basic Research Program

The SARA legislation of 1986 established a university-based program of basic research and training within the NIEHS. This program, funded through

15

Table 2. Chemicals on Test by the National Toxicology Program, under
Superfund Support from ATSDR

Acetone	13-week dose study
Aniline	Genetic toxicology
Barium chloride	2-year chronic study
Benzene	Deferred from further study
Bromoform/Chloroform	Genetic toxicology, Reproductive
Carbon tetrachloride	Genetic toxicology
Cresols	Prechronic study
Cupric sulfate	Prechronic study
1,2-Dichloroethane	Prechronic study
1,2-Dichloroethylene	Prechronic study
Dichloromethane	Behavioral teratology
1,2-Dichlorpropane	Testing decision pending
Ethyl benzene	Genetic toxicology, Prechronic
Ethylene glycol	Chronic study
Hexachloro-1,3--butadiene	Chemical disposition, Prechronic
Hexachlorocyclopentadiene	Chronic study
n-Hexane	Cytogenetics, Reproductive, Prechronic
Manganese sulfate	Chronic study
Methyl ethyl ketone	Teratology
Pentachlorobenzene	Dose setting, Reproductive
Phthalates	Reproductive, Teratology, Prechronic
Sodium cyanide	Prechronic study
Sodium selenate	Prechronic study
1,2,4,5-Tetrachlorobenzene	Prechronic, Reproductive
1,1,2,2-Tetrachlorethane	Prechronic study
Toluene	Chronic study
1,1,1-Trichloroethane	Prechronic, Teratology, Disposition
Trichloroethylene	Reproductive, Dose finding
Xylenes	Dropped from testing

research grants, complements the existing activities at the ATSDR and the
EPA. The SARA legislation specifically mandates that the NIEHS program
include topics such as:

°   Methods and technologies to detect hazardous substances in the
    environment.

°   Advanced techniques for the detection, assessment, and evaluation
    of the effects on human health of hazardous substances.

°   Methods to assess the risks to human health presented by hazardous
    substances.

°   Basic biological, chemical, and physical methods to reduce the
    amount and toxicity of hazardous substances.

In initiating the Superfund Basic Research and Training Program, the NIEHS chose to identify a variety of research opportunities that would be appropriate to address the mandate of the legislation. These were detailed in the Program Announcement (NIEHS, 1987) and included research topics not only in biomedical areas and epidemiology, but also in engineering, ecology, and hydrogeology.

As a result of two solicitations, the NIEHS currently supports programs at nine institutions. These programs, characterized as multicomponent, interdisciplinary efforts linking basic biomedical research with related ecologic, hydrogeologic, and engineering studies, contain about 78 separate diverse projects. In this respect, the Program represents a unique and powerful combination of science being brought to bear on the problem of health effects of hazardous wastes.

The number and diversity of the research projects preclude a comprehensive description of them in this article. Selected examples, however, illustrate the Program's research. There is considerable emphasis on the development of biomarkers of exposure, using a variety of extremely sensitive methods such as analytical chemistry, immunoassays, and spectra of changes in DNA. In many cases, there are plans to extend these studies in future epidemiologic investigations. Investigators are also developing state-of-the-art methods for assessing the completeness of combustion of toxic waste products. These engineering efforts are combined with a toxicologic assessment of the resulting combustion products. Finally, several grantees are working on new approaches to the biodegradation of toxic compounds.

In summary, the focus of the NIEHS Program is on human health and potential ecological damage, and a major emphasis is on the use of advances in basic research to improve the sensitivity and specificity of techniques for detecting exposure and injury to humans or in ecological systems. Knowledge gained in such research will be subjected to a careful risk assessment with improved techniques (also being developed within the Program) to evaluate the possible impact and extent of the danger to human populations and the environment.

INFORMATION AND EDUCATION

The development, maintenance, and dissemination of information about hazardous substances is required of ATSDR by CERCLA. Health professionals have found that such information bases are critical to their ability to assess and manage health hazards at Superfund sites. The requirements for a useful toxics data base are several. First, it must be accurate, relevant, and contemporary. Second, it must be accessible to its users. And, third, the cost of using the information data base must not be prohibitive. To meet these criteria, the National Library of Medicine (NLM), National Institutes of Health, developed the Hazardous Substances Data Base under ATSDR sponsorship, and incorporated it into NLM's TOXNET computer data base. The TOXNET is a system of integrated toxicological data bases that can be accessed by personal computer and telephone modem.

The SARA also authorizes an ATSDR program of education for health care providers regarding hazardous substances. In particular, ATSDR must develop and provide materials and training for primary care physicians, medical educators, and emergency care providers. In response, ATSDR has developed a series of Tox Quizzes for physicians who desire continuing medical education credits. The content of the Quizzes is based on the Agency's toxicological profiles. Additionally, training materials are being developed for health care providers.

SUMMARY

Yesterday's waste has become today's health concern. Yet the magnitude of the public health problem awaits better measurement. The Superfund law provides a comprehensive response to public and congressional concern over hazardous waste in the environment and a means to address public health concerns. Uncontrolled waste sites and emergency releases of hazardous substances present a potential threat to public health and to the environment. Effective remedies for protecting human health and the environment require sound toxicological knowledge and reasoned judgment. Through its various authorities encompassing applied public health services, basic and applied research, and public education, Superfund provides the engine for effective actions against health and environmental consequences of uncontrolled waste in the environment.

REFERENCES

ATSDR, 1987, Annual Report for Fiscal Year 1987, Agency for Toxic Substances and Disease Registry, Atlanta, Georgia.

ATSDR, 1988, National Exposure Registry, Agency for Toxic Substances and Disease Registry, Atlanta, Georgia.

ATSDR and EPA, 1987, Notice of the first priority list of hazardous substances that will be the subject of toxicological profiles, Federal Register, 52(74): 12866-12874.

ATSDR and EPA, 1988, Hazardous substances priority list, toxicological profiles, Federal Register, 53(203): 41280-85.

Binder, S., Forney, D., Kaye, W., and Paschal, D., 1987, Arsenic exposure in children living near a former copper smelter, Bull. Environ. Contam. Toxicol., 39:114-21.

Hoffman, R.E., Stehr-Green, P.A., and Webb, K.B., 1986, Health effects of long-term exposure to 2,3,7,8-tetrachlorodibenzo-p-dioxin, J. Am. Med. Assoc., 255:2031-2038.

Johnson, B.L., 1987, Health risk communication at the Agency for Toxic Substances and Disease Registry, Risk Analysis, 7:409-12.

Johnson, B.L., 1988, Health effects of hazardous waste: The expanding role of the Agency for Toxic Substances and Disease Registry, Environ. Law Reporter, 18:10132-10140.

NIEHS, 1987, Announcement--Superfund Hazardous Substances Basic Research Program, Phase II. National Institute of Environmental Health Sciences, Research Triangle Park, North Carolina.

Patterson, D.G., Jr., Hampton, L., and Lapeza, C.R., Jr., Belser, W.T., Green, V., Alexander, L., and Needham, L.L., 1987a, High-resolution gas chromatographic/high resolution mass spectrometric analysis of human serum on a whole-weight and lipid basis for 2,3,7,8-tetrachlorodibenzo-p-dioxin, Anal. Chem., 59:2000-2005.

Patterson, D.G., Jr., Holler, J.S., Belser, W.T., Boozer, E.L., Lapeza, C.R., Jr., and Needham, L.L., 1987b, Determination of 2,3,7,8-tetrachlorodibenzo-p-dioxin (TCDD) in human adipose tissue on whole-weight and lipid bases, Chemosphere, 16:935-6.

Patterson, D.G., Holler, J.S., and Lapeza, C.R., Jr., 1986, High-resolution gas chromatograph/high-resolution mass spectrometric analysis of human adipose tissue for 2,3,7,8-tetrachlorodibenzo-p-dioxin, Anal. Chem., 58:705-13.

APPENDIX

The ATSDR/EPA List of the 200 Most Hazardous Superfund Substances.
Group 1 chemicals have been assigned a higher priority than those in
Group 2, etc.

Group 1

Arsenic
Benzene
Benzo(a)anthracene
Benzo(b)fluoranthene
Benzo(a)pyrene
Beryllium
Bis(2-ethylhexyl)phthalate
Cadmium
Chloroform
Chromium
Chrysene
Cyanide
Dibenz(a,h)anthracene
Dieldrin/aldrin
1,4-Dichlorobenzene
p-Dioxin
Heptachlor/heptachlor epoxide
Lead
Methylene chloride
Nickel
N-nitrosodiphenylamine
PCB-1260, 54,48,42,32,21,1016
Tetrachloroethene
Trichloroethene
Vinyl chloride

Group 2

Benzidine
BHC-1,2,3,4
Bis(chloromethyl)ether
Bis(2-chloroethyl)ether
Bromodichloromethane
Carbon tetrachloride
Chlordane
Chloroethane
4,4'-DDE, DDT, DDD
3-3'-Dichlorobenzidine
1,2-Dichloroethane
1,1-Dichloroethene
1,1-Dichloropropane
2,4-Dinitrotoluene
Isophorone
Mercury
N-nitrosodimethylamine
N-nitrosodi-n-propylamine
Pentachlorophenol
Phenol
Selenium
1,1,2,2-Tetrachloroethane
Toluene
1,1,2-Trichloroethane
Zinc

Group 3

Acrolein
Acrylonitrile
Ammonia
Bromoform
Chlorobenzene
Chlorodibromomethane
Chloromethane
Copper
Di-N-butyl phthalate
1,1-Dichloroethane
1,2-trans-Dichloroethene
2,6-Dinitrotoluene
1,2-Diphenylhydrazine
Endrin aldehyde/endrin
Ethylbenzene
Hexachlorobenzene
Indeno (1,2,3-cd) pyrene
Naphthalene
Nitrobenzene
Oxirane
Silver
Toxaphene
1,1,1-Trichloroethane
2,4,6-Trichlorophenol
Xylenes (total)

Group 4

Aniline
Benzoic acid
Bromomethane
2-Butanone
Carbon disulfide
p-Chloro-m-cresol
1,2-Dichlorobenzene
1,3-Dichlorobenzene
Dichlorodifluromethane
2,4-Dichlorophenol
Diethyl phthalate
2,4-Dinitrophenol
Dimethyl phthalate
2,4-Dimethylphenol
4,6-Dinitro-2-methylphenol
1,4-Dioxane
Fluoranthene
Fluorotrichloromethane
Hexachlorobutadiene
Hexachloroethane
4-Methyl-2-pentanone
Phenanthrene
2-Methylphenol
Thallium
1,2,4-Trichlorobenzene

Group 5

Anthracene
Antimony
Asbestos
Barium and compounds
Boron and compounds
Cobalt and compounds
Cresote
trans-1,3-Dichloropropene
Endosulfan (alpha, beta, sufate)
Fluorides/fluorine/hydrogen fluoride
Hexachlorocyclopentadine
2-Hexanone
Hydrazine
Manganese
Mechloroethamine
Plutonium
Pyrene
Radium and compounds
Radon and compounds
Styrene
Thorium and compounds
Tin
2,4,6-Trinitrotoluene
Uranium and compounds
Vinyl acetate

Group 6

Acenaphthene
Acenaphthylene
Acetone
Atrazine
Benzo(g,h,i)perylene
Benzo(k)fluoranthene
4-Chloroaniline
Chlorodibenzodioxins
2-Chloroethyl vinyl ether
Dibenzofuran
1,2-Dibromoethane
cis-1,2-Dichloroethylene
Di-n-octyl phthalate
Disulfoton
Fluorene
Formaldehyde
2-Methylnaphthalene
Mustard gas
Nitrates/nitrites
2-Nitrophenol
Parathion (DNTP)
Phosgene
Sulfur dioxide
1,2,3-Trichloropropane
Vanadium

Group 7

Benzyl alcohol
Bromochloromethane
Butylbenzyl phthalate
Chlorodibenzofurans
Chlorodifluromethane
2-Chlorophenol
2,4,5-D, salts, and esters
Dibromochloropropane
1,2-Dichloroethylene
Dimethyl formamide (DMF)
Hexane
Hydrogen sulfide
Methoxychlor
4,4'-Methylene-bis-(2-chloroaniline)
Mirex
o-Nitroaniline
2-Nitrophenol
n-Pentane
Polybrominated biphenyls
RDX (cyclonite)
Sevin (carbaryl)
2,4,5-T
Toluene diisocyanate
2,4,5-TP acid (silvex)
2,4,5-Trichlorophenol

Group 8

Aramite
1-Bromo-4-phenyoxy benzene
bis(2-Chloroethoxy)methane
4-Chlorophenyl phenyl ether
Cresols
Cyclohexanone
cis-1,3-Dichloropropene
Ethylene glycol
Heptane
Malathion
Methanol
Methyl methacrylate
4-Methylphenol
Molybdenum
m-Nitroaniline
Nitrophenol
Octane
Pentachlorobenzene
Strontium
Sulfuric acid
Tetrahydrofuran
1,1,2-Trichloro-1,2,2-trifluroethane
1,3,5-Trinitrobenzene
Trinitrophenylmethylnitramine
Tritium

BIOLOGICAL ASSESSMENT AND ECOLOGICAL RISK ASSESSMENT

NEW TOOLS FOR CLEAN-UP DECISIONS AT HAZARDOUS WASTE SITES

S.B. Norton[1], J. Benforado[1], C. Zamuda[1]
A. Mittleman[2], I. Diwan[2], and R. Fulton[2]

[1]U.S. Environmental Protection Agency
Washington, DC

[2]Technical Resources, Inc.
Rockville, Maryland

INTRODUCTION

The goal of EPA's Superfund program is deceptively straightforward--
find the nation's worst abandoned hazardous wastes sites and clean them
up. Achieving this goal in a logical and defensible manner has proven to
be a much more difficult task. The initial strategy of the program has
emphasized monitoring and testing the toxicity of individual chemicals.
However, there has been renewed interest in testing site-specific complex
mixtures and organisms. Biological assessment techniques are particularly
useful for these purposes and hence have great potential for supporting
decisions made at Superfund and other hazardous waste sites.

A wide variety of tests and measurements are referred to as bioassessment techniques. They include the following:

° conducting toxicity tests on contaminated environmental media
brought into the laboratory;

° using biochemical characteristics of organisms (biomarkers) to
estimate exposure or toxic response;

° conducting toxicity tests at the study location (in situ) using
laboratory-raised organisms; and

° comparing characteristics or organisms at study location to a
reference location.

In this paper, we will briefly discuss how bioassessment techniques are
currently used in the Superfund program. This will be followed by a more
general discussion of how bioassessments might be more extensively used to
support decisions. Finally, we provide some recommendations for further
research that will make bioassessment a more effective tool for decision
making.

In Situ Evaluations of Biological Hazards of Environmental Pollutants
Edited by S. S. Sandhu *et al.*
Plenum Press, New York, 1990

THE CURRENT USE OF BIOASSESSMENT TECHNIQUES AT SUPERFUND SITES

The use of bioassessment techniques at uncontrolled hazardous waste sites is at a relatively early stage of development. The use of these techniques is not specifically required by the Comprehensive Environmental Response Compensation, and Liability Act (CERCLA), amended by the Superfund Amendments and Reauthorization Act of 1986 (SARA). However, Section 104 of CERCLA does authorize investigations, monitoring, surveys, testing, and other information gathering as necessary or appropriate to identify the existence and extent of a release or threatened release, and the extent of danger to the environment.

Bioassessment techniques may be useful at many stages of the Superfund process. Specifically, the studies may be used as a monitoring tool to detect the presence of hazardous materials to determine if remedial action is needed, to gauge the real extent of required remedial action, to evaluate alternative remedial actions, and to assess and characterize the risks posed by releases from the waste sites. Additionally, bioassessments can be used both during and after remedial action to evaluate the efficacy of clean-up activities.

Two examples of how bioassessment techniques have been used at Superfund sites are discussed in the remainder of this section. More examples of bioassessment applications at hazardous waste sites can be found in U.S. EPA, 1989a,b,c, and d. These examples are described here to provide a foundation for the more general discussion of bioassessment and risk assessment in the following sections of the paper.

The first example is work done at the Rocky Mountain Arsenal near Denver (U.S. EPA, 1987). This site was used principally by the U.S. Army and by Shell Chemical Co. to manufacture, test, and dispose of toxic chemicals, including pesticides and chemical warfare agents. Preliminary testing indicated that an area referred to as Basin A was contaminated, and suggested that a gradient in contamination extended north-south from a trench that drains Basin A. Therefore, a sampling design consisting of four parallel transects extending south from the trench was used. Soil samples were collected and a battery of toxicity tests were conducted using the soil and elutriates of the soil.

Toxic effects were exhibited primarily in lettuce seed germination and lettuce root elongation tests. Little toxic effect was seen in _Daphnia_, Microtox, or algal toxicity tests using the elutriates, indicating the toxicants probably do not leach readily from the soil. Based on lettuce seed mortality tests, contour maps were developed for the 0- to 15-cm soil fraction (shown in Figure 1) and also for the 15- to 30-cm soil fraction.

Based on these results, a 30% lettuce mortality was selected as a criterion for cleanup of the Basin A site. The 30% criterion was selected because it was two standard deviations above the mean control mortality. Using this criterion, the shaded areas below the heavy black line in Figure 1 would be selected for clean-up.

Another example of the use of bioassessment techniques at a hazardous waste site is a study conducted at a wood treatment plant in Mississippi that was known to be contaminated with creosote and other wood-preserving materials (U.S. EPA, 1987). An exploratory study was designed to determine whether toxicity tests could be used to detect creosote contamination in stream sediments and water and to define the boundaries of contaminated zones on the site.

22

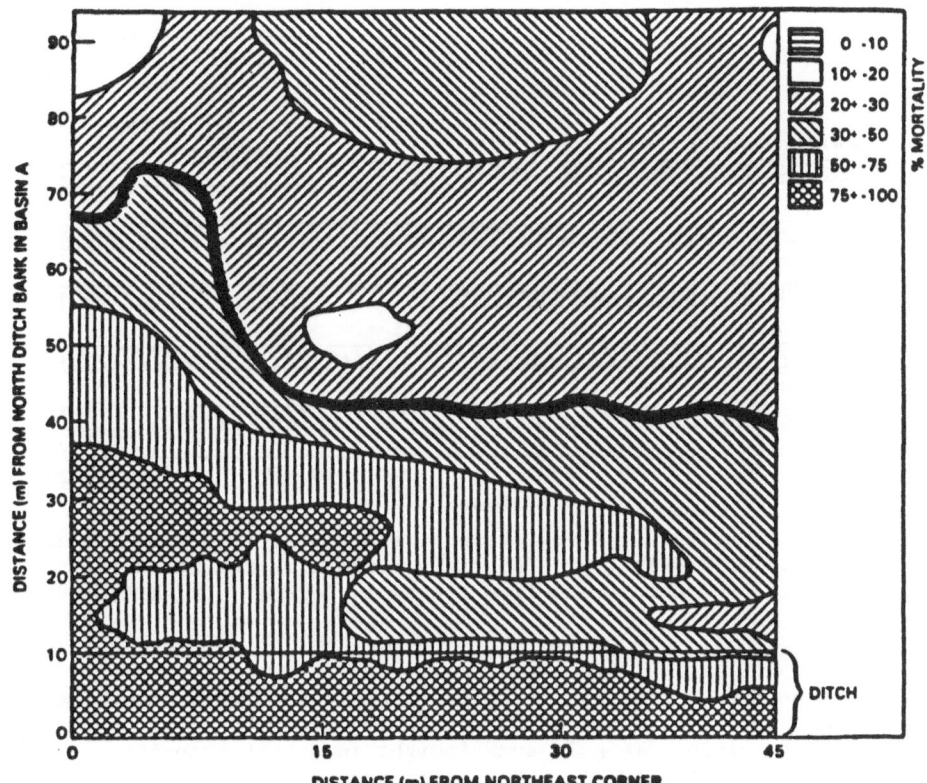

Figure 1. Estimated lettuce seed mortality (based on kriging) for the 15-
to 30-cm soil fraction from the Rocky Mountain Arsenal.
Source: U.S. EPA, 1987.

Sediment and water samples were collected for use in bioassays on the
site upstream and downstream of a visibly contaminated zone, with an additional control sample upstream of the site. Figure 2 shows the results
from sediment toxicity tests using four different organisms as well as chemical analysis of creosote concentrations. The chemical analysis showed the
highest creosote concentration in the furthest downstream portion of the
creek study area. Although the toxicity tests showed somewhat variable results, the algae, Daphnia, and Microtox bioassays showed the greatest
adverse responses in samples taken from upstream of the highest creosote
concentrations. These results indicated that contaminants other than creosote may be present in the stream sediments. The Microtox bioassay, and to
a lesser extent the Daphnia bioassay, also exhibited toxic effects in the
area of detectable creosote contamination. Based on the results of the
toxicity tests, it was recommended that clean-up should begin about 220 m
west of the initial point and continue to about the 660-m area.

As can be seen in the previous examples, bioassessment can be a very
useful tool for evaluating complex wastes. Although it is beyond the scope
of this paper, there are many available techniques that are either readily
available or easily adaptable for routine use (e.g., U.S. EPA, 1983; Greene
et al., 1988; U.S. EPA, 1989). However, bioassessment still has not made

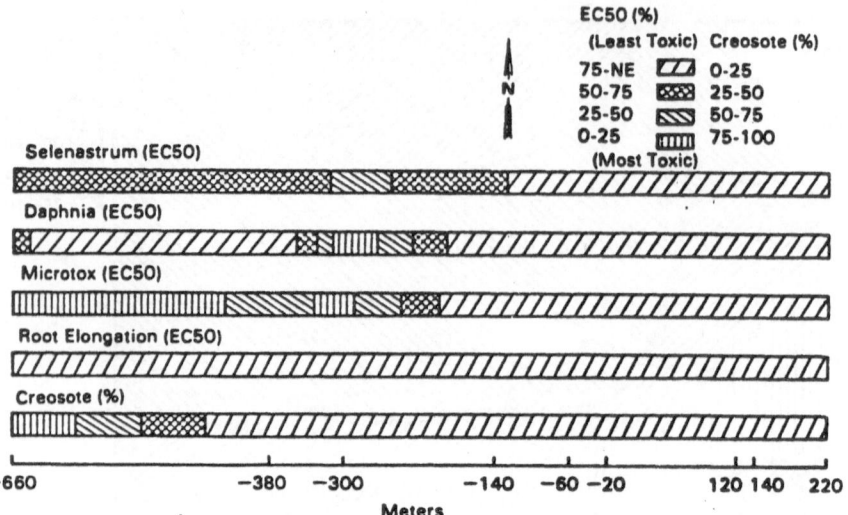

Figure 2. Bioassay results from sediment elutriates at a wood treatment site in Mississippi. Negative numbers represent samples collected downstream from the site. Source: U.S. EPA, 1987.

the transition from a research technique that is used only under special circumstances to a tool that can be routinely used to support decisions at hazardous waste sites. To gain some insight into this transition, it is worth examining how the Agency makes decisions.

APPLICATIONS OF BIOASSESSMENT IN RISK ASSESSMENT

The Agency makes many of its decisions aimed at protecting human health and the environment using a framework called risk assessment (Figure 3). EPA's risk assessment process consists of a series of steps: (1) identify the adverse effects caused by a material (hazard identification), (2) characterize the relationship between exposure to the material and the extent of adverse effects (dose- or exposure-response assessment), (3) assess the extent of exposure to the material, and (4) combine the results of the first three steps to characterize the probability of adverse effects occurring in the environment (U.S. EPA, 1984). The first two steps of this process, hazard identification and exposure-response assessment, are often referred to together as the toxicity assessment component of risk assessment.

Risk assessments are commonly conducted for individual chemicals. There are some distinct advantages to assessing risk on a chemical-specific basis. Chemical analyses are essential for identifying a specific pollutant, such as an EPA priority pollutant, or in conforming to an established chemical-specific regulatory standard. Chemical analyses may be more cost-effective when characterizing simple wastes containing few chemicals. Finally, many treatment systems are developed specifically to manage a particular chemical. Chemical-specific analyses provide a common currency that can be easily communicated to the engineers and chemists who operate the facility or design the treatment system.

Bioassessment techniques, however, offer some advantages over chemical analysis. Because bioassessment can directly measure effects of a contaminated medium on biota, it incorporates a measure of bioavailability (i.e.,

24

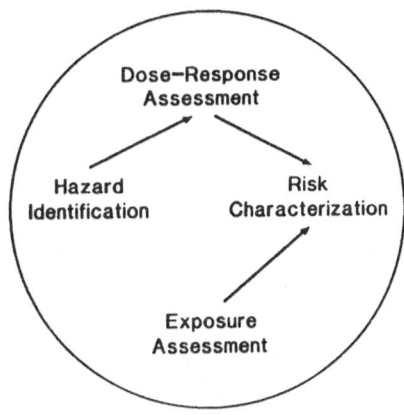

Figure 3. Risk Assessment

desorption from the medium and subsequent uptake by the organism). In addition, bioassessment techniques evaluate the combined toxicity of all components of a complex waste mixture or medium. This is a definite advantage in that some toxic chemicals may not be detected by routine chemical analyses. While chemical analyses can be conveniently compared with guidelines or standards, such benchmarks are not available for all toxic chemicals potentially present in hazardous wastes or effluents. Finally, for complex wastes, conducting a select group of bioassessments may be more cost-effective than completely characterizing chemical content.

Bioassessment techniques can be readily applied to all of the different components of the risk assessment paradigm (Figure 4). Currently, however, bioassessment techniques are used most frequently for two purposes in site assessment. First, they are used at the beginning of the process when identifying potential problems. In particular, bioassessment has been used to define the extent of contamination at sites, as illustrated by the bioassays conducted at the Rocky Mountain Arsenal site. Second, as illustrated by the creosote example, bioassessment techniques have been used to directly characterize risk by measuring adverse effects in organisms on the site or exposed (in a laboratory) to media from the site.

TOXICITY ASSESSMENT AND REMEDIAL DECISIONS

Bioassessment techniques have been used less often as tools for the exposure and toxicity assessment components of risk assessment. It is especially important to develop bioassessment techniques for the toxicity assessment component or risk assessment. While the risk characterization part of risk assessment indicates whether remedial action is necessary at all, it is the toxicity assessment that provides more specific guidance for risk reduction decisions (Figure 5).

As discussed previously, toxicity assessment consists of two steps, hazard identification and exposure-response assessment. the adverse effects potentially caused by a chemical are identified in the hazard identification step. The definition of what constitutes an adverse effect is not necessarily an easy task. An adverse effect is what the decision maker is trying to protect against--for example, increased cancer in humans, or a decrease in a population of sportfish. Suter (in U.S. EPA, 1989e) has called the adverse effects important to decision makers "assessment endpoints." Unfortunately, it is often difficult or impossible

25

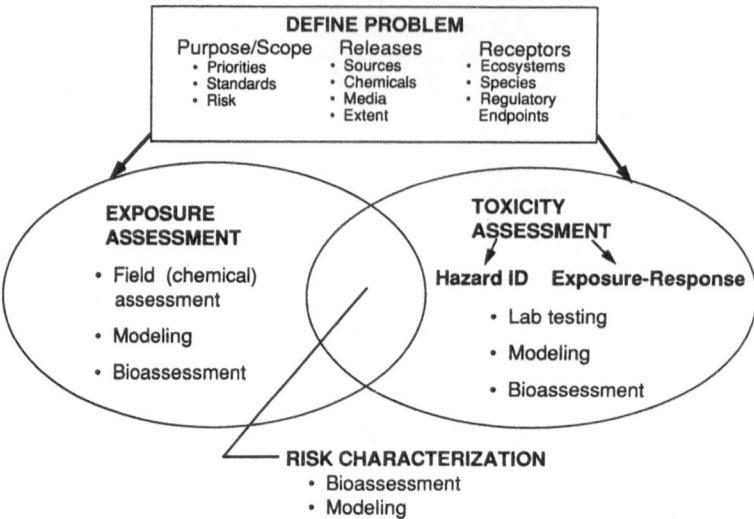

Figure 4.   Ecological risk assessment

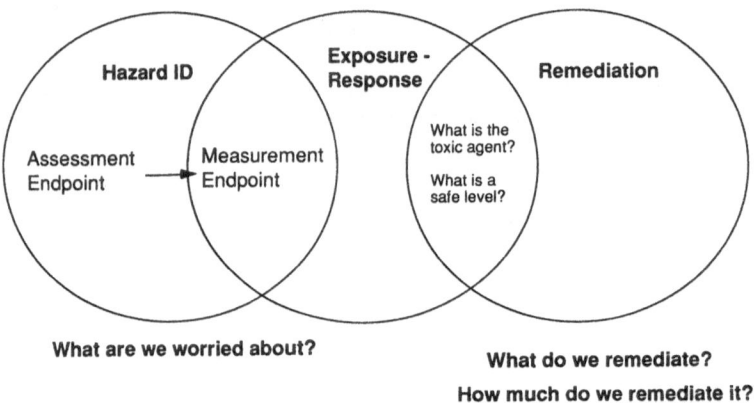

Figure 5.   Toxicity assessment and remedial decisions

to measure changes in assessment endpoints. Hence, the scientist document-
ing adverse effects will measure changes in "measurement endpoints," for
example, reproductive impairment in fish or cancer incidence in laboratory
animals. The measurement endpoint is the "response" in the exposure-
response characterization step of risk assessment. The "exposure" part of
the exposure-response characterization defines what toxic agent is being
considered, for example, a chemical, a mixture, or a contaminated medium.
This is an important step in that it identifies what will be targeted for
remediation. The exposure-response curve itself provides the pivotal
information that is used to identify acceptable levels of exposure.

Many other factors are considered in making a risk reduction decision,
including engineering constraints and cost-benefit trade-offs. However, it
is the toxicity assessment part of the risk assessment process that answers
the health-based questions: What are we worried about? What do we remediate? and How much do we remediate it?

RECOMMENDATIONS

Some recommendations for future directions and emphases in bioassessment research can be made based on the previous discussion. One recommendation is to expand techniques for identifying toxic agents. Currently, toxic agents are most often characterized by chemically analyzing the medium concurrently with bioassessment. Conducting bioassays on fractions of waste or effluent is another way to focus on the toxic components. The important consideration here is to communicate what exactly needs to be remediated to the engineers and scientists designing treatment systems or remedial alternatives. For example, biomarkers may be a powerful way to measure exposure to a contaminant. However, if exposure as measured by a biomarker cannot be related to something that can be treated, the biomarker will be of limited use in choosing remedial options.

A second recommendation is to expand techniques for obtaining exposure-response relationships. For example, media-based exposure- response curves can be constructed by diluting the medium being tested to provide a range of contaminant concentration. Another way to obtain a range of concentration is to locate a gradient on a site. An exposure- response curve could also be obtained by conducting tests sequentially during the course of remediation. As bioassessment techniques are used more frequently to evaluate the efficacy of remedial activities, tests conducted before and during remediation will provide a needed baseline for comparison.

One final recommendation is to continue to solidify links between measurement endpoints and assessment endpoints. Unless we can connect what EPA is mandated to protect (e.g., environmental health) with what we can measure, risk reduction actions based on bioassessment will be difficult to support.

Measurement and assessment endpoints can be linked both qualitatively and quantitatively. Qualitatively, connections can be made using our knowledge of environmental systems. While these relationships may be common sense for ecologists, they must be communicated effectively to the people making the decisions. For example, while decision makers may not base a remedial decision on possible adverse effects to _Daphnia_ per se, they may base a decision on possible adverse effects to the food web of an important community. The idea of using certain species as sentinels or indicator organisms is certainly not new; we have used canaries in coal mines to monitor air quality and trout in cold-water streams to monitor water quality. However, careful reasoning may be required to base the selection of a costly remedial alternative on, for example, mortality in earthworms. Quantitatively, we can clarify connections between assessment and measurement endpoints by continuing efforts in ecosystem modeling. For example, models developed at the Oak Ridge National Laboratory use information of effects on individual species to predict effects at the community level (e.g., SWACOM in Barnthouse et al., 1986).

Bioassessment techniques will undoubtedly be used to a greater degree in the future as a tool in environmental research. Unfortunately, the development of methods for bioassessment has outpaced the ability of many decision makers to easily interpret the implications of their results. It is important for scientists to continue developing and refining bioassessment techniques. However, it is also essential that we continue to search for ways to make bioassessment more interpretable, applicable, and useful in supporting environmental decisions.

REFERENCES

Barnthouse, L.W., Suter, G.W., Bartell, S.M., Beauchamp, J.J., Gardner, R.H., Linder, E., O'Neill, R.V., and Rosen, A.E., 1986, User's Manual for Ecological Risk Assessment, Environmental Sciences Division Publication No. 2679, Oak Ridge National Laboratory, Oak Ridge, TN.

National Research Council (NRC), 1983, Risk Assessment in the Federal Government: Managing the Process, National Academy Press, Washington, D.C.

U.S. Environmental Protection Agency (U.S. EPA), 1983, Protocol for Bioassessment of Hazardous Waste Sites, EPA 600/2-83-054, Environmental Research laboratory, Corvallis, OR.

U.S. Environmental Protection Agency (U.S. EPA), 1984, Risk Assessment and Management: Framework for Decision Making, EPA 600/9-85-002, PB85-170157.

U.S. Environmental Protection Agency (U.S. EPA), 1987, Role of Acute Toxicity Bioassays in the Remedial Action Process at Hazardous Waste Sites, EPA 600/8-87/044.

U.S. Environmental Protection Agency (U.S. EPA), 1988, Protocols for Acute Toxicity Screening of Hazardous Waste Sites, Final Draft, Environmental Research Laboratory, Corvallis, OR.

U.S. Environmental Protection Agency (U.S. EPA), 1988a, Summary of Ecological Risks, Assessment Methods, and Risk Management Decisions in Superfund and RCRA, EPA 230/03-89-046.

U.S. Environmental Protection Agency (U.S. EPA), 1988b, The Nature and Extent of Ecological Risks at Superfund Sites and RCRA facilities, EPA 230/03-89-043.

U.S. Environmental Protection Agency (U.S. EPA), 1989c, Ecological Risk Assessment Methods; a Review and Evaluation of Past Practices in the Superfund and RCRA Programs, EPA 230/03-89-044.

U.S. Environmental Protection Agency (U.S. EPA), 1989d, Ecological Risk Management in the Superfund and RCRA Programs, EPA 230/03-89-045.

U.S. Environmental Protection Agency (U.S. EPA), 1989e, Ecological Assessment of Hazardous Waste Sites, EPA 600/3-89/013, Environmental Research Laboratory Corvallis, OR.

FIELD STUDIES: AQUATIC SYSTEMS

BIOASSESSMENT METHODS FOR DETERMINING THE HAZARDS OF DREDGED MATERIAL

DISPOSAL IN THE MARINE ENVIRONMENT

J.H. Gentile[1], G.G. Pesch[2], K.J. Scott[3]
W. Nelson[2], W.R. Munns[3] and J.M. Capuzzo[1]

[1]Woods Hole Oceanographic Institution
Woods Hole, Massachusetts 02543
[2]Environmental Protection Agency
Narragansett, Rhode Island 02882
[3]Science Applications International Corporation
Narragansett, Rhode Island 02882

INTRODUCTION

Approximately 325 million m^3 of sediment are dredged annually for navigation purposes in the United States. Of this, 46 million m^3 are disposed of annually in the ocean (Peddicord, 1987). Decisions regarding the ocean disposal of dredged material result, in large part, from bioassessment-based estimates of contaminant exposure and ecological impacts (U.S. EPA/COE, 1977). Predictions of impacts for an individual dredging project are estimated from laboratory determinations of the magnitude, bioavailability, bioaccumulation, and hazards (toxicity) of dredged material contaminants. Disposal site management of individual and multiple dredging projects requires monitoring for contaminant transport, availability and accumulation in biota, and the hazards to ecologically and commercially important populations. Because of their importance, suites of bioassessment methods representing several levels of biological organization have been proposed for predicting and assessing the hazards resulting from the ocean disposal of dredged material (Gentile and Scott, 1987; Gentile et al, 1988c).

Hazard assessment methods can be used for either of two purposes: to detect a contaminant-induced biological change, and to infer or predict the ecological or ecosystem impacts from such a change. Hazard methods that simply detect or measure a contaminant-induced change in biological response are useful for mapping the spatial extent of environmental impact, ranking the relative toxicity of different dredged materials, monitoring temporal changes in toxicity at a disposal site, inferring contaminant bioavailability, and providing insight into the potential mechanisms of toxic action. More commonly, changes detected in responses measured at different levels of biological organization are used to predict complex ecological and ecosystem impacts (Underwood and Peterson, 1988). Because the number of compensatory and adaptive mechanisms increases with system complexity, considerable uncertainty exists in predicting population and community impacts from measures of cellular response (Capuzzo, 1981). To address the problems of extrapolation, conceptual and analytical models have been proposed that link multiple levels of biological organization (McIntyre and

In Situ Evaluations of Biological Hazards of Environmental Pollutants
Edited by S. S. Sandhu *et al.*
Plenum Press, New York, 1990

Pearce, 1980; Sheehan et al., 1984; Kooijman and Metz, 1984; Bayne et al., 1988; Capuzzo et al., 1988).

The purpose of this paper is to examine the utility of using a suite of bioassessment methods to predict, verify and assess the hazards associated with dredged material disposal. Evidence is presented in support of two hypotheses implicit in both laboratory predictions of environmental hazards and predictions of ecological impacts from field assessments. The first hypothesis states that measures of direct toxicity in the laboratory are analogous to those measured in the field (field verification). The hypothesis assumes that the same responses can be measured successfully in both environments, that the change in response is proportional to exposure in both the laboratory and field, and that laboratory and field exposures are of similar duration and intensity. The second hypothesis states that responses measured at the subcellular, cellular, and organism levels of biological organization can be used to predict population and community level responses. Predictions of this type are frequently employed in files assessments because the direct measurement of population and community properties is complex, lengthy, and expensive. Because of the limitations and uncertainties associated with these types of extrapolations, suites of responses with complementary properties are recommended for assessing environmental hazard.

CASE STUDY

The ocean disposal of contaminated dredged material provided a unique opportunity to conduct a "holistic" research program that included both prospective, predictive laboratory studies and retrospective field assessments of environmental impacts. A risk assessment strategy was chosen as the integrating framework for this program (Bierman et al., 1986). This facilitated the evaluation of a suite of biological responses, ensured the temporal compatibility of laboratory and field exposures, and permitted the field verification of laboratory estimates of risk and impact.

Approximately 55,000 m^3 contaminated material was dredged from Black Rock Harbor (CT) disposed of (May, 1983) in 25 m of water at the Central Long Island Sound (CLIS) disposal site located 15 km southeast of New Haven, CT (Figure 1). Field stations, sampled for one year pre- and three years post-disposal, were located on a 1,000-meter transect from the mound apex along the primary axis (east-west) of the total current flow. The dredged material was a fine grained (93.6% fines) anoxic harbor sediment containing 72% water and 5.6% organic carbon. Contaminant concentrations (dry weight) included 6.4 ug/g PCB, 3.9 ug/g benzo(a)pyrene, 142 ug/g total PAHs, 2.9 mg/g cooper, and 1.5 mg/g chromium (Rogerson et al., 1985).

The biological responses used to assess hazard included: sister chromatid exchange, a measure of genotoxicity that was used to infer bioavailability and detect the hazard of mutagenic and carcinogenic contaminants; histopathological changes in principal organ systems; biochemical assessment of adenylate energy charge; physiological measures of energetics; measures of somatic growth and reproduction; long-term population growth rates projected from demographic models; and recruitment, recolonization, and succession in benthic communities.

Hazard Assessment: Methods and Verification

Adenylate energy charge (AEC). Adenine nucleotides and AEC are of interest in measuring stress effects because of their central role in energy transformation and their importance as regulators of metabolic processes. The metabolic costs of stress may result either from diversion of assimilated energy to nonproductive functions such as increased respiration and

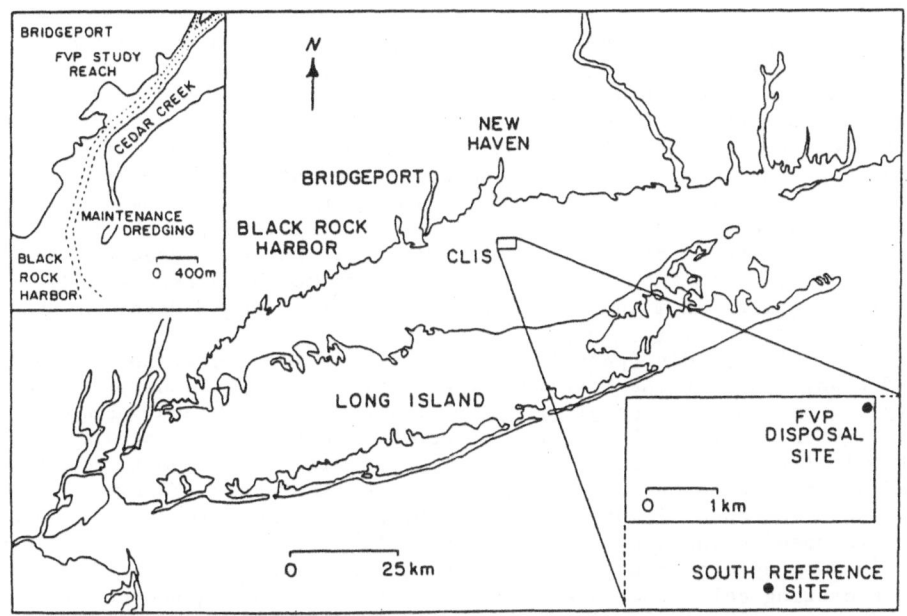

Figure 1.   Central Long Island Sound disposal site and Black Rock Harbor
            dredging site.

cell repair, or from a decrease in the efficiency of energy transfer reactions through toxic effects on enzyme systems, changes in membrane potent-
ials or genetic damage (Vetter and Hodson, 1984).   Either mechanism increases the dissipation of assimilated energy and decreases the energy
available for growth and reproduction.

Adenine nucleotides were measured in the adductor muscle tissue of
Mytilus edulis and the whole body of Nephtys incisa from laboratory and in
situ exposed organisms.   None of the nucleotides examined, ATP, ADP, AMP,
AEC, or the total pool showed a quantitative or predictable response to
dredged material in laboratory or field exposed organisms (Zaroogian et
al., 1988).   Vetter and Hodson (1984) caution that since virtually all metabolic energy flows through ATP, the measurement of static quantities or
pool sizes will convey little information about energy flow.   Consequently,
AEC was neither a useful in situ indicator of stress nor laboratory predict-
or of potential hazard (Gentile et al., 1988a).   Combining adenylate measurements with measures of energy reserves (lipid and glycogen) will improve detection of chronic stress and provide information on possible mechanisms of toxic action.

Sister chromatid exchange (SCE).   The SCE response has been recommended
for environmental application by the U.S. Environmental Protection Agency
Gene-Tox Program (Latt et al., 1981).   Several studies have shown that SCE
is a more sensitive method for detecting mutagens and carcinogens than the
traditional chromosome and chromatid observations (Perry and Evans, 1975;
Solomon and Bobrow, 1975; and Bloom, 1978).   The application of SCE to polychaete worms (Pesch et al., 1981) and mussels (Dixon and Prosser, 1986) has
created a practical tool for studying genetic problems in marine environments.

The SCE technique was applied to N. incisa, an infaunal polychaete dominant in the benthic community at the CLIS disposal site.   Differences in
SCE response among three laboratory replicate tests and the inability to
obtain a reproducible dose response relationship suggest that one or more
factors important in SCE induction under laboratory conditions were not

33

controlled (Pesch et al., 1987). In the field, the frequency of SCE increased in organisms exposed to dredged material, declined when the worms were held in the laboratory in clean sediments, but increased to field levels upon re-exposure to dredged sediments. The positive correlation between increased frequency of the SCE response and the presence of dredged material suggests the presence of bioavailability of mutagenic contaminants. The magnitude of the SCE response in N. incisa to contaminated sediment was comparable in both the laboratory and field. Although the use of SCE in quantitative hazard assessments must await further research (Gentile et al., 1988a), the method can be used to determine the presence and infer the bioavailability of mutagens and carcinogens in dredged material.

Histopathology. Histopathological changes were detected in laboratory exposed populations of the bivalve mollusc Mytilus edulis, the polychaetes Nephtys incisa and Neanthes arenaceodentata, and the amphipod, Ampelisca abdita (Yevich et al., 1986). The incidence of pathology involving the gastrointestinal tract and gills in M. edulis was directly proportional to the dredged sediment exposure concentration in the laboratory (Figure 2). Degeneration of parapodial muscles and metaplasia of the epidermis in both polychaete species was also directly proportional to the intensity of exposure. In N. arenaceodentata, histological changes were reported in the mucous secreting cells that are essential for facilitating the movement of this species over sediment surfaces. Several well-defined patterns of histological change were detected in A. abdita exposed to dredged sediments. These included necrosis of gill epithelium and lamellae, loss of normal gill architecture, and atrophied and vacuolated mucous cells and mucous tube glands.

Histological changes were not detected in transplanted M. edulis exposed in situ during and after disposal of dredged material because the magnitude and duration of exposure was below the response threshold for these effects as determined from laboratory studies (Figure 2). Histological changes detected in the epidermis N. incisa collected from the disposal site, however, were analogous with those reported in the laboratory. Generally, the type and incidence of histopathological changes detected in the field were consistent with laboratory predictions and field exposures.

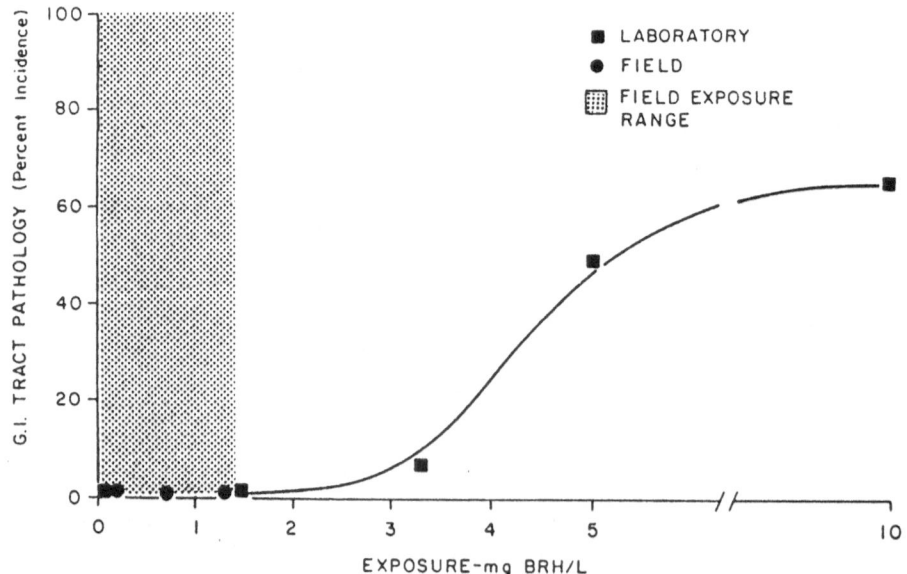

Figure 2. Percent incidence of gastronintestinal tract pathology in M. edulis measured in the laboratory and field.

The value of pathology as an in situ measure of pollutant impact has been well demonstrated (Sindermenn, 1980; Patton and Couch, 1984; Auffret, 1988; and Yevich et al., 1988). It is suggested that future studies improve response quantification and focus on determining the casual links between tissue pathology and genetically induced and biochemically mediated responses to contaminant stress (McMahon et al., 1988; Stegeman, 1987; and Stegeman and Kloepper-Sams, 1987).

Energetics. The energetics of Mytilus edulis was measured using the scope for growth index (SFG). This index represents an instantaneous assessment of energy balance in an organism for that set of environmental conditions under which it is measured. As applied in this study, SFG, was used to measure relative differences resulting from exposure to dredged material.

A quantitative and reproducible exposure-response relationship was exhibited in M. edulis exposed to suspended dredged material in the laboratory (Nelson et al., 1987). SFG and shell growth decreased 45% and 80% at exposure concentrations of 1.5 and 3.3 mg/L, respectively. The SFG index showed a clear response in the field, during and 1 month after disposal, when in situ exposures were estimated to be 1.4 mg/L (Figure 3). These data demonstrate that the laboratory and field exposure-response relationships were comparable and that SFG and shell growth are useful in situ measures of stress.

The changes noted in SFG were due primarily to a depression in clearance rates. Absorption efficiencies, respiration rates and ammonia excretion rates were not significantly related to dredged material exposure. Widdows and Johnson (1988) also report that decline in scope for growth of mussels exposed to copper and aromatic hydrocarbons was due primarily to a reduction in clearance (feeding) rate. These results would lead one to conclude that clearance rate, which is easily measured, could be used in place of SFG when the intent is simply to rank or compare laboratory or field exposures. However, if the goal is to make long-term projections of M. edulis population dynamics one must analyze all the components of the energy budget in the presence and absence of gametogenesis.

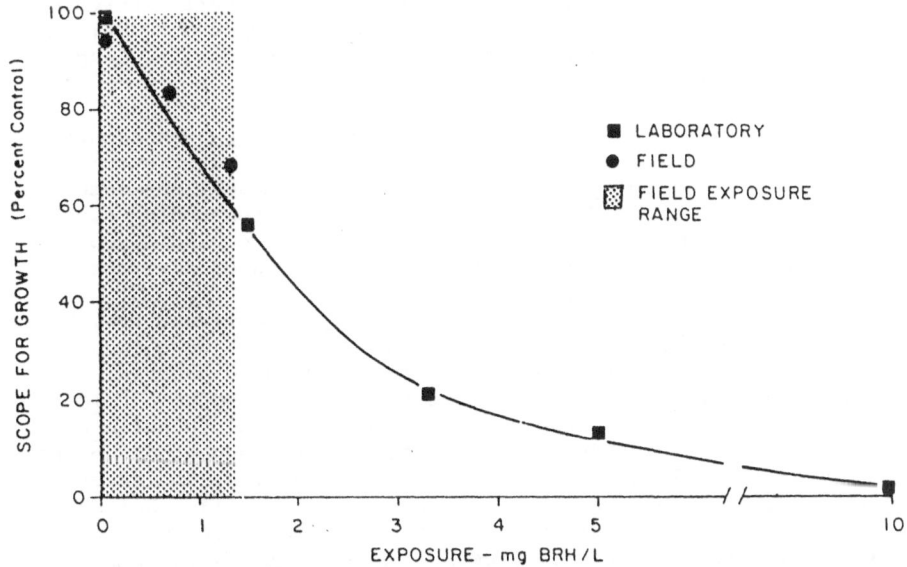

Figure 3. Exposure-response relationship for SFG in M. edulus measured in the laboratory and field.

Physiological responses measured in Nephtys incisa juveniles included tissue production, respiration, ammonia excretion rate, the cumulative energy for production, and net growth efficiency. In the laboratory, all the responses showed a quantitative and reproducible exposure-response relationship to dredged material exposure. Laboratory respiration rates were dependent upon whether N. incisa were exposed to bedded or suspended sediments. In bedded sediment exposures, N. incisa were continuously exposed to contaminated sediment, which decreased burrowing, feeding, and respiration rates and ultimately led to starvation (Johns et al., 1985). In suspended sediment exposures N. incisa were placed in bedded reference sediment and exposure occurred only when they were in contact with suspended sediment at the sediment-water interface. In this case, normal burrowing and feeding activity was maintained in the bedded reference sediment while the observed increase in respiration represents a response to the contaminated suspended sediments (Johns and Gutjahr-Gobell, 1988).

In the field, respiration and excretion rates measured in N. incisa collected on the disposal mound were significantly lower than those in worms collected from the reference stations during seasons when the seawater temperatures were between 12 and 21°C (Johns and Gutjahr-Gobell, 1988). Therefore, respiration rates measured in N. incisa collected from the disposal mound correspond with decreased respiration rates in laboratory bedded sediment exposures. Excretion rates in the laboratory and field decreased significantly with increasing exposure to dredged material. Because of the inability to conduct a complete energy budget in the field and the limited utility of individual measures of respiration and excretion, energetic responses in N. incisa are not recommended for in situ assessments.

Bioaccumulation. Laboratory-derived contaminant bioaccumulation patterns and kinetics were used as an indirect measure of field exposure for water column and benthic compartments, and as a measure of contaminant bioavailability. PCBs were chosen as a tracer for the dredged material because of their persistence, particle affinity, equilibrium-partitioning, and kinetic properties. Studies with M. edulis and N. incisa show that the PCB distributions and patterns were quite similar whether exposure was from suspended or bedded sediments (Lake et al., 1988). The patterns of uptake and lipid-normalized PCB tissue residues in M. edulis increased proportionally with increasing exposure to suspended dredged material in the laboratory and agreed with values obtained from analogous field exposures (Figure 4).

An exposure-residue relationship for PCB in N. incisa was constructed from both suspended and bedded dredged material exposures (Figure 5). PCB tissue residues were directly proportional to dredged sediment concentrations and agreed with field derived residues from corresponding exposures. Thus laboratory exposure-residue relationships could be used in conjunction with field exposures to estimate tissue residue values of field organisms.

Growth, reproductive and population responses. Survival, growth, reproduction, and intrinsic rate of population increase, "r", were measured in Mysidopsis bahia and Ampelisca abdita exposed to dredged material in the laboratory. All responses showed a significant and reproducible impairment of function with increasing exposure to dredged material (Gentile et al., 1988b; Scott and Redmond, 1989). Somatic growth, reproduction, and population growth rate, calculated from age-specific survival and fecundity data, were the most sensitive responses measured (Table 1). Neither of A. abdita nor M. bahia were obtainable from the disposal site which precluded verification of these responses. Growth of Nephtys incisa, exposed to dredged material in the laboratory, was significantly reduced and was among the most sensitive responses measured in this species (Table 2). Reduced

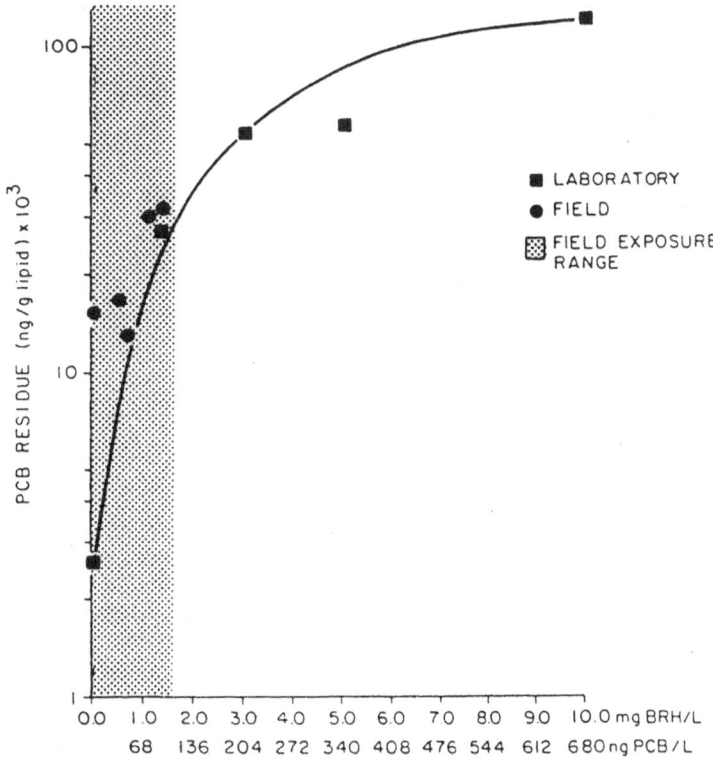

Figure 4.   Field verification of tissue residues (PCB) in M. edulis.

shell growth in M. edulis was also directly proportional to dredged material exposure in the laboratory and was directly correlated with decreased scope for growth (Table 2).

The long-term population growth rate, size class distribution, and reproductive value were measured in N. incisa field populations from the reference and four experimental stations for 1 year before and 3 years after disposal. The population patterns revealed an absence of older cohorts on the disposal mound. This could be due to: (1) reduced migration of larger worms, (2) the inability of larger worms to survive at depth in the contaminated dredged sediment, (3) impaired somatic growth that predicted from laboratory studies, or (4) differential size migration within a cohort on the disposal mound. The decreased size of females on the mound resulted in decreased fecundity (fecundity is directly proportional to size), which was reflected in a 93% decrease in the projected long-term growth rate, on the mound compared to the reference station (Zajac and Whitlach, 1989). Recruitment failure was not distinguishable as a disposal effect since it occurred throughout Long Island Sound during this study. As a result, the recovery of N. incisa populations on the mound apex did not follow the typical recovery pattern. However, although the magnitude and direction of the projected change in N. incisa population growth rate on the disposal mound may be significant when compared to the reference station, this change fell within the range of natural variation for Long Island Sound populations observed during the study period (Zajac and Whitlach, 1989). The population dynamics at the mound apron (400 m) were indistinguishable from those at the reference site after 4 to 6 months and no population effects were detected at 1000 m. Understanding the temporal and spatial variability in population patterns is important for interpreting the significance of experimental results.

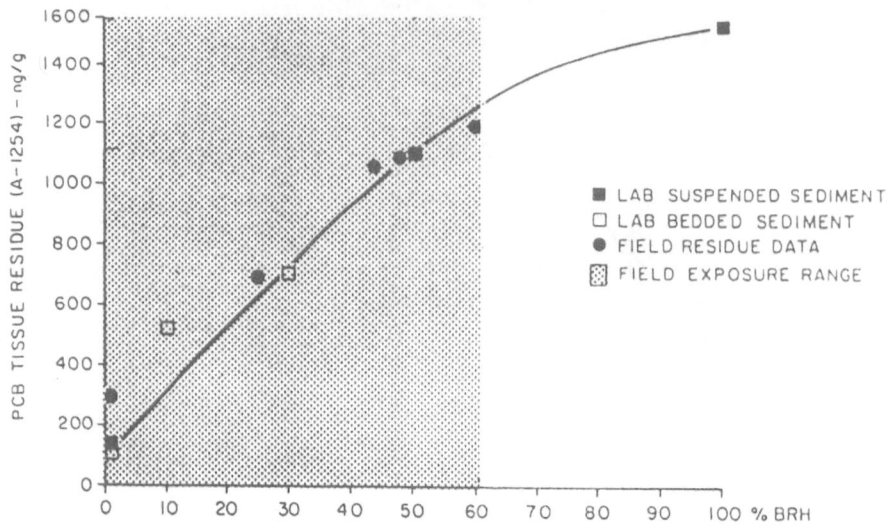

Figure 5. Field verification of tissue residues (PCB) in N. incisa.

Table 1. Hazard Responses for Ampelisca abdita and Mysidopsis bahia Exposed to Suspended Black Rock Harbor Dredged Material

Response	A. abdita	M. bahia
Acute mortality	80[1]	262
Chronic mortality	11	100
Gill pathology	10	N.A.
Mucous gland pathology	10	N.A.
Somatic growth	2.2	25
Reproduction	2.2	18
Population growth rate	1.1	8

[1]Suspended sediment exposures (mg/1).

Benthic recolonization and community structure. Recolonization of the disposal mound and convergence with predisposal conditions were determined from: species numbers, abundances of numerically dominant species, degree of infaunalization (successional stage), and depth of biogenic mixing of the bottom sediments. In addition to analyzing the infaunal populations from sieved sediment grab samples, a vertical imaging technique, REMOTS,

Table 2. Hazard Responses (EC50) for Nephtys Incisa and Mytilus Edulis Exposed to Suspended Exposed to Black Rock Harbor Dredged Material

Response	Nepthys Incisa	Mytilus Edulis
Acute mortality	>200	>10
Adenylate energy charge	>200	>10
Adenylate pool	>200	>10
Pathology	>200	---
Pathology (G.I. Tract)	---	5
Pathology (Gill)	---	1.5
Excretion rates	250	---
Respiration rates	65-150	---
Cumulative energy for production	29-55	---
Net growth efficiency	23-65	---
Scope for growth	---	1.5
Shell growth	---	1.5
Somatic growth	100	---

[1]Suspended sediment exposures (mg/1).

was used as a rapid reconnaissance method for mapping benthic mosaics, resolving fine structure of sedimentary fabric, and characterizing successional patterns (Rhoads and Germano, 1986). The temporal pattern of recolonization consisted of two separate processes operating at different time scales. First was the immediate recolonization of the dredged material mound, which occurred during the first 6 months following disposal. This phase of recolonization was not unlike that seen for other disturbed sites within Long Island Sound and elsewhere (Rhoads et al., 1977). Enhanced species and individual abundances on the disposal mound were not unexpected since many early colonizers thrive in disturbed and defaunated habitats where neither competition for space nor the biologically mediated geochemical conditions of the sediment pose problems for recruitment (McCall, 1977).

The second component of the recovery process was the progressive development of subsurface bioturbation associated with reestablishment of the long-lived species (1-3 year time scale). Maldanid polychaetes and other "head-down" deposit feeders were not observed on the disposal mound until 19 months after disposal even though the major frequency mode of the biological mixing depth (BMD) had converged with that of the reference station within 1 year (Rhoads and Germano, 1986). The recolonization pattern (based upon density) for the bivalve Yoldia limatula and the polychaete Nephtys incisa showed the disposal mound converging with the reference site (Scott et al., 1987). However, closer examination of the age and size distributions of these species indicated that they were in fact growing slower and were limited in distribution to the surficial (0-4 cm) sediment layers. The protobranch mollusc, Nucula proxima did not recolonize the mound in significant numbers and reflects the broader failure of the ambient Stage III assemblage (Nephtys, Nucula, and Yoldia) to become established on the mound 2 years after disposal (Figure 6).

The impairment of recolonization could be due to differences in granulometry between the disposal site and the surrounding sea floor but could also be due to resource limitations (feeding) due to the toxic effects of deeper sediments. The presence of key members of the Stage III assemblage, in low densities, suggests that they were recruited and grew in the upper (0-4 cm), less contaminated surficial sediment. When a critical size was reached requiring feeding depths to penetrate the sub-surface contaminated silts, feeding activity, growth, and survival may have been impaired. Laboratory data on burrowing behavior in Yoldia (Rogerson et al., 1985) and burrowing behavior, bioenergetics, and growth in Nephtys (Johns and Gutjahr-Goebell, 1988) support such as interpretation.

The vertical imaging reconnaissance method (REMOTS) is ideally suited for rapid surveys of the horizontal and vertical distribution of contaminated sediments, mapping benthic assemblages, characterizing successional patterns, defining organism sediment relations, and identifying areas for intensive study. However, to forecast the successional recovery patterns in disturbed habitats requires time series information on the age/size structure and reproductive properties of dominant species.

Hazard Assessment: Response Relationships

The goal of hazard assessment is to predict and document the effects of wastes on populations and communities. To achieve this goal, causal linkages must be defined and quantified for integrating and interpreting the changes detected at various levels of biological organization (Capuzzo et al., 1988). The following examples are presented to illustrate the relationships between responses measured at different levels of biological organization in Mytilus edulis, Nephtys incisa, Ampelisca abdita and Mysidopsis bahia resulting from exposure to dredged material.

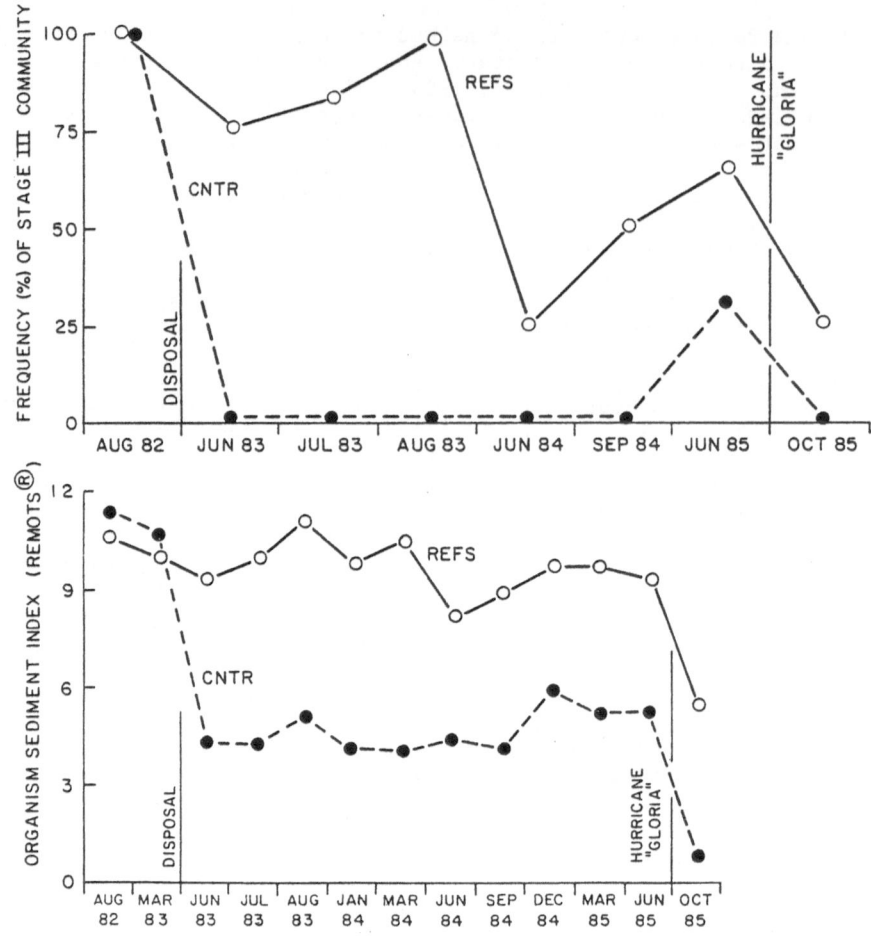

Figure 6.   Temporal and spatial patterns of the community parameters,
             successional stage, and organism sediment index at the FVP
             disposal site.

Examination of the biological responses measured in M. edulis indicate
that the energetic index, scope for growth and gill pathology, and shell
growth showed similar sensitivity to dredged material exposure (Table 2).
Each response was correlated with the intensity and duration of dredged
material exposure, tissue residues of selected lipophilic contaminants and
with each of the other responses (Nelson et al., 1987).

The relationships between gill pathology and physiological measures of
energetics in M. edulis provide insight into the probable causes for the
observed alterations in organism shell growth. For example, pathology of
the gill architecture might be expected to alter both respiratory and filtration functions Examination of the individual responses comprising the
scope for growth index showed that respiration rate was unaffected. Filtration rate, however, was significantly impaired and was responsible for the
observed decreases in scope for growth (Nelson et al., 1987). Filtration
rate also could have been affected by exposure to the narcotizing effects
of soluble, low molecular weight polycyclic aromatic hydrocarbons that were
present in the dredged material (Cappuzzo and Kester, 1987). Collectively,
these observations suggest that, initially, gill function was impaired by
contaminant narcotization and later by structural changes in the gills.
Both effects on the gills led to decreased filtration rate and a reduction
in available energy for growth.

In the field, M. edulis populations showed a significant change in scope for growth when dredged material exposure reached the response threshold found in the laboratory (Figure 3). In addition, scope for growth was correlated inversely with tissue contaminant concentrations. Gill pathology was not detected in the field as would have been predicted from laboratory results (Nelson et al., 1987). One possible explanation for these results is that field exposures were neither continuous nor of the same intensity and duration as those that elicited structural changes in the laboratory. Consequently, the observed changes in field scope for growth were probably due to gill narcotization which altered filtration rate.

Laboratory results with N. incisa showed that pathology of epidermis, physiological measures of energetics, behavioral response (burrowing) and somatic growth were among the most sensitive responses measured (Table 2). Individual responses were correlated with each other, with exposure to dredged material, and tissue reside concentrations of lipophilic contaminants.

The following example illustrates how burrowing behavior in N. incisa can be linked to somatic growth through energetics. Johns et al. (1985) determined that the length and depth of burrows in N. incisa were significantly reduced by the presence of contaminated dredged material. If the primary energy source for this species is carbon from ingested sediment, then curtailing burrowing activity has the same effect as starving the individual. While the decreased burrowing activity reduced maintenance costs, it forced the organism to catabolize tissue to satisfy the energy requirements of routine metabolism (Johns et al., 1985). In the absence of a net input of assimilated energy from sediment carbon, energy reserves would be expected to become depleted and unavailable for new somatic growth. In fact, the cummulative energy available for growth decreased significantly in the presence of dredged sediment as did the actual increases in tissue weight (Johns et al., 1985).

The size and patterns of abundance of N. incisa populations collected from the disposal mound and reference sites were monitored for three years after disposal. N. incisa did not recruit or colonize the mound where the depth of dredged material exceeded 10 cm until the upper 2-4 cm layer of dredged material had become oxidized and diluted with ambient sediment. The sizes of the N. incisa from the disposal mound population were smaller than those from the reference site population. One explanation for these results is that the worms become resource limited after reaching a certain size because of the limited depth of available sediment (2-4 cm) for burrowing. To continue to grow and avoid predation, the worms would have to use the deeper contaminated and anoxic sediments. However, laboratory studies have demonstrated that N. incisa do not burrow and ingest contaminated sediments (Johns et al., 1985). Consequently, their inability to obtain adequate nutrition would reduce the energy available for growth which, in fact, was reflected in decreased respiration, excretion, and size of worms on the mound.

Growth and reproduction of the field populations were used in a size-dependent demographic model to project the long-term growth rate of N. incisa populations at the disposal and reference sites (Zajac and Whitlach, 1989). Because fecundity in N. incisa is a function of size, the decreased growth and size of disposal mound organisms resulted in a 93% decrease in the projected long-term population growth rate when compared to reference site organisms. These results clearly demonstrate the quantitative relationship between organism level responses (e.g., feeding behavior, growth, and reproduction) and projected long-term population growth rates.

Laboratory studies with the amphipod, A. abdita, showed pathological changes in gill architecture and mucous cell glands used in tube construction (Yevich et al., 1986), impaired tube building (Rogerson et al., 1985), acute and chronic mortality, and decreased somatic growth, reproduction, and intrinsic rate of population growth (Scott and Redmond, 1989). Understanding the relationships between responses provides insight into the possible causes for the observed effects. For example, pathology of the gills provides an explanation for the observed decrease in growth and reproduction through alteration of energetics. Pathology of the mucous cell glands interferes with tube building, which affects feeding, resulting in energetic changes that lead to decreased growth and reproduction. Organism level changes in survival and reproduction in laboratory populations of A. abdita were linked to population growth rates through the use of demographic models (Scott and Redmond, 1989).

In addition to the uncertainties associated with intraspecific response extrapolations, it is necessary to mention the importance of interspecific differences in response sensitivity. This is illustrated by the comparative toxicity data for the crustaceans, A. abdita and M. bahia (Table 1). Interspecific comparisons of the same response shows that, in general, A. abdita is 8-10 times more sensitive than M. bahia. The intraspecific range of sensitivity for all responses is 73 for A. abdita and 33 for M. bahia. Thus, the intraspecific range of all responses is greater than the interspecific differences for an individual response. The order of response sensitivity, however, is identical in both species.

DISCUSSION

The field verification of laboratory toxicity data is an implicit assumption in predictions of environmental hazard. The results from this study clearly demonstrate that specific hazard responses can be measured successfully in the same species in both environments and that the exposure proportional response determined in the laboratory applies in the field. Because the types of continuous time variable data necessary to quantitatively described field exposures were unavailable, discrete data on water and sediment chemistry and contaminant tissue residues were used to estimate upper and lower boundaries for field exposures. Nevertheless, when similar laboratory and field exposures occurred, the magnitude of hazard was comparable in both environments indicating that the laboratory measures of direct toxicity are applicable in the field.

The data obtained in this study clearly demonstrate that knowledge of the relationships between responses can provide valuable insight into mechanisms of toxic action which can then be used to explain the toxic responses observed at other levels of organization. Alterations in energetic responses clearly were related to the pathology of respiratory and feeding organs as well as feeding behavior itself. Energetic changes were also correlated with decreased organism growth and reproduction that were responsible for changes in population growth. It is apparent, therefore, that understanding the reproductive and developmental processes at the organismal level provides a critical link between bio-energetic responses at the suborganismal level and population growth and stability (Capuzzo et al., 1988).

However, the ability to predict environmental impacts with a known degree of certainty requires a rigorous statement of causality and quantitative models for linking responses from different levels of biological organization. Quantifying the linkages between responses can be accomplished either through correlation analyses or through the development of mechanistic models. In the human health field, bio-chemical and physiological changes can be statistically correlated with the incidence of specific diseases because of the extensive clinical data base on the human population.

These types of statistical correlations are often both disease and stressor specific, require large data bases, provide limited understanding of the mechanisms causing the response, and therefore have limited predictive value. An alternative approach is to develop a mechanistic and analytical framework for quantifying linkages between responses. For example, in laboratory studies on Daphnia magna, individual growth and reproduction was quantitatively linked to population growth rate and to the size-dependent assimilation through the partitioning of energy into maintenance, growth, and reproductive functions (Kooijman and Metz, 1984). Unfortunately, these types of multicomparment models have not been widely applied to marine systems, which highlights the need for quantitative paradigms for linking subcellular and cellular responses to the whole organism. On the other hand, quantitative changes in organism development, growth and reproduction can be used in demographic models to project quantitative changes in population growth and stability (Caswell, 1982). The degree to which such projections actually predict environmental effects on populations is a function of the model's ability to account for density- dependent interactions, and external recruitment.

Until such time as the problems with extrapolation are resolved, suites of responses that measure structural and functional properties at the subcellular and cellular level should be conducted in concert with organismic responses that are more readily interpretable and that can be related to population and community effects (Widdows, 1982; Gentile and Scott, 1987; Capuzzo et al., 1988). To further minimize extrapolation uncertainties, selection of responses must be tailored to the objectives and needs of the assessment and take advantage of the unique properties of responses measured at different levels of organization (NAS, 1971; Capuzzo, 1981; Widdows, 1982). For example, cellular and subcellular responses (biomarkers) often are characterized by high contaminant specificity, discrimination, sensitivity and rapid response to specific pollutant classes (Figure 7). These properties are helpful in determining causality from multicontaminant exposures, providing insight into mechanisms of toxic action, and measuring the extent and magnitude of contaminant exposure. Organism and population level responses, on the other hand, are general rather than specific, require longer time periods to be expressed, and integrate both pollutant and environmental variables making them useful indicators of long-term ecological effects (Figure 7). Rather than dealing with the extrapolation uncertainties associated with using cellular and subcellular responses to predict long-term ecological impacts, a more productive approach would be to measure ecological responses directly, or to develop methods for rapidly estimating the critical attributes and properties of populations, communities, and ecosystems. Conversely, a short-term biomarker may be a more appropriate method for determining the spatial or temporal distribution of a carcinogenic contaminant than a whole organism tumorgenic response that may require several months for expression. Therefore, a well-chosen suite of biological responses can be effective in predicting and assessing environmental impacts by providing information on both the exposure and hazards of waste disposal in marine environments.

CONCLUSION

A suite of methods that measures responses from several levels of biological organization was used to predict and determine the magnitude and extent of hazard resulting from the ocean disposal of dredged material. The results show that tissue and organ pathology, bioenergetics, growth, reproduction, and population growth rate were excellent measures of hazard both in the laboratory and the field. In addition, sediment profile photography (REMOTS), which relates organism-sediment relationships to the dynamical aspects of benthic community successional stage, proved to be an excellent reconnaissance method for field assessments.

43

The ability to extrapolate laboratory results to the field is an implicit assumption in hazard assessments. The field verification of laboratory hazard responses tested the hypothesis that when measuring direct toxicity, laboratory exposure-response relationships were not different from those in the field. For pathology, energetics, behavior, bioaccumulation, growth and population growth there was general concurrence between laboratory and field patterns of response. Lack of laboratory-field concurrence for sister chromatid exchange and adenylate energy charge was due to limitations with the respective methodologies.

Extrapolations between the hazard responses studied in this program provided explanations for the observed toxicity, insight into the possible mechanisms and causes of toxicity, and an appreciation for the uncertainties associated with these types of extrapolations. In general, organism growth, development, and fecundity could be quantitatively coupled to the acquisition and partitioning of energy resources that were qualitatively linked to behavior and pathology. Alterations of organism survival, growth, and reproduction were used in demographic models to project quantitative changes in population growth. However, because these models do not account for density-dependent reproductive compensation or external recruitment they may not provide accurate predictions of field population dynamics.

The data and observations from this study both confirm existing and suggest additional hypotheses for investigation. Regarding the former, the results clearly illustrate that using suites of biological responses in hazard assessments is a necessity of and that organism developmental and reproduction functions provide a critical link to both suborganismal responses and to populations. However, it is also apparent that the development of an interpretive framework for qualitatively and quantitatively linking subcellular and cellular responses to impacts on the survival, growth, and reproduction of the organism is necessary if these responses are to have predictive value in hazard assessments. Further, we need to use the unique and diverse properties of responses measured at different levels of biological organization. Given these differences, it is unlikely that a single method will have such properties as high contaminant specificity and discrimination, rapid response time, and a high degree of ecological relevance. However, if the need for different requirements is recognized in the program design, they can be satisfied by the judicious selection of suites of methods with the desired complementary properties. Finally, these studies on

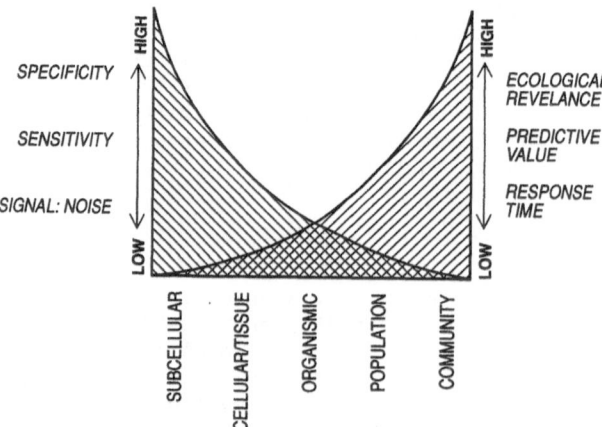

Figure 7. Properties of assessment methods and levels of biological organization.

dredged material have contributed to an understanding of the uncertainties associated with extrapolating hazard results from laboratory to the field and between levels of biological organization.

REFERENCES

Auffret, M., 1988, Histopathological changes related to chemical contamination in Mytilus edulis from field and experimental conditions, Mar. Ecol. Prog. Ser. 46:1.

Bayne, B.L., Clarke, K.R., Gray, J.S., eds., 1988, Biological effects of pollutants, a practical workshop, Mar. Ecol. Prog. Ser. 46:1-5.

Bierman, V.J., Jr., Gentile, J.H., Paul, J.F., Miller, D.C., and Brungs, B.A., 1986, Research strategy for ocean disposal: conceptual framework and case study, in: "Environmental Hazard Assessment of Effluents," H.L. Bergman, R.A. Kimerle, and A.W. Maki, eds., Pergamon Press, New York, pp.313-329.

Bloom, S.E., 1978, Chick embryos for detecting environmental mutagens, in: "Chemical Mutagens: Principles and Methods for their Detection," A. Hollaender, and F.J. Serres, eds., Plenum Press, New York, Vol. 5, pp.203-232.

Caswell, H., 1982, Stable population structure and reproductive value for populations with complex life cycles, Ecology, 63:1223-1231.

Capuzzo, J.M., 1981, Predicting pollution effects in the marine environment, Oceanus, 24(1):25-33.

Capuzzo, J.M., and Kester, D.R., 1987, in: "Biological effects of waste disposal: experimental results and predictive assessment," J.M. Capuzzo and D.R. Kester, eds., Ocean Processes in Marine Pollution, Vol. 1, Biological Processes and Wastes in the Ocean, R.E. Krieger Publishing Co., Malabar, FL, pp. 1-15.

Capuzzo, J.M., Moore, M.N., and Widdows, J., 1988, Effects of toxic chemicals in the marine environment: predictions of impacts from laboratory studies, Aquat. Toxicol., 11:303-311.

Dixon, D.R., and Prosser, H., 1986, An investigation of the genotoxic effect of an organotin antifouling compound (bistributyl tin oxide) on the chromosomes of Mytilus edulis., Aquat. Toxicol., 8:185-195.

Gentile, J.H., and Scott, K.J., 1987, in: "The application of hazard assessment strategy to sediment testing: issues and case study," K.L. Dickson, A.W. Maki, and W.A. Grungs, eds., Fate and effects of sediment-bound chemicals in aquatic systems, Pergamon Press, New York, pp.167-182.

Gentile, J.H., Pesch, G.G., Lake, J., Yevich, P.P., Zaroogian, G., Rogerson, P., Paul, J., Galloway, W., Scott, K.J., Nelson, W., Johns, D.M., and Munns, W., 1988a, Synthesis of research results: applicability and field verification of predictive methodologies for aquatic dredged material disposal. Technical Report D-88-5, prepared by the U.S. EPA, Narragansett, RI, for the U.S. Army Engineer Waterways Experiment Station, Vicksburg, MS.

Gentile, J.H., Scott, K.J., Lussier, S., Redmond, M.S., 1988b, The assessment of Black Rock Harbor dredged material impacts on laboratory population responses. Technical Report D-87-2, prepared by the U.S. EPA, Environmental Research Laboratory, Narragansett, RI, for the U.S. Army Engineer Waterways Experiment Station, Vicksburg, MS.

Gentile, J.H., Pesch, G.G., Dillon, T., 1988c, in: "Urban Wastes in Coastal Marine Environments," The application of a hazard assessment research strategy to the ocean disposal of dredged material: an overview, D.A. Wolfe and T.P. O'Connor, eds., Oceanic Processes in Marine Pollution: Vol. 5, pp.115-122, R.E. Krieger Publishing Co., Malabar, FL.

Johns, D.M., Gutjahr-Gobell, R., Schauer, P., 1985, Use of bioenergetics to investigate the impact of dredged material on benthic species: a laboratory study with polychaetes and Black Rock Harbor dredged material,

Technical Report D-85-7, prepared by the U.S. EPA, Narragansett, RI, for the U.S. Army Engineer Waterways Experiment Station, Vicksburg, MS.

Johns, D.M., and Gutjahr-Gobell, R., 1988, Bioenergetic effects of Black Rock Harbor dredged material on the polychaete Nephtys incisa: a field verification, Technical Report D-88-3, prepared by the U.S. EPA, Narragansett, RI, for the U.S. Army Engineer Waterways Experiment Station, Vicksburg, MS.

Kooijman, S.A.L.M., Metz, J.A.J., 1984, On the dynamics of chemically stressed populations: the deduction of population consequences from effects on individuals, Ecotoxicol. and Environ. Saf. 8:254-274.

Lake, J., Galloway, W., Hoffman, G., Nelson, W., Scott, K.J., 1988, Comparison of field and laboratory bioaccumulation or organic and inorganic contaminants from Black Rock Harbor dredged material, Technical Report D-87-6, prepared by the U.S. EPA, Narragansett, RI, for the U.S. Army Engineer Waterways Experiment Station, Vicksburg, MS.

Latt, S.A., Allen, J., Bloom, S.E., Carrano, A., Falke, E., Kram, D., Schneider, E., Schreck, R., Tice, R., Whitfield, B., and Wolff, S., 1981, Sister chromatid exchange: a report of the gene-tox program, Mutat. Res. 87:17-62.

McCall, P.L., 1977, Community patterns and adaptive strategies of the infaunal benthos of Long Island Sound, J. Mar. Res., 35:221-226.

McIntyre, A.D., and Pearce, J.B., eds., 1980, Biological effects of marine pollution and problems of monitoring, Rapp. P. V. Reun. Cons. Int. Explor. Mer., Vol. 179, pp.346.

McMahon, G., Huber, L.J., Stegemen, J.J., and Wogan, G.N., 1988, Identification of a c-Ki-ras oncogene in a neoplasm isolated from winter flounder, Marine Environ. Res., 24:345-350.

National Academy of Sciences (NAS), 1971, Suggested research programs for understanding man's effects on the oceans, The effects on marine organisms, National Academy of Sciences, Washington, D.C., pp.63-81.

Nelson, W.G., Galloway, W., Phelps, D., 1987, Effects of Black Rock Harbor dredged material on the scope for growth of the blue mussel, Mytilus edulis, after laboratory and field exposures, Technical Report D-88-7, prepared by, Environmental Research Laboratory, U.S. EPA, Narragansett, RI, prepared for the Environmental Laboratory, U.S. Army Engineer Waterways Experiment Station, Vicksburg, MS.

Patton, J.S., and Couch, J.A., 1984, Can tissue anomalies that occur in marine fish implicate specific pollutant chemicals, H. White, ed., in: "Concepts in Marine Pollution Measurements," College Park, MD: Maryland Sea Grant Publication, pp. 511-538.

Peddicord, R.K., 1987, Overview of the influence of dredged material disposal on the fate and effects of sediment-associated chemicals, in: "Fate and effects of sediment-bound chemicals in aquatic systems," K.L. Dickson, A.W. Maki, and W.A. Brungs, eds., Pergamon Press, New York, pp. 167-182.

Perry, P., and Evans, J.H., 1975, Cytological detection of mutagen-carcinogen exposure by sister chromatid exchange, Nature, 258:121-125.

Pesch, G., Pesch, C.E., Malcolm, A.R., 1981, Neanthes arenaceodentata, a cytogenetic model for marine genetic toxicology, Aquat. Toxicol. 1:301-311.

Pesch, G., Mueller, C., Pesch, C., Rogerson, P.F., 1987, Sister chromatid exchange in marine polychaetes exposed to Black Rock Harbor sediment, Technical Report D-87-5, prepared by the U.S. EPA, Narragansett, RI, for the U.S. Army Engineer Waterways Experiment Station, Vicksburg, MS.

Rhoads, D.C., Allen, R.C., and Golhaber, M., 1977, The influence of colonizing benthos on physical properties and chemical diagenesis of the estuarine sea floor, in: "Ecology of the Marine Benthos," B.C. Coull, ed., Belle Baruch Library in Marine Sciences, University of South Carolina Press, Columbia, SC, pp.113-138.

Rogerson, P.F., Schimmel, S.C., and Hoffman, G., 1985, Chemical and biological characterization of Black Rock Harbor dredged material, Technical

Report D-85-9, prepared by the U.S. EPA, Narragansett, RI, for the U.S. Army Engineer Waterways Experiment Station, Vicksburg, MS.

Scott, K.J., and Redmond, M., 1989, The effects of a contaminated dredged material on laboratory populations of the tubicolous amphipod, Ampelisca abdita, Symposium on Aquatic Toxicology and Hazard Assessment, American Society of Testing Materials, In press.

Scott, K.J., Rhoads, D.C., Pratt, S., Rosen, J., and Gentile, J.H., 1987, Impact of open-water disposal of Black Rock Harbor dredged material on benthic recolonization at the FVP site, Technical Report D-87-4, U.S. EPA, Narragansett, RI, prepared for the U.S. Army Engineer Waterways Station, CE, Vicksburg, MS.

Sheehan, P.J., Miller, D.R., Butler, G.L., Bourdeau, P., and Ridgeway, J.M., eds., 1984, Effects of pollutants at the ecosystem level, SCOPE 22, Wiley Press, Chichester, England.

Sindermann, C.J., 1980, The use of pathological effects of pollutants in marine environmental monitoring programs, Rapp. P.-V. Reun. Cons. Int. Explor. Mer. 179:129-134.

Solomon, E., and Bobrow, M., 1975, Sister chromatid exchanges: a sensitive assay of agents damaging human chromosomes, Mutat. Res. 30:273-278.

Stegemen, J.J., 1987, Polynuclear aromatic hydrocarbons and their metabolism in the marine environment, in: "Polycyclic Hydrocarbons and Cancer, Vol. 3," H. V. Gelboin, and P.O.P. Ts'O eds., Academic Press, New York, p. 1-60.

Stegemen, J.J., and Kloepper-Sams, P.J., 1987, Cytochrome P-450 isozymes and monooxygenase activity in marine animals, Environ. Health Perspect. 71:87-95.

Underwood, A.J., and Peterson, C.H., 1988, Towards an ecological framework for investigating pollution, Mar. Ecol. Prog. Ser. 46:227-234.

U.S. EPA/Army Corps Engineers, 1977, "The Ecological Evaluation of Proposed Discharge of Dredged Material into Ocean Waters: Implementation Manual for Section 103 of PL 92-532," Engineer Waterways Experiment Station, Vicksburg, Mississippi.

Vetter, R.D., and Hodson, R.E., 1984, Metabolic indicators of sublethal stress: changes in adenine nucleotides, glycogen, and lipid. in: "Concepts in Marine Pollution Measurements," H. White, ed., Maryland Sea Grant Publication, College Park, MD, pp.471-498.

Widdows, J., 1982, Field measurement of the biological impacts of pollutants, in: "Proceeding of a Pacific Regional Workshop on the Assimilative Capacity of the Oceans for Man's Wastes," J.C. Su, and T.C. Jung, Scope/ICSU Academia Sinica, Taipei, Republic of China, pp.111-129.

Widdows, J., and Johnson, D., 1988, Physiological energetics of Mytilus edulis: scope for growth, Mar. Ecol. Prog. Ser., 46:113-121.

Yevich, P.P., Yevich, C., Scott, K.J., Redmond, M., Black, D., Schauer, P. and Pesch, C., 1986, Histopathological effects of Black Rock Harbor dredged material on marine organisms, Technical Report D-86-1, prepared by the U.S. EPA, Narragansett, RI, for the U.S. Army Engineer Waterways Experiment Station, Vicksburg, MS.

Yevich, P.P., Yevich, C., Pesch, G., and Nelson, W., 1988, Effects of Black Rock Harbor dredged material on the histopathology of the blue mussel Mytilus edulis, and polychaete worm Nephtys incisa after laboratory and field exposures, Technical Report D-88-8, prepared by the U.S. EPA, Environmental Research Laboratory, Narragansett, RI, for the U.S. Army Engineer Waterways Experiment Station, Vicksburg, MS.

Zajac, R.N., and Whitlach, R.B., 1989, Natural and disturbance induced demographic variation in an infaunal polychaete, Nephtys incisa, Mar. Ecol. Prog. Ser., In press.

Zaroogian, G.E., Rogerson, P.F., Hoffman, G., Johnson, M., Johns, D.M., and Nelson, W.G., 1988, A field and laboratory study using adenylate energy charge as an indicator of stress in Mytilus edulis and Nephtys incisa treated with dredged material, Technical Report D-88-4, prepared by the U.S. EPA, Narragansett, RI, for the U.S. Army Engineer Waterways Experiment Station, Vicksburg, MS.

USE OF HERRING EMBRYOS FOR IN SITU AND IN VITRO MONITORING OF MARINE

POLLUTION

Richard M. Kocan and Marsha L. Landolt

School of Fisheries
University of Washington
Seattle, Washington 98195

INTRODUCTION

Developing embryos are frequently very sensitive to the adverse effects of environmental contaminants (Rosenthal and Alderdice, 1976; McKim, 1985). Aquatic toxicologists have successfully employed sea urchin and oyster embryos in bioassays of pure chemicals and complex mixtures (Hose et al., 1983; Kobayashi, 1972; Roberts, 1980), and have used several species of fish embryos to study the carcinogenic and teratogenic potential of numerous compounds (Horning and Weber, 1985; Kocan and Landolt, 1984, 1987; Liguori and Landolt, 1985; Linden, 1978; Longwell and Hughes, 1980; Westernhagen et al., 1979). These systems have proven to be suitable laboratory models, but they have not been used for in situ monitoring. We will describe the utility of herring embryos for both in situ and in vitro testing of environmental contaminants.

Herring (Clupea harengus) are widely distributed, circumpolar marine organisms that serve as forage for many species of predatory birds and fishes. They spawn over a period of several months during late winter and early spring, producing hydrated eggs that measure 1.2 to 1.5 mm and that hatch in 10 to 14 days, depending upon temperature. The eggs are sticky and normally adhere to eelgrass, kelp, pilings and rocks in near-shore areas. This adhesive quality is especially useful for experimental purposes, because the eggs can be spawned directly onto an artificial substrate (e.g., a microscope slide), fertilized, and placed in the field for in situ monitoring, or manipulated in the laboratory for in vitro testing. If the eggs are attached to the substrate in regular arrays, it is possible to recognize individual organisms and to maintain accurate measurements of fertilization rate, embryo mortality, etc. Other attractive features of herrings eggs include the fact that they are extremely hardy and can be transported or handled without fear of causing damage to embryos, and that their size and transparency permit visual monitoring of events such as embryonic development, eye formation, heartbeat, and movement.

METHODS

Spawning Techniques

Fish collection. Sexually mature herring can be captured with drifting gill nets or purchased from commercial bait dealers and fishermen. If salt-

In Situ Evaluations of Biological Hazards of Environmental Pollutants
Edited by S. S. Sandhu *et al.*
Plenum Press, New York, 1990

49

water holding facilities are available, the fish should be kept alive until spawning. If such facilities are not available, the fish should be placed in a chilled container (avoiding direct contact with ice) and transported to the laboratory within two hours of capture. All of the following procedures should be carried out at 10-11°C, preferably in a cold room.

Spawning procedures. Ripe female fish are selected and the area near the genital pore is cleansed with an alcohol-soaked gauze pad. By gently squeezing the abdomen, eggs can be extruded onto the surface of a clean glass slide. The eggs should be deposited in a single plane, since three dimensional deposition results in developmental retardation for those embryos that are surrounded on two or three sides by other eggs, presumably due to competition for oxygen (Figure 1). If eggs are deposited too densely, they can be gently removed with fine tipped forceps to obtain the desired number and arrangement. The slide is then immersed in seawater of the required salinity and the eggs are fertilized with sperm obtained from a ripe male.

Sperm is collected in a manner similar to that described for eggs, except that the thick sticky milt is deposited onto a glass rod and then stirred into a beaker of seawater to obtain a homogeneous mixture consisting of approximately 1 mL of sperm per 500 mL of seawater. To ensure adequate fertilization, sperm should be collected from at least two males. The sperm suspension is poured over the eggs and allowed to remain in contact for 60 minutes. The slides are then washed with clean seawater to remove extraneous sperm and examined with a dissecting microscope to determine whether

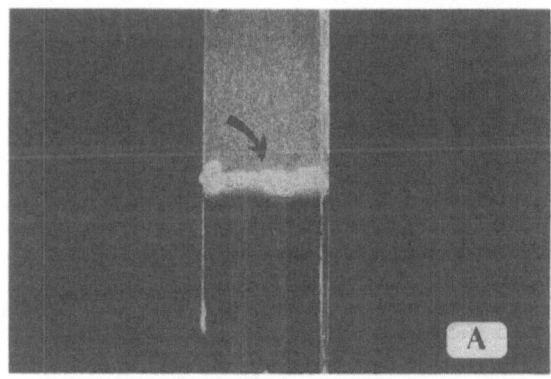

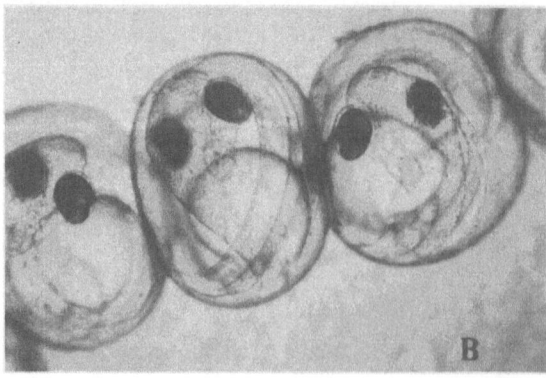

Figure 1. Herring eggs on glass slide. (A) Eggs aligned across slide for in vitro exposure at water surface. (B) Fully developed herring embryos visible through the chorion at 20X magnification.

fertilization has occurred. Fertilized eggs take up water (hydrate) and can be recognized by the presence of a perivitelline space between the yolk and the chorion. Unfertilized eggs lack such a space. After the fertilization rate has been recorded, the slides can be placed into incubation vessels for laboratory exposure or transported to field sites for in situ testing.

Embryo Rearing Techniques

In vitro testing. Figure 2a shows the type of incubation vessel we have successfully used for in vitro studies. The eggs are arranged horizontally across the slide and submerged in seawater so that they are located at the meniscus. This placement allows for direct exchange of oxygen from the air and eliminates the need for aeration during the early stages of development. Once eye formation occurs, it becomes necessary to exchange the air over the eggs at least three to four times per day. This is due because the metabolic rate increases dramatically during the last half of development, resulting in embryos producing large quantities of CO_2 which, being denser than air, layers over the eggs and causes suffocation. Oxygen exchange can be accomplished by attaching tubing to an aquarium pump and periodically passing a stream of air over the vials. Embryos can be reared from fertilization through hatching in this way, or can be removed at any time for other uses, such as in situ exposures.

In situ testing. For in situ exposures, slides containing eggs at the desired stage of development are removed from the fertilization or incubation vessels and placed in glass slide holders of the type used for staining blood films (Figure 2b). For our studies we normally place five slides slides (20 to 25 eggs/slide) into each holder. The number of fertile eggs per slide is recorded prior to transport to the field site so that accurate assessments of mortality and nonspecific loss can be determined. Embryos can be exposed in situ for any desired length of time, depending on the design of the experiment, and returned to the laboratory (prior to hatching) for final evaluation. To prevent predation by marine invertebrates and/or

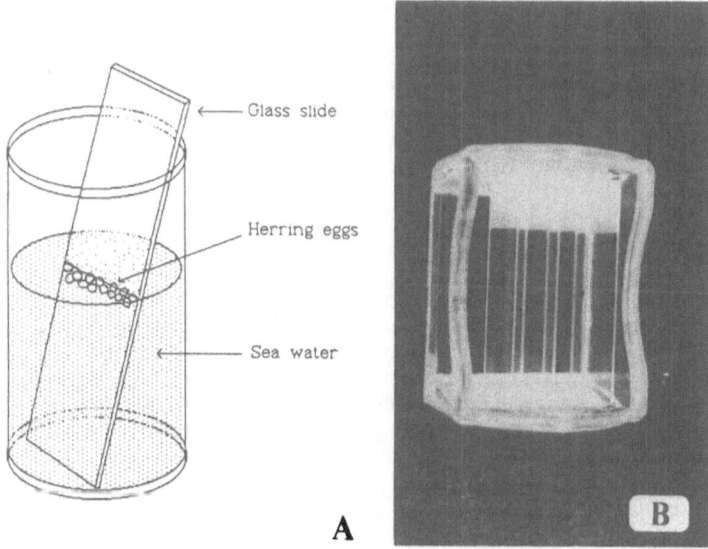

Figure 2. (A) Diagram of in vitro exposure vessel containing eggs on glass slide of 5 ml seawater, (B) Glass slide holder used for in situ exposure.

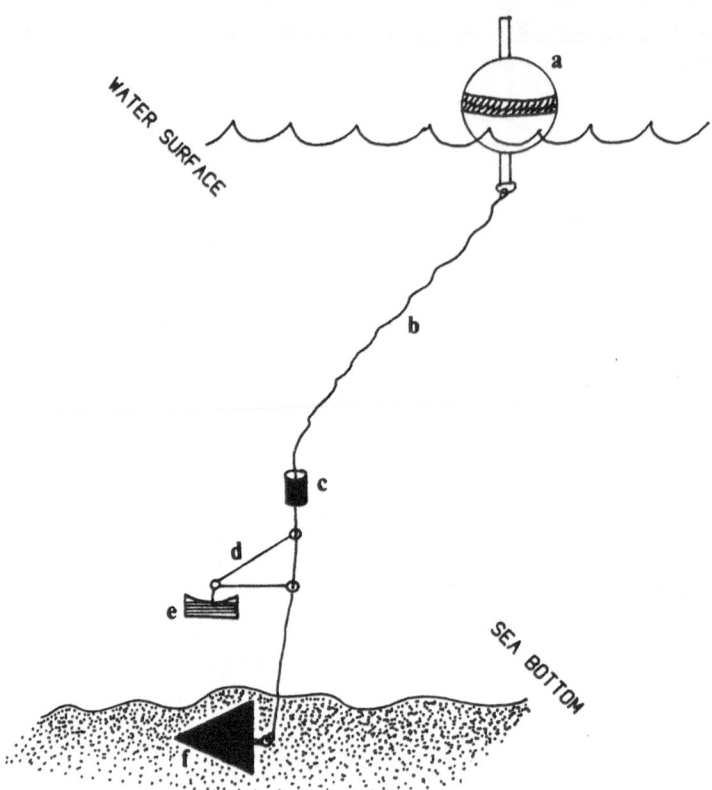

Figure 3. Diagram of mooring apparatus used to expose herring eggs in situ
           at a fixed distance from the sea bottom: (a) Mooring and re-
           trieval buoy; (b) mooring/retrieval line; (c) secondary buoy
           used to maintain eggs at constant distance from bottom; (d) wire
           support to hold glass slide holder in place; (e) glass slide
           holder with eggs; and (f) anchor.

small fish, the glass slide carrier can be wrapped with a fine mesh net
that allows the free transfer of water and particulates to the eggs. The
procedure allows one to place embryos at multiple depths at a single site,
at a single depth at multiple sites, etc. Figure 3 shows the design of
equipment we have used successfully in the field for multiple site
exposures.

Assessment Techniques

     Several types of observations and measurements can be made to assess
the response of embryos to toxin exposure. Embryo mortality can be documented throughout development through daily examination of the eggs. This
procedure will reveal not only how many embryos die, but also the stage of
development at which mortality occurs. Chromosome damage can be determined
by collecting embryos at various stages of development, preserving them in
methanol/acetic acid, making a squash preparation and staining the cells
with aceto-orcein dye (Longwell and Hughes, 1980). Figures 4a and b show
mitotic cells obtained from herring embryos at the blastodisc stage of development. Hatching success (total live hatch) can be determined by counting
the number of larvae that emerge alive. Similarly, teratogenesis rates can
be estimated by determining the percentage of live-hatched larvae that have

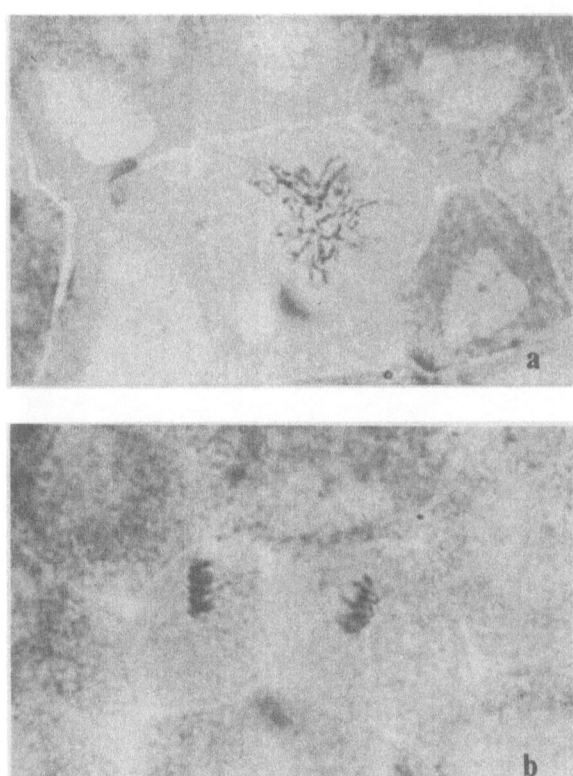

Figure 4. (A) Metaphase and (B) late anaphase/early telophase chromosomes in mitotic cells from herring embryo blastula.

grossly visible defects (e.g., spinal deformities) (Figure 5a,b). More subtle embryopathies can also be detected by special staining techniques or electron microscopy (Figure 6a,b). Modifications in <u>hatching time</u> (hatching dynamics) can be determined following toxin exposure. These often appear as extended time periods for complete hatching, delays in peak hatching time, early hatching or bimodal hatching. These types of modifications have been reported for several fish species following exposure to environmental toxins (Kocan and Landolt, 1984; 1987).

FIELD TEST

The in situ and in vitro rearing methods described above were field tested in 1987 at Port Gamble Bay (Puget Sound, Washington). This bay was selected for study because it is known to be a herring spawning area and because it is a site that frequently experiences unusually high embryo mortality rates.

Water samples were collected from three locations within Port Gamble Bay. These sites have been designated by the Washington State Department of Fisheries as Sites 25, 30 and 33 (Pentilla et al., 1985). At the time of our study, mortality rates among natural herring spawn were 30% (Site 33), 80% (Site 30) and 90-100% (Site 25) (Yake and Norton, 1987). The water samples were collected in glass containers, transported to the University of Washington School of Fisheries and stored at 9°C until used as incubation medium for newly fertilized herring eggs. A reference water

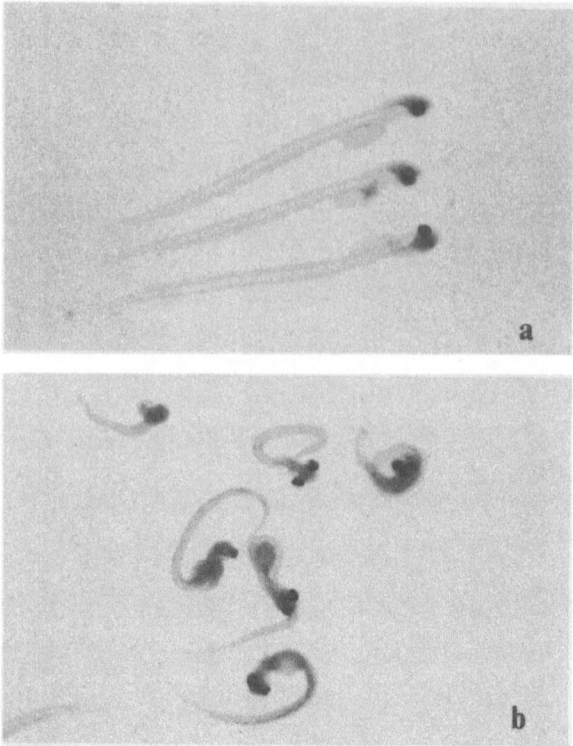

Figure 5. (A) Normal herring fry just hatched in 16 ppt sterile North Sea water; and (B) deformed herring fry reared in sterile seawater containing 200 ppt petroleum hydrocarbons extracted with hexane from North Sea Water.

sample was collected from a site (Bywater Bay) well outside Port Gamble Bay for comparison with the samples taken within the affected area.

Sediment samples (0.5 kg) were collected from each of the four sites using a Van Veen core sampler. Two hundred grams of each sample was extracted by vigorous shaking with 1 L of synthetic seawater (27 ppt). The extract was then sterilized by passage through a 0.45 um filter and stored at 9°C until used as an incubation medium for laboratory spawned herring embryos. Samples from sites 25 and 30 were pooled during collection and were subsequently treated as a single sample.

Sexually mature herring, obtained from a commercial bait dealer, were used as a source of eggs and sperm. Slides were prepared in the manner described above. Eggs were examined every other day to determine mortality rates and hatching success. Live hatched larvae were collected daily, anesthetized in methane tricaine sulfonate (MS 222), preserved in 10% neutral buffered formalin and examined microscopically for the presence of physical defects--primarily defects affecting spine, eye and skull development (Rosenthal and Alderdice, 1976) (Figures 5, 6).

Field exposures. From March 27-March 31 (Exposure I) and from March 31-April 3 (Exposure II) a minimum of 100 newly fertilized eggs were suspended 2 to 3 feet from the bottom at each site. This was accomplished by placing five slides into glass slide carriers, covering them with fine mesh screen, and attaching the carriers to a mooring line that bore an anchor at

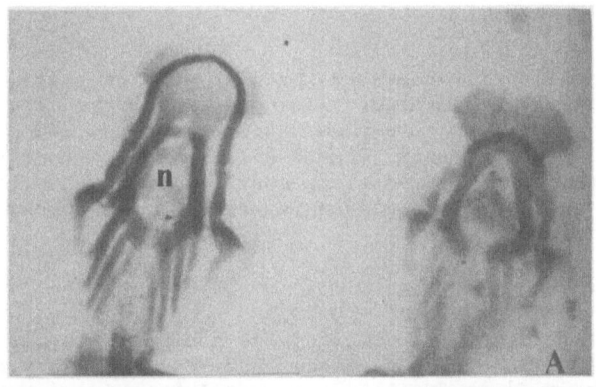

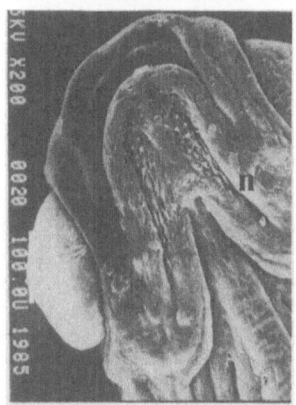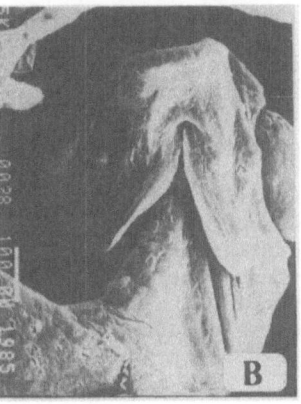

Figure 6. Normal and abnormal herring fry following exposure to 200 ppm
petroleum hydrocarbons extracted from North Sea water with
hexane. (A) Cartilage development in normal (n) and abnormal
fry; alcian blue stain. (B) Normal (n) and abnormal jaw and
gill development from same group of fry; scanning EM.

one end and a retrieval buoy at the other (Figure 3). Eggs that were maintained in synthetic seawater served as a laboratory control that could be
compared with field and reference samples.

Following the exposure period, the slides were retrieved and returned
to the laboratory for evaluation of survival rates. This was done by examining each slide under 60 x magnification and calculating embryo mortality
(%). Eggs from Exposure II were maintained for an additional period of
time in the laboratory and then evaluated for hatching success and developmental defects.

Laboratory exposures to water and sediment extract. Five slides, each
containing 20 to 25 newly fertilized eggs, were placed into 5 mL of seawater sample or sediment extract for incubation (100 + eggs/exposure). The
eggs were observed microscopically prior to hatching to determine the mortality rate, and the hatched larvae were counted for calculations of hatching success. Following hatching, the larvae were also examined for the
presence of physical defects.

RESULTS OF FIELD TEST

Field Exposures

In Exposure I, significantly elevated embryo mortality was recorded
from Sites 25, 30 and 33 immediately after retrieval from the field. In

Exposure II, significantly elevated mortality at the time of retrieval was
recorded only from Site 33 (Figure 7); however, embryos from all three
sites ultimately experienced reduced hatching success and elevated deformity rates relative to controls (Figure 8). There was no statistical difference between the laboratory and field controls, even though there was a
slight decrease in live hatch and an increase in the number of abnormal
larvae in the field controls.

Laboratory Exposures to Water

Water from all three Port Gamble sites and from the control site produced no difference in hatching success (93-98%) for embryos exposed continuously from 24 hours post fertilization until hatching. Water from Sites
30 and 33 did, however, produce a significant increase in larval abnormalities (63% and 42%, respectively) compared to Site 25 and the field control
(15% and 10%, respectively). Figure 9a summarizes the results of the laboratory exposures to water.

Laboratory Exposures to Sediment Extract

Seawater extracts of sediment from Sites 25/30, Site 33 and the field
control site produced no difference in embryo survival (95-99%) following
continuous exposure form 24 hours post fertilization until hatching. Sediment extracts from Sites 25/30 and Site 33 did, however, produce a significant (P<.01) increase in abnormality rates (26% and 22%, respectively) as
compared to the field control (3%). Figure 9b summarizes the results of
the laboratory exposures to sediment extract.

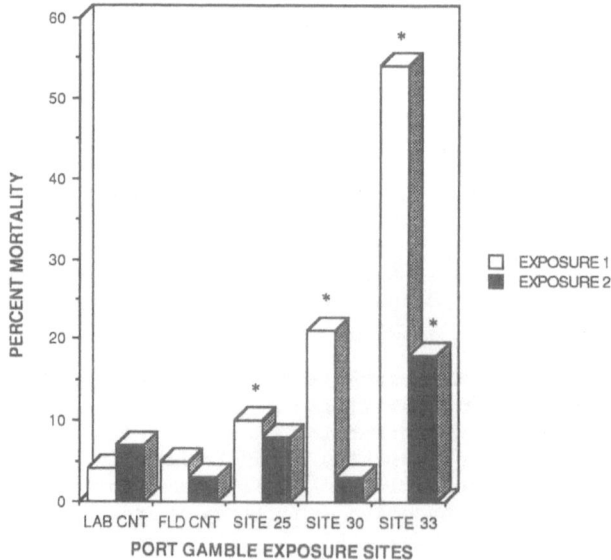

Figure 7. Mortality of field-exposed embryos 24 hours after return to the
laboratory. Exposure I (96 hr) resulted in significantly elevated mortality rates from three sites. Exposure II (72 hr) resulted in similar mortalities at only one site, while filed and
laboratory controls were not significantly different from each
other. (* = P<0.01; X^2.) Field exposures were in Port Gamble
Bay, Washington; the field control site was Bywater Bay,
Washington.

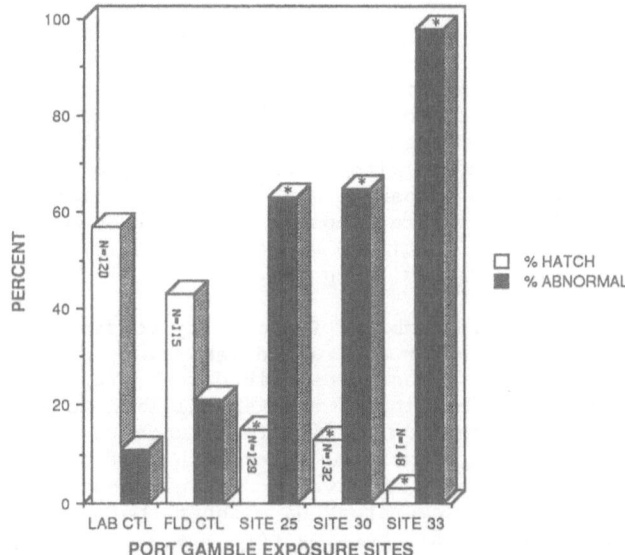

Figure 8. Percent of hatching success and larval abnormalities in field-exposed herring embryos. Hatching success (% based on total number of viable eggs at end of exposure) was significantly reduced at three sites. The frequency of abnormal live larvae was also increased at these same sites. (* = P<0.01; X^2).

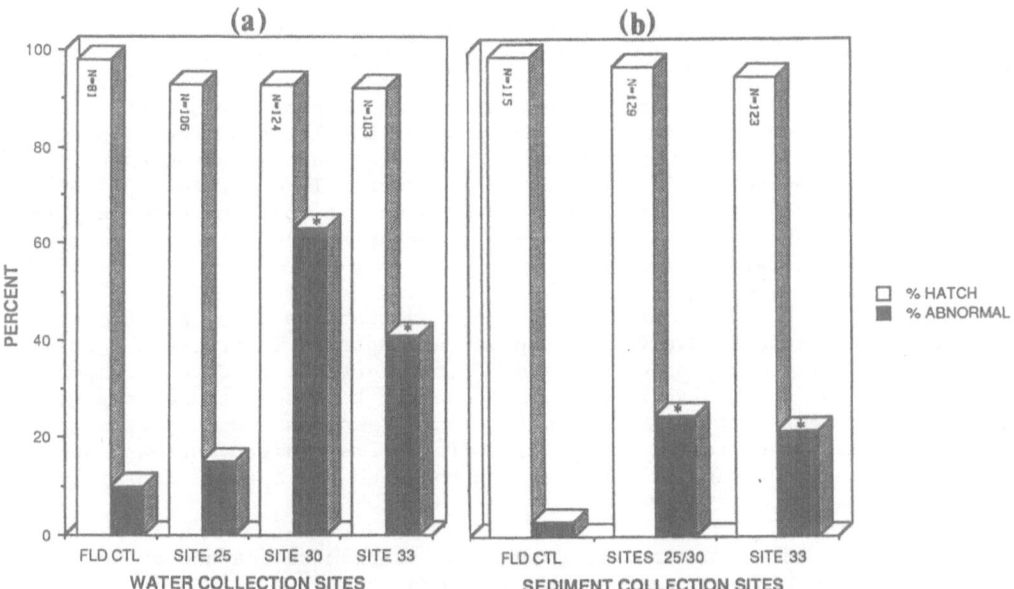

Figure 9. (a) Percentage of hatching success and larval abnormalities in herring embryos exposed in the laboratory to water from three Port Gamble sites as compared to a control site. Physical defects increased significantly in live larvae that hatched from eggs incubated in water from two of the sites. (b) Percentage of hatching success and larval abnormalities in herring embryos exposed in the laboratory to seawater extract (27 ppt) of sediment collected from Port Gamble Bay and a control site. Exposure resulted in a significant increase in the number of physically defective larvae as compared to field control extracts. (* = P<0.01; X^2).

DISCUSSION

Herring eggs are well suited for use as in situ monitors of environmental contamination. Because the eggs are adherent, they can be placed onto artificial substrates and transported between the laboratory and the field. The eggs can also be handled without causing adverse effects to the developing embryos, so comparisons can be made between laboratory exposures, field exposures, multiple field site exposures or other combinations. Because individual embryos can be identified and observed daily throughout development, it is possible to study the pathogenesis of lesion development.

The utility of herring embryos in toxicity studies depends upon their being carefully spawned under controlled laboratory conditions. By spawning the eggs onto glass slides as described here, it is possible to control the number, density and spacing of eggs in relation to each other, thus allowing for direct comparison of different exposure groups. Previous studies have shown that the spatial relationship of herring eggs to each other influences prehatching survival and developmental rate. Toxicity studies conducted in our laboratory have shown that data cannot be compared to exposure groups unless all herring eggs are exposed to identical physical conditions. These conditions include water/egg ratio, water depth, number of eggs in contact with each other, and air circulation at the water surface. For our studies we normally place 20-25 eggs on each slide and incubate the slide in 5 mL seawater. The eggs are arrayed in two closely spaced rows and positioned at the air/water interface (meniscus). This protocol allows a maximum number of eggs to be exposed in a minimum volume of water without affecting normal development.

The data obtained from the field test demonstrate that it is possible to detect biological activity arising from the presence of water soluble or particulate born toxicants. Both our in situ studies and our in vitro studies yielded patterns of embryo mortality and teratogenesis that were similar to those observed in naturally spawning populations (Pentilla et al., 1985). Unfortunately, mortality and skeletal deformity represent nonspecific responses to physical or chemical insult. In order to implicate specific substances as the cause of such changes, it would be necessary to conduct more extensive chemical analyses of water and sediment and to conduct targeted laboratory studies.

The ability to conduct on-site toxicity evaluations using a resident species offers obvious advantages when one attempts to interpret the results of toxicological studies. One can circumvent the need to extrapolate results from one species to another, and one can test only those components of the environment that naturally contact the organisms. On the other hand, use of undomesticated species may offer certain disadvantages such as limited seasonal availability of eggs and the need to capture free-living adult spawners.

The purpose of the present study was not to promote the use of herring eggs per se, but rather to encourage the use of "sticky" or adherent fish eggs of any species. Relatively little information has been published regarding piscine gametes. Out of 20,000 to 25,000 species of fish known to exist today, the eggs of only 746 species have been described (Breder and Rosen, 1966). Based on these descriptions most freshwater fish species produce adherent eggs that could be used for aquatic toxicity testing. In contrast, relatively few saltwater species that produce adherent eggs. This paucity should not pose an insurmountable obstacle to marine toxicologists, since suitable candidates are known to exist within major saltwater taxa (e.g., herring, sardines, anchovies, smelt, scuplins, pochards, gunnels), and research might reveal the existence of others. Some of the species with adherent eggs are relatively easy to rear in captivity; others have not been extensively studied. It has been our experience that numerous intertidal species adapt readily to laboratory conditions and will spawn in captivity (e.g., sculpins and gunnels).

In summary, herring embryos can be used as sensitive detectors of toxic substances in the marine environment. They can be used in the laboratory as well as in the field, provided that they are spawned in such a way as to ensure uniformity in egg density, spatial arrangement and egg/water ratio. The major drawback to their use is their limited availability. The proven utility of herring embryos suggests that attempts should be made to identify other species of fish that produce adherent eggs and that adapt to rearing under laboratory conditions. By using eggs that can be manipulated in the laboratory or transferred to field sites, one can study environmental problems under both in situ and in vitro conditions.

ACKNOWLEDGMENTS

This work was supported in part by grants from the U.S. Environmental Protection Agency (R811348) and the National Institutes of Environmental Health Sciences (ES 02190 and ES 04696). Contribution No. 802, School of Fisheries University of Washington, Seattle, WA 98195.

REFERENCES

Breder, C.M., Rosen, D.E., 1966, Modes of reproduction in fishes, The American Museum of Natural History, The Natural History Press, Garden City, N.Y. p.941.

Horning, W.B., and Weber, C.I., 1985, Short-term methods for estimating the chronic toxicity of effluents and receiving waters to freshwater organisms, U.S. EPA Publication EPA/600/4-85/014, Office of Research and Development, Cincinnati, OH.

Hose, J.E., Puffer, H.W., Oshida, P.S., and Bay, S.M., 1983, Development and cytogenetic abnormalities induced in the purple sea urchin by environmental levels of benzo(a)pyrene, Arch. Environ. Contam. Toxicol. 12:319-325.

Kobayashi, N., 1972, Marine pollution bioassay using sea urchin eggs in the Inland Sea of Japan (The Seto-Naikai), Publ. Seto Mar. Biol. Lab. 19:359-363.

Kocan, R.M., and Landolt, M.L., 1984, Alterations in patterns of excretion and other metabolic functions in developing fish embryos exposed to benzo(a)pyrene, Helgol. Wiss. Meeresunters, 37:493-504.

Kocan, R.M., and Landolt, M.L., 1987, Toxicity of sea-surface microlayer: Effects of hexane extract on Baltic herring (Culpea harengus) and Atlantic cod (Gadus morhua) embryos, Mar. Environ. Res. 23:291-305.

Liguori, V.M., Landolt, M.L., 1985, Anaphase aberrations: An in vivo measure of genotoxicity, in: "Short-Term Bioassays in the Analysis of Complex Environmental Mixtures IV," by M.D. Waters, S.S. Sandhu, J. Lewtas, L. Claxton, G. Strauss and S. Nesnow eds., New York:Plenum Press, pp.87-98.

Linden, O., 1978, Biological effects of oil on early development of the Baltic herring, Clupea harengus membras, Mar. Biol. 45:273-283.

Longwell, A.C., and Hughes, J.B., 1980, Cytologic, cytogenetic and developmental state of Atlantic mackerel eggs from sea surface waters of the New York Bight, and prospects for biological effects monitoring with ichthyoplankton, Rapp. P., v. Reun. Cons. Int. Explor. Mer. 179: 275-291.

McKim, J.M., 1985, Early life stage toxicity tests, in: "Fundamentals of Aquatic Toxicology, Methods and Applications," G.M. Rand and S.R. Petrocelli, eds., New York:Hemisphere Publishing Corp., pp.58-95.

Pentilla, D., Burton, S., and Gonyea, G., 1985, Summary of 1985 Herring Recruitment Studies in Puget Sound, Washington Dept. of Fisheries Progress Report No. 223, Olympia, WA.

Roberts, M.H., Jr., 1980, Flow-through toxicity testing system for molluscan larvae as applied to halogen toxicity in estuarine water, in: "Aquatic Invertebrate Bioassays," A.L. Buikema, Jr., and J. Cairns, Jr., eds., ASTM STP 715, American Society for Testing and Materials, Philadelphia, pp.131-139.

Rosenthal, H., and Alderdice, D.F., 1986, Sublethal effects of environmental stressors, natural and pollutional, on marine fish eggs and larvae, J. Fish. Res. Board Can. 33:2047-2065.

Westernhagen, H., Dethlefsen, V., Rosenthal, H., 1979, Combined effects of cadmium, copper and lead on developing herring eggs and larvae, Helgol. Wiss. Meeresunters, 32:257-278.

Yake, W., and Norton, D., 1987, Port Gamble Bay: A Reconnaissance Survey of Sediment Quality, Washington Dept. Ecology, Segment NO. 25-00-03, 07-15-03, Olympia, WA, p.17.

NEW METHODS FOR ON-SITE BIOLOGICAL MONITORING OF EFFLUENT WATER QUALITY

Henry S. Gardner[1], William H. van der Schalie[1],
Marilyn J. Wolfe[2], and Robert A. Finch[1],

[1]U.S. Army Biomedical Research and Development Laboratory
Frederick, MD, and [2]Experimental Pathology Laboratories
Inc., Herndon, VA.

INTRODUCTION

The presence of contaminated water and wastewater at industrial and
hazardous waste sites is a costly problem for the military. The Department
of Defense (DOD) operates 500 to 700 domestic wastewater treatment plants
and 100 to 200 industrial wastewater treatment plants in the continental
United States (Anon, 1985). In addition, DOD has 400-800 waste disposal
sites requiring remedial actions that are conservatively estimated to cost
$5-10 billion over the next 5-10 years. Many of these sites contain com-
plex mixtures of chemicals and an estimated 35% contain military-unique
materials (Naugle, 1986).

The actual cost of treatment required for contaminated water or wastewater depends in part upon the degree of clean-up required, and this in
turn is based upon toxicity estimates. Toxic hazard is usually determined
by comparing the concentrations of individual chemical constituents in
water or wastewater with available data on their toxicity to mammals. Unfortunately, there may be little or no toxicity data on many of the compounds present in the contaminated water, and even when toxicity data are
available, their application to a specific contaminated water may be difficult because of interactions between chemicals present in complex mixtures
and large variations in the concentrations of chemicals over time. When
faced with such uncertainty, target concentrations for clean-up usually are
set using generous (and very costly) safety factors.

One way to reduce high costs associated with treatment of contaminated
aquatic media would be to obtain direct, on-site measurements of toxicity
using traditional mammalian test species. This approach is technically
impractical and would be very costly if used. The U.S. Army Biomedical
Research and Development Laboratory (BRDL) has, for several years, been
working on an alternative approach: the development of fast, relatively
inexpensive, non-mammalian toxicity assessment techniques that can be used
not only in the laboratory but also at field sites having potentially contaminated water or wastewater. There is certainly precedent for using
biological endpoints for evaluating (and regulating) toxicity. Aquatic
toxicity tests of wastewater effluents are an integral part of the Federal
Clean Water Act of 1987.

This paper describes two of the non-mammalian toxicity testing techniques that have been utilized at the BRDL for on-site toxicity assessment

In Situ Evaluations of Biological Hazards of Environmental Pollutants
Edited by S. S. Sandhu *et al.*
Plenum Press, New York, 1990

61

in aqueous media, and discusses the installation and preliminary testing of these techniques in a mobile biomonitoring facility designed for on-site toxicity evaluations at Army sites. Both test methods use fish: one uses the computer-monitored fish ventilatory patterns to check water or wastewater continuously for developing acutely toxic conditions, while the other method uses fish to test for the presence of carcinogenic substances in the water. Both methods have been installed in a mobile biomonitoring facility to facilitate on-site toxicity determinations.

The advantages of using the computer-monitored responses of aquatic organisms to continuously monitor for acute toxicity have been previously described (Cairns and van der Schalie, 1982) and the uses of such systems in field applications have been detailed (van der Schalie, 1986). Automated biomonitoring systems can be useful to the Army both for rapid detection of accidentally or purposefully introduced toxicants in drinking water supplies and for monitoring toxicant spills at Army wastewater treatment facilities. The BRDL biomonitor detects developing toxic conditions by monitoring changes in the ventilatory and movement patterns of fish. Fish are held individually in small chambers through which test water flows continuously. As each fish passes water over its gills, it generates an electrical signal in the water that is picked up by a pair of electrodes in the test chamber and sent to a microcomputer for analysis. While most other biomonitoring systems monitor only one parameter, the BRDL biomonitoring system monitors three ventilatory parameters as well as body movement. Use of multiple parameters improves the speed and sensitivity of the system response, since the relative responses of the individual parameters to toxicants can vary considerably. In this study, we evaluated the response of the automated field biomonitoring system to a simulated spill of the munitions-related material, 1,3,5-trinitrobenzene (TNB).

Fish can respond not only to acutely toxic conditions in the water but also to the presence of chemical carcinogens. Fish have been shown to develop neoplasms in response to many known mammalian carcinogens (Hoover, 1984; Couch and Harshbarger, 1985). There are many advantages to using fish for carcinogenicity assessments (Black, 1984; Masahito et al., 1988; Hawkins et al., 1988). Fish can provide direct, continuous assessment of the effects of complex, varying chemical mixtures. When small fish species are used, large numbers of individuals can be exposed in a relatively small space. Finally, carcinogenicity tests using fish are quick and inexpensive relative to tests with mammals and, when it comes to aqueous media, fish can be used for monitoring in ways that cannot be duplicated using mammals. One approach, used by Grizzle and co-workers (1984), is to monitor caged fish placed directly into a wastewater pond. At the BRDL we have developed a bioassay system incorporated into the biomonitoring facility that allows us to expose a sensitive fish species (the medaka, Oryzias latipes) to water or wastewater at a field site. In the studies reported here, we continuously exposed medaka in the biomonitoring facility to effluent at a wastewater treatment plant and checked for the presence of tumors after exposure.

METHODS AND MATERIALS

Biomonitoring Facility

All toxicity tests were done in a 24-foot biomonitoring trailer that was situated at a 1.5 million gallon per day trickling filter-type sewage treatment plant located on the Monocacy River in Frederick County, Maryland. Incoming wastewater originated from both residential and laboratory facilities. Effluent wastewater from the plant was chlorinated, then dechlorinated prior to discharge into the river. Submersible pumps were used to deliver both Monocacy River water (upstream and downstream of the effluent discharge point) and pure effluent (after dechlorination) to the trailer. River water was filtered to remove sediment particles larger than 80 microns to prevent excessive fouling of test equipment. Solenoid-type proportional diluters were used to deliver different concentrations of test solutions using river water as a diluent for both the acute toxicity and carcinogenicity studies. Effluent from the TNB tests was filtered through activated carbon to remove any residual TNB prior to discharge.

Standard Toxicity Assessments

Wastewater from the sewage treatment plant was tested using standard mutagenicity and aquatic toxicity tests to provide reference points for the new toxicity assessment techniques. Composite samples (24 h) of effluent were evaluated for mutagenic potential using the Ames bacterial mutagenic- ity assay (Jagannath, 1987a,b). Tests were done using both direct effluent samples as well as XAD-2 resin extracts of effluent samples. The proce- dures followed those recommended by the New Jersey Department of Environ- mental Protection for effluent water samples. Two strains of Salmonella typhimurium were used (TA-98 and TA-100), and the assays were done both with and without metabolic activation. Seven-day fathead minnow and Ceriodaphnia tests were also conducted to assess aquatic toxicity (Goodfellow, 1987). Methods were similar to those described by Horning and Weber (1985). A TNB flow-through acute toxicity test with bluegills was conducted to facilitate comparisons between the results of the field venti- latory study with those from previously conducted laboratory tests. Meth- ods followed those reported by van der Schalie et al. (1988) except that the acute field test was conducted at the ambient river water temperature (6°C), not the 22°C used in the laboratory study.

Acute Toxicity Monitoring System

The acute toxicity monitoring system was situated behind a partition in the rear of the biomonitoring facility to minimize disturbance of the fish. A total of 26 bluegill (Lepomis macrochirus) were held in individual chambers through which test solution flowed continuously. Four groups of six or seven fish each were first held in Monocacy River water for four days to establish normal baseline levels for the ventilatory and movement parameters, then one group each was exposed for six days to mean measured TNB concentrations of < 0.025 (Monocacy River water), 0.052, 0.173, or 0.564 mg/L. Electrical signals generated by ventilatory or movement activ- ity were detected by remote electrodes in each chamber, amplified and filtered to remove 60 cycle electrical noise, and sent to a small computer for analysis of ventilation rate, average depth of ventilation, gill purge (cough) rate, and body movement rate. Data from each fish were printed out every 15 minutes. The computer also gathered the data from sensors that monitored the pH, dissolved oxygen concentration, temperature, conductiv- ity, and turbidity of the river water every 30 minutes. Methods for venti- latory data collection and analysis as well as TNB analysis have been previously described (van der Schalie et al., 1988). Changes in the maxi- mum and minimum ventilatory and movement parameters from the pre-exposure to the TNB exposure periods were calculated for each group of fish. Comparisons were made to determine which of the TNB-exposed groups were significantly different from the controls.

Carcinogenicity Assessment System

Medaka used in the carcinogenicity studies were obtained from labora- tory cultures. Newly hatched medaka were raised for 14 days in laboratory well water, then divided into two groups. One group was exposed in the laboratory to 10 mg/L of a routinely used mammalian initiating chemical (diethylnitrosamine or DEN) for 48 hours, while the other group was kept in well water. Unexposed fry were used to monitor water for the presence of complete carcinogens. DEN-exposed fish were used to assay for the presence of carcinogen promoters; substances that encourage the growth and development of cells transformed by the initiating agent but that do not, by themselves, cause neoplasms (the 10 mg/L-DEN exposure is routinely insufficient, by itself, to result in the formation of liver neoplasms in the medaka.) One set of 50 unexposed and one set of 50 DEN-exposed medaka were placed in tanks of clean laboratory well water, and similar sets of tanks were placed in the biomonitoring facility that received wastewater treatment plant effluent. Approximately five tank volumes per day of test solution were provided to each tank. Medaka at the biomonitoring facility were exposed to the effluent for three months, at which time half the surviving animals in each tank were sacrificed and fixed in Bouin's solutions. Five step sections were cut in duplicate on a sagittal plane through each fish. These steps included two right paramedian sections, one midsagittal section and two left paramedian sections. A total of ten sections for each fish were stained with hematoxylin and eosin and evaluated for lesions. Thirty tissues were accounted for in each fish. The remaining animals were held in the laboratory for an additional three months. These remaining animals were then sacrificed and processed in a manner identical to the earlier sacrifice.

RESULTS AND DISCUSSION

Acute Toxicity Monitoring System

The 96-hour median lethal concentration (LC50) of TNB to bluegills in river water at the test temperature of the ventilatory study (6°C) was 1.27 mg/L, which is about twice the 96-hour LC50 value of 0.57 mg/L reported from a laboratory test with bluegills and TNB conducted at 22°C (van der Schalie et al., 1988). A similar temperature-toxicity trend was reported by Liu et al. (1984), who found that the toxicity of another nitroaromatic chemical, 2,4-dinitrotoluene, dropped by a factor of three as temperature decreased from 24.5 to 17.3°C.

Ventilatory responses of bluegills to TNB are shown in Table 1. As evidenced by the negative values, maxima for all parameters showed decreases in the controls from the pre-exposure to the exposure periods. Maximum levels of ventilatory rate, ventilatory depth, and cough rate increased significantly during exposure to 0.564 mg/L, although none of these responses was evident when data analysis was restricted to the first four hours after the start of exposure. Changes in ventilatory depth maxima were most sensitive to TNB, with effects occurring as low as 0.172 mg/L, but no significant effects on movement were found.

As with the parameter maxima, minima for all parameters (except movement) also showed decreased values during the exposure period (Table 1). The movement values are all zero since there were 15-minute periods where no movement was observed during both the pre-exposure and exposure periods. Significant effects were noted at 0.564 mg/L TNB for ventilatory rate and at both 0.564 and 0.172 mg/L for cough rates. The changes in both ventilatory and cough rates were rapid, and effects were noted within the first four hours after the initiation of TNB exposure at 0.564 mg/L. Strong increases in ventilatory depth minima and movement maxima at the top TNB concentration were not statistically significant due to high variability in the responses of the exposed fish.

In summary, TNB caused changes in bluegill ventilatory patterns at concentrations as low as 0.172 mg/L (about 14% of the 96-hour LC50). Significant changes in ventilatory and cough rates occurred within the

64

Table 1. Changes in Ventilatory and Movement Response Over 6 days Exposure to TNB

Parameter	Concentration of TNB (mg/L)	Change in: Maximum Values[a]	Change in: Minimum Values[a]
Ventilatory Rate (per min)	< 0.025	-2.78	-1.88
	0.053	-2.66	-2.33
	0.172	-6.95	-2.51
	0.564	27.0[b]	6.50[b,c]
Ventilatory Depth (volts)	< 0.025	-44.1	-25.6
	0.053	-30.1	-50.6
	0.172	-15.5[b]	-13.8
	0.564	415.3[b]	95.6
Cough Rate (per min)	< 0.025	-0.09	-0.02
	0.053	-0.03	-0.05
	0.172	-0.50	0.05[b]
	0.564	5.15[b]	0.68[b,c]
Movement (%)	< 0.025	-0.02	0.00
	0.053	-0.13	0.00
	0.172	1.72	0.00
	0.564	8.23	0.00

[a]Mean difference between the parameter maxima (or minima) from the pre-exposure to the exposure periods.
[b]Significantly different from the controls (p < 0.05).
[c]Significant effects occurred within the first four hours of exposure.

first four hours of exposure to 0.564 mg/L. While dramatic changes in the ventilatory signals from these fish could be observed visually in much less than four hours, the statistical method used in this study employed 4-hour moving average technique that prevented analysis of any time interval prior to the first four hours of exposure. A new on-line statistical procedure now being added to the ventilatory monitoring system will obviate this shortcoming and will allow much more rapid detection of developing toxic conditions.

The data from this field exposure to TNB at 6°C may be compared to the results of a similar ventilatory test with bluegills done under laboratory conditions at 22°C (van der Schalie et al., 1988). Besides the difference in temperature, the field test had more fish per concentration (six or seven) than the laboratory test (five; three at the top concentration). Although TNB was less acutely toxic in the colder water of the field study (the 96-hour LC50 was 1.27 mg/L versus 0.57 mg/L at 22°C) the sensitivity of the biomonitoring system to TNB was not appreciably changed. The lowest concentrations of TNB causing changes in the ventilation or movement responses were 0.17 and 0.13 mg/L in the field and laboratory tests, respectively. Rapid biomonitor responses to toxicant concentrations that approach acutely toxic levels were found in both laboratory and field tests.

Carcinogenicity System

Chemical analysis of a grab sample of the effluent tested in the long-term fish carcinogenicity study demonstrated the presence of low-level chlorinated organics consistent with the chlorination treatment of the effluent. Chloroform (10 ug/L) and bromodichloromethane (9.5 ug/L) were the only volatile organics present above the detection limit (Cudnason et al., 1987). No acid/base/neutral extractables, metals, or pesticides were found in this sample. By traditional measures of toxicity, the sewage treatment plant effluent was nontoxic. Standard 7-day toxicity tests with fathead minnows and Ceriodaphnia showed no toxic effects at concentrations of up to 100% effluent (Goodfellow, 1987).

Direct Ames mutagenicity assays of a composite effluent sample showed no mutagenic activity in either TA98 or TA100 Salmonella strain irrespective of S-9 metabolic activation of the samples (Jagannath, 1987a). On the other hand, a XAD-2 resin extract of the effluent was mutagenic, in a dose-dependent fashion, to Salmonella strain TA-98 after metabolic activation, but was not mutagenic to strain TA100 (Jagannath, 1987b). Without metabolic activation, strain TA98 had only a very weak response that was insufficient to conclude that positive mutagenic activity was shown. TA100 results were negative.

Exposure of fish to various concentrations of effluent and/or Monocacy River water pumped from upstream and downstream of effluent the discharge point for 13 weeks resulted in degenerative and/or neoplastic changes in the liver of both male and female fish (see Table 2 for neoplastic summary). Hepatocellular adenomas and one hepatocellular carcinoma were identified only in fish exposed to effluent and/or river water and among both fish pretreated and not pretreated with DEN. Cystic degeneration of the liver occurred in males and females among treated and control fish, but the lesions were, in general, more severe in the treated fish. Other non-neoplastic lesions occurred randomly in various organs and tissues and were considered to be incidental and unrelated to treatment. The significance of a thymic neoplasm of non-lymphoid origin in one fish pretreated with DEN and exposed to 10% effluent in upstream water is not known.

Fish exposed to various concentrations of effluent and/or water from various segments of the Monocacy River for 13 weeks and then kept in laboratory well water for an additional 13 weeks had treatment-related changes in the liver (see Table 3 for neoplasia summary). Cholangiocarcinomas and one hepatocellular carcinoma were identified only in fish exposed to effluent and/or river water and not among fish pretreated with DEN. Cystic degeneration of the liver occurred in males and females among treated and control fish, but the incidence of the lesion was greater in the treated groups than in the control groups. Other non-neoplastic lesions occurred randomly in various organs and tissues and were considered to be incidental and unrelated to treatment. The significance of non-hepatic neoplasms is not known. The significance of intestinal carcinoma in two treated fish and

Table 2. Pathology Incidence (# Animals) for 13 Weeks (Wolfe, 1988)

Treatment	DEN	#	FCA+	Adenoma	Carcinoma	Non-Hepatic Neoplasm
Control	+	24	-	-	-	-
Lab Water	-	20	-	-	-	-
Upstream	+	9	-	-	-	-
Of Discharge	-	20	2	1	-	-
Downstream	+	24	-	-	-	-
Of Discharge	-	23	-	-	-	-
1% Effluent	+	22	-	-	-	-
	-	15	-	-	-	-
10% Effluent	+	22	-	-	-	1*
	-	16	-	-	-	-
100% Effluent	+	23	2	1	1	-
	-	15	-	1	-	-

+Focus of cellular alteration
*Thymoma of reticular cells

Table 3.   Pathology Incidence (# Animals) for 26 Weeks (Wolfe, 1988)

Treatment	DEN	#	FCA+	Adenoma	Carcinoma	Non-Hepatic Neoplasm
Control	+	21	4	-	-	-
	-	20	1	-	-	1++
Upstream	+	2	-	-	-	-
	-	25	-	-	3*	-
Downstream	+	22	-	-	-	-
	-	13	-	-	-	-
1% Effluent	+	2	-	-	-	-
	-	15	-	-	1	1++
10% Effluent	+	22	1	-	-·	-
	-	14	1	-	-	2**
100% Effluent	+	24	1	-	-	-
	-	12	-	-	1	-

+Focus of cellular alteration
++Carcinoma of intestine
*Two cholangiocarcinomas, one hepatocellular carcinoma
**One neurogenic neoplasm, one carcinoma of intestine

one control fish was unclear. The lesion was independently diagnosed as a carcinoma by three pathologists but was felt to be essentially the result of a reactive process by two other pathologists who failed to diagnose the lesion as a neoplasm. For the purpose of this study the lesion was counted as a neoplasm but should be viewed as somewhat problematic.

The conclusion of this field bioassay is that the effluent is very mildly mutagenic and possibly weakly carcinogenic but that this effect is not remarkably different from the water of the laboratory or the receiving river. In future tests of both effluent and ground water, the exposure and holding times will be extended and these tests will include additional mutagenicity and chemical characterization of all waters to which fish are exposed. These latter data are useful in providing a "snapshot" of the water at one point in time but are not used as confirmatory evidence for carcinogenicity potential.

CONCLUSIONS

Preliminary evaluations were made of the performance of two toxicity assessment systems incorporated into a mobile biomonitoring facility situated at a wastewater treatment plant. An automated toxicity monitoring system was able to identify rapidly developing acutely toxic conditions in a simulated spill of 1,3,5-trinitrobenzene. Additional development of this system will permit continuous, on-line evaluation of toxic changes in water or wastewater at Army sites.

Incorporation of a fish carcinogenicity assay into the biomonitoring facility also was found to be feasible. Both carcinogenicity and carcinogen promotion have been evaluated. Future research will focus on the assessment of other potentially contaminated effluents and groundwaters and on cross-species extrapolation to allow the use of fish bioassay data in human hazard assessment.

On-site assessment techniques can provide direct and cost-effective evaluations of the toxicity of both surface and groundwaters. More realistic estimates of toxicity at contaminated sites can help ensure that any remedial actions are appropriate to the degree of hazard involved. This is an important consideration, given the high cost of cleaning up contaminated aquatic resources.

ACKNOWLEDGMENTS

The authors acknowledge the technical assistance of Mr. Tommy R. Shedd, Mr. Robert C. Bishoff, and SPC David L. Harvey.

The opinions or assertions contained herein are the private views of the author(s) and are not to be construed as official or as reflecting the views of the Department of the Army or the Department of Defense.

In conducting the research described in this report, the investigators adhered to the "Guide for the Care and Use of Laboratory Animals" as promulgated by the Committee on the Guide for Laboratory Animal Facilities and Care of the Institute of Laboratory Animal Resources, National Research Council.

REFERENCES

Anon, In Defense of Treatment, 1985, Water Pollut. Control. Fed. Highlights, 22(4):3.

Black, J.J., 1984, Aquatic Animal Neoplasia as an Indicator for Carcinogenic Hazards to Man, J. Sarena, ed., Hazard Assessment of Chemicals, Academic Press, Orlando, FL, pp.181-232.

Cairns, J. Jr., and van der Schalie, W.H., 1982, Biological Monitoring Part I - early warning systems, in: "Biological Monitoring in Water Pollution," J. Cairns, Jr. ed., Pergamon Press, New York, pp. 1179-1199.

Couch, J.A., and Harshbarger, J.C., 1985, Effects of Carcinogenic Agents on Aquatic Animals: An environmental and experimental overview, Environ. Carcinog. Rev., 3(1):63-185.

Goodfellow, W.L., 1987, Results of Chronic Toxicity Tests Conducted on Effluent, EA Aquatic Toxicology Report 12-2-87-165, EA Engineering, Science, and Technology, Inc., Sparks, MD.

Grizzle, J.M., Melius, P., and Strength, D.R., 1984, Papillomas on Fish Exposed to Wastewater, J. Nat. Cancer Inst., 73(5):1133-42.

Gudnason, H.M., Huynh, M., Harris, C., Allen, G.W., and Orth, D., 1987, Comprehensive Water quality Analysis Report, Rockville, MD, Biospherics Inc.

Hawkins, W.E., Overstreet, R.M., and Walker, W.W., 1988, Small fish models for identifying carcinogens in the aqueous environment, Water Res. Bull., 24(5):941-949.

Hoover, K.L., ed., 1984, The Use of Small Fish Species in Carcinogenicity Testing, NCI Monograph No. 65.

Horning, W.B., II, and Weber, C.I., 1985, Short-Term Methods for Estimating the Chronic Toxicity of Effluents and Receiving Waters to Freshwater Organisms, EPA-600/4-85-014, U.S. Environmental Protection Agency, Cincinnati, OH.

Jagannath, D.R., 1987a, Mutagenicity Test on Water Sample 87-307-1 in the Ames Salmonella/microsome Reverse Mutation Assay, Hazleton Laboratories America, Inc., Kensington, MD.

Jagannath, D.R., 1987b, Mutagenicity Test on an Extract of the Water Sample 87-307-1 in the Ames Salmonella/microsome Reverse Mutation Assay, Hazleton Laboratories America, Inc., Kensington, MD.

Liu, D.H.W., Spanggord, R.J., Bailey, H.C., Javitz, H.S., and Jones, D.C.L., 1984, Toxicity of TNT Wastewaters to Aquatic Organisms, Volume II, Acute Toxicity of Condensate Wastewater and 2,4-Dinitrotoluene, SRI International, Menlo Park, CA.

Masahito, P., Ishikawa, T., and Sugano, H., 1988, Fish Tumors and Their Importance in cancer research, Gann, 79:545-555.

Naugle, D., 1986, "Report of the DOD/EPA/DOE Working Group to Explore Hazardous Waste Technology Cooperative Efforts," PEER Consultants, Inc., Rockville, MD, p.13.

van der Schalie, W. H., 1986, Can biological monitoring early warning systems be useful in detecting toxic materials in water? in: "Aquatic Toxicology and Environmental Fate: Ninth Volume, ASTM STP 921," T. M. Poston and R. Purdy, eds., American Society for Testing and Materials, Philadelphia, PA, pp.107-121.

van der Schalie, W.H., Shedd, T.R., and Zeeman, M.G., 1988, Ventilatory and movement responses of bluegills exposed to 1,3,5-trinitrobenzene, in: "Aquatic Toxicology and Hazard Assessment, Tenth Volume, ASTM STP 971," W. Adams, G. Chapman, and W.G. Landis, eds., American Society for Testing and Materials, Philadelphia, PA, pp.307-315.

Wolfe, M.J., 1988, Utilization of Fish to Evaluate the Carcinogenic Potential of Army Wastewaters: Pathology Report, Experimental Pathology Laboratories, Inc., Herndon, VA.

MICRONUCLEUS TEST USING PERIPHERAL RED BLOOD CELLS OF AMPHIBIAN

LARVAE FOR DETECTION OF GENOTOXIC AGENTS IN FRESHWATER POLLUTION

A. Jaylet, L. Gauthier, and C. Zoll

Centre de Biologie du Developpement
UA No. 675 (CNRS) Universite Paul Sabatier
118 route de Narbonne, 31062 Toulouse Cedex, France

INTRODUCTION

Increased environmental pollution can be attributed to a variety of factors resulting from new industrial and agricultural technologies together with changes in our way of life. Moreover, the nature of the pollution itself has become more diverse. Whatever the origin of the pollution it tends to find its way into the aquatic environment. Genotoxic pollutants affect the aquatic ecosystem, and their presence in the water can also have repercussions on non-aquatic species via food chains or simply from drinking the polluted water. Man is not exempt from this risk; much drinking water is derived from surface water that is more or less polluted. Before distribution to the consumer, this water is thus subjected to a number of treatments. Ironically, if the water is rich in organic matter some disinfectant treatments may in fact lead to the formation of genotoxic compounds that were not present in the initial water source. One should therefore be aware of the hidden risks stemming from potential genotoxins in the aquatic environment. Moreover, a considerable time may elapse between the action of the mutagenic agent and the outward signs of its effects. The relationship between cause and effect may thus become obscured.

Mutagenic agents may exert their action on any cell in the organism. Mutation in a somatic cell may trigger a process leading to carcinogenesis.

Mutagenic agents also exert their action on germ cells. If the toxicity is severe and many cells are affected, there may be a lowered or even temporary loss in fertility. If a gamete with a genetic anomaly contributes to the formation of a zygote, the disorder created in the hereditary material may be serious enough to lead to the death of the embryo. Although a genetic anomaly that is compatible with survival of the organism frequently leads to an immediately obvious anomaly in the phenotype, the effect may be deferred and only become apparent in future generations. This is often the case with equilibrated chromosomal re- arrangements and recessive mutations.

The mutagenic risk is particularly apparent in prokaryotes, and readily discernible in plants and animals with a rapid rate of reproductions, although it is often not very perceptible in plants and animals (including humans) with a slower reproduction rate. It should be remembered that the manufacture and use of aggressive mutagenic substances is too recent to be able to judge effects over the relevant number of generations.

In Situ Evaluations of Biological Hazards of Environmental Pollutants
Edited by S. S. Sandhu *et al.*
Plenum Press, New York, 1990

The mutagenicity of an unknown substance is usually evaluated by putting it in contact with a living system, which is then examined for genetic damage. It is generally agreed that it is difficult to extrapolate results obtained from one living system to another, or even from animal to man. Nevertheless, the commonly used tests, at least as a first step, are based on bacteria. The main advantages are that such tests can be carried out rapidly and are low in cost. One of the main drawbacks of these bacterial tests for the detection of genotoxins in water is that they are relatively insensitive, and in general they cannot be used on unconcentrated water samples.

Ideally, one should evaluate the biological hazards of environmental genotoxic pollutants in situ. In vivo mutagenicity tests applied to unconcentrated water samples represent a step in this direction (Chouroulinkov and Jaylet, 1989; Jaylet and Zoll, 1989; Sandhu and Lower, 1987). They give an indication of the overall genotoxic potential of the water under testing. One example is the micronucleus test adapted to larvae of amphibians that we developed over the last few years. The larvae can be reared not only in containers filled with unconcentrated water samples (laboratory conditions) but also in running water of various sources (effluents of factories river water or even drinking water taken directly from the tap.

Three main categories of mutation are usually distinguished: gene mutations, chromosomal mutations and genomic mutations. In gene mutations, the changes are localized to a single gene. The anomaly, whatever its extent, is too small to be detected by microscopic observation. In chromosome mutations some chromosome segments are lost or displaced. In the case of genome mutations the structure of the gene or chromosome is unaltered, and only the number of chromosomes is affected. Mutagenic agents essentially differ from one another by the relative frequency with which they produce the different categories of mutation. However, in the majority of cases there is no absolute specificity. A given mutagen that essentially leads to gene mutations can also give rise to chromosome mutations and vice versa. In the case of genome mutations, the DNA remains unchanged and the variations in the number of chromosomes are due to a disruption of chromosomal migration during cell division. Substances that influence the formation of the mitotic spindle or the interactions between centromere and spindle fibers are likely to lead to this type of aberration.

In most eukaryotes, chromosome and genome mutations result in the formation of micronuclei. These micronuclei are small intracytoplasmic masses of chromatin resembling small nuclei. They are formed from chromosome fragments or complete chromosomes which have not migrated to a spindle pole during anaphase. Therefore, the formation of micronuclei stems either from chromosome fragmentation, or a malfunction of the mitotic apparatus. In the former case, micronuclei correspond to chromosome fragments that, having lost the centromere, have been unable to connect with the spindle fibers. In the latter case, they arise from complete chromosomes lagging at anaphase due to spindle abnormalities. Clastogenic compounds and spindle poisons both lead to an increase in the number of micronucleated cells.

Evans et al. (1959) was the first to suggest counting cells containing micronuclei as a method for the evaluation of cytogenetic damage. Since then, induction of micronuclei has been widely used for genotoxicity testing. A detailed description of the micronucleus test using bone marrow polychromatic erythrocytes from small mammals is given by Schmid (1976). Results from the micronucleus test and recommendations for its practical application have been reviewed by Heddle et al. (1983).

In the animal kingdom, micronucleus formation has been studied mainly in mammals. In aquatic vertebrates, a micronucleus test using a fish (Umbra pigmea) has been developed by Hooftman and de Raat (1982). However, in our hands a similar test, on red blood cells (RBCs) with two other fish species (Brachydanio rerio and Cyprinus carpio) did not lead to statistically significant results (Jaylet and Deparis, unpublished results). In 1987, Krauter et al. demonstrated micronuclei formation in peripheral erythrocytes of Rana catesbeiana tadpoles after irradiation. They proposed this animal for in vivo genotoxicity studies.

We have implemented a micronucleus test in three amphibian species Pleurodeles waltl, Ambystoma mexicanum and Xenopus laevis (Siboulet et al., 1984; Grinfeld et al., 1986; Jaylet et al., 1986a,b; Jaylet et al., 1987; Fernandez et al., 1988; Zoll et al., 1988; Van Hummelen et al., 1989). In the larvae of these three amphibians, the RBCs are nucleated cells that divide actively in the bloodstream. Micronuclei in these cells can be readily detected on blood smears. Initially, the optimal larval stage and duration of treatment were determined for each test system. Then a concentration-response curve was established for well-known clastogens. The procedure currently used in our laboratory and the results obtained so far are summarized below.

PROCEDURE OF THE AMPHIBIAN MICRONUCLEUS TEST

Animals

We have established the test procedure in two species of urodel: Pleurodeles waltl (the pleurodele) and Ambystoma mexicanum (the axolotl), as well as in the Anuran Xenopus laevis (the South African toad or xenopus). These three amphibians are widely used in research in genetic and developmental biology. They are abundant egg-layers: a female of pleurodele or axolotl can lay up to 1,000 eggs at once, while 2,000 or 3,000 is not rare in xenopus. Their rearing and development are now well described. The details of the test procedure have been described in the publications mentioned above, and for urodele a documentation sheet has been published by the French Standards Institute (AFNOR 1987).

Rearing

Urodele. Similar methods are used to rear both axolotl and pleurodele larvae. After they are laid, the eggs are placed in an aquarium. The water is normal tap water filtered through active carbon or "ultrapure water" reconstituted with salts.

The young larvae eat only live food. After hatching, the animals are fed on freshly hatched artemia (Artemia salina) or daphnia (any species). When the food is switched to chironoma larvae. Generally this latter food is subsequently used throughout the experiments.

The temperature of the water can range from 12°C to 20°C. Within this range and increase in temperature increases the growth rate of the animals. In this way it is possible to change the rate of development of subjects depending on requirements. Several groups at different stages of development can therefore be studied at the same time. However, the temperature dependence of the mitotic index must be taken into account. All treatments are carried out at 20°C. They are always carried out after a period of 8 days' habituation at this temperature.

Xenopus. Xenopus tadpoles are fed with dehydrated aquarium fish food. The temperature of the water can range from 18°C to 23°C; the growth rate also depends on the rearing temperature. All treatments are carried out at 22°C after a 6 days' habituation period at this temperature.

Treatment Stage

The larvae must be of sufficient size to enable blood samples to be taken easily, and also they must be at a stage of intense erythropoiesis with a large number of divisions of RBCs in circulating blood. As the rate of growth is a function of temperature, and as treatment should be carried out when the larvae are in a identical physiological state, age cannot be used as a reference. An accurate morphological marker is required. For the axolotl, we have found that treatment must start when the hind limb buds of the larvae exhibit slight indentation (onset of formation of the two first digits). For pleurodele, treatment is started when the hind limbs present four well-formed digits with an outline of the fifth. In both cases, the mean size of the larvae is around 30 mm (about 6 weeks after laying). They reach 40 mm within the next 10 days.

The growth of xenopus tadpoles is quicker, and the test can begin 2 weeks after laying. At this time one must choose larvae at stage 50 of the chronological table of Nieuwkoop and Faber (1956) (hind limb bud longer than broad, constricted at base).

Experimental Design

The treatment procedure is basically the same for all three animal species. They are treated in groups of twenty in 5-liter flasks filled with 2 liters of the sample. Control groups are reared in purified water. The glass containers are placed in a large water tank in order to keep them at the same temperature. The media are renewed and food is added every 24 hours. At the end of the treatment period (generally 12 days for pleurodele and xenopus and 10 days for axolotl), the animals are anesthetized with 0.02% tricaine methane sulfonate. Blood samples are taken by cardiac puncture into heparinized micropipettes (20% solution at 5,000 IU/ml). Blood smears are made, then fixed in methanol and stained with hemalun solution. Slides are examined under the microscope with an immersion lens (X 100). For each animal the number of micronucleated RBCs per 1,000 cells is determined irrespective of the number of micronuclei per RBC. Proportions of the RBCs containing one, two, three and four (or more) micronuclei may be recorded as additional data.

Significance of the Test

For a given sample of treated larvae, the levels of micronucleated cells are scattered, hence the need for a large number of animals per group. Fifteen seems to be an adequate number. The absolute minimum is nine. The choice of smear is based solely on the spreading properties of the cells. The values in each sample (levels per thousand) are not necessarily normally distributed and so median values and quartiles are calculated instead of means. A suitable statistical method must be chosen in consequence. In the absence of statistical calculations, a result is considered positive if the two following conditions are satisfied: a) the median of the treated animals is twice that of controls; b) the lowest quartile of the treated animals is above the highest quartile of the controls. A positive results can be taken for any duration of treatment, but in general 4 to 8 days is long enough for a strongly clastogenic substance used at maximal concentration (MC); MC is defined as half the lethal concentration (LC 50 within a period of 6 days), or, in certain cases, as the maximal concentration compatible with survival and normal food intake during 2 weeks.

A negative result can be accepted only if the chemical has been tested at as high a concentration as possible (MC). If the MC has been accurately determined it is also necessary to determine that the animals have ingested as much food as the controls, and that the substance does not inhibit cell division. Treatment must have been continued for a reasonable length of time (12 days seems adequate for pleurodele and xenopus, and 10 days for axolotl).

Comments on Specific Points of the Method

It seems worthwhile to emphasize specific points that would be of value to potential users of this method. A the larval stage employed in the test, the young urodele larvae feed exclusively on live material. Daphnia of any species, Chironomus larvae or tubifex can be used. If either of these cannot be obtained, then Artemia (Artemia salina) hatched less than 24 hours previously in salt water can be used. This latter food leads to a slightly slower development than the others, and there are consequently fewer mitoses. The level of micronucleated cells will be correspondingly less, and the sensitivity of the test will also be reduced.

The larvae are given food every 24 hours when the water is changed. If the food is daphnia, chironomus or tubifex, it should be given in sufficient quantity for some to be left 7-8 hours later), since this species does not survive for more than a few hours in fresh water.

For urodele, as for xenopus, it is important not to give the animals more food than they can eat. The compound tested may bind to the non-ingested food, producing inaccurate results.

CURRENT RESULTS WITH THE AMPHIBIAN MICRONUCLEUS TEST

The Micronucleus Test in Pleurodele

Studies with X-Rays and Chemicals. Initially, the sensitivity and dose-response of the micronucleus test in pleurodele were evaluated with X-ray irradiation, a well-known physical clastogenic agent. The measurements were made 6 days after X-ray irradiation. A dose of 6 rad (relatively weak) lead to a significant effect. The dose-response is approximately linear up to 150 rad, after which the slope falls and the maximal effect is reached at 600 rad.

Results on 19 organic compounds have been published elsewhere (Fernandez et al., 1988). Aroclor 1254, butylated hydroxy-anisole, phenobarbital and 12-0-tetradecanoyl-phorbol-13-acetate produced negative results, while acridine orange, benzo(a)pyrene, e-caprolactam, cyclophosphamide, diethyl sulfate, epichlorhydrin, ethidium bromide, ethyl methanesulfonate, ethyl dibromide (dibromoethane), N-ethyl-N'-nitro-N-nitrosoguanidine, N-ethyl-N-nitrosourea, hexa-methylphosphoramide, 3-methyl cholanthrene, pyrene and o-toluidine gave positive responses.

The cytogenetic effects of mercury compounds have been widely studied in plants, drosophila and tissue culture cells, but to our knowledge they have not been evaluated in vertebrates in vivo. Pleurodele larvae were raised in water containing low concentrations of methyl mercuric chloride (CH3HgCL) or mercuric chloride (HgCl2) (Zoll et al., 1988). It should be noted that a low concentration of the two compounds (12 ppb) gave a positive result, and that at equivalent concentrations in the water of both CH3HgCl and HgCl12 led to similar levels of micronucleated RBCs. The test gives positive results for concentrations below those often found in samples of contaminated water (Giraud and Guillet, 1972). Both chromosome

aberrations and abnormalities in cell division were observed in cells from
animals treated with these two substances. The micronuclei could thus have
derived from chromosome fragments lacking a centromere or whole chromosomes
that had not migrated during an abnormal anaphase. Bioaccumulation of both
compounds was also evaluated by determination of mercury levels in the
larvae. After 12 days of treatment, concentration factors (concentration
in the organism/concentration in the water) of 1,200 and 600 were found for
CH_3HgCl and $HgCl_12$ respectively. These findings can be compared to those
for benzo(a)pyrene (Grinfeld et al., 1986). The concentration of benzo(a)-
pyrene in unfed pleurodele larvae was 200-fold of that in the water after
12 hours.

The marked bioaccumulation potential of newt larvae partially explains
why it is not necessary to concentrate mutagenic micropollutants in samples
of natural or drinking water to detect genotoxic effects.

These results demonstrate the sensitivity and the reliability of the
test for known genotoxic agents experimentally added to the rearing water.
It was important to find out whether the pleurodele test test was also applicable to in situ conditions.

In Situ Studies. Samples of river water contaminated by factory effluents, as well as those taken at the input and output of a water purifying
plant, increased the level of micronucleated RBCs (unpublished results).
The tap water supplying the laboratory was also found to yield positive
results (Jaylet et al., 1987). In this last case, groups of larvae were
reared in tap water that had been filtered over sand and active carbon to
remove micropollutants. Seven separate tests carried out between October
1985 and May 1986 all gave positive results of varying degree depending on
the time of the year. This test is therefore able to detect clastogens in
normal drinking water. It could be used for quality control of drinking
water during the various stages in the treatment of raw water without any
requirement for prior extraction or concentration of micropollutants (work
in progress, unpublished results).

Validation of the Assay. A collaborative study was carried out to verify the reproducibility of the results of the newt micronucleus test. One
Belgian and five French laboratories simultaneously evaluated the clastogenic effects of benzo(a)pyrene. Despite the fact that experimental conditions differed somewhat between the laboratories (age of the larvae,
rearing conditions) all laboratories found a dose-response relationship.
The results have been published as a French Standard (AFNOR, 1987). In
addition, for a better evaluation of the specificity of the micronucleus
test, non-carcinogens need to be tested. We are currently investigating
the mechanisms leading to the formation of clastogenic metabolites in the
newt using the well-known mammalian and human promutagen benzo(a)pyrene,
involved both in cancer initiation and promotion processes. Results of
investigations of its metabolism (Marty et al., 1988) together with
previous data (Fernandez and Jaylet, 1987) indicate that the clastogenic
effects of benzo(a)pyrene involve, at least in part, similar mechanisms in
both the newt and mammals.

The Micronucleus Test in Axolotl

In order to investigate the generality of the micronucleus test in
pleurodele, larvae from another urodele, the axolotl, were reared in water
containing either of the compounds: benzo(a)pyrene (BaP) or ethyl methanesulfonate (EMS). The level of micronucleated erythrocytes on blood smears
was compared with control samples from larvae reared in fresh water. The
optimum larval stage for this test system was determined. The effects of
the indirect mutagen (BaP) and the direct mutagen (EMS) were found to

76

depend on both dose and exposure to the clastogen. For BaP, positive results were obtained after 8 days of treatment at a concentration of 0.025 ppm. After 10 days of treatment at a concentration of 0.1 ppm, numerous micronuclei were seen (< 250%/00). Positive results were also obtained with EMS after 8 days of treatment at a concentration of 24 ppm. At 62 ppm, positive results were found after 6 days, while at 124 ppm positive results were found after only 4 days. The results with both these agents show that the axolotl also holds promise as an _in vivo_ test system for the detection of low concentrations of clastogens in the aquatic environment. This is not very surprising, since the axolotl is morphologically and biologically similar to the pleurodele at both the embryonic and larval stages, although not of the same genus and of different geographical origin.

The Micronucleus Test in Xenopus

The third species used was xenopus. It differs in a number of respects from the previous two species, including its feeding behavior.

Three different variables: temperature, stage of larval development and frequency of renewal of the test substance were investigated using ethyl methanesulfonate as the clastogenic compound. In addition, a dose-response curve was established for benzo(a)pyrene (BaP) in order to determine the limits of sensitivity of the test (Van Hummelen et al., 1989).

With BaP, the lowest concentration (0.03 ppm) gave a negative response. From 0.06 ppm up to 0.50 ppm, an approximately linear increase in median value of cells with micronuclei was observed. This linear response indicates that the test is reliable, although the lower frequencies of cells with micronuclei at doses above 0.50 ppm, are probably accounted for by a lower rate of growth of the larvae exposed to high doses of BaP. In fact, above BaP concentrations of 1 ppm the larvae eat less and grow more slowly than the larvae in the other samples. The mitotic index is thus lower, and since the production of micronuclei depends on cell division, a fall in the mitotic index results in a decrease in the number of RBCs with micronuclei. It is worth noting that a similar phenomenon occurs, for the same concentrations, in axolotl and pleurodele with BaP.

The effects of other organic products, different kinds of polluted water, an mercuric salts are under investigation. With respect to mercury compounds, it should be noted that they are strongly accumulated by xenopus tadpoles. For example, after 12 days the test is positive with only 2.5 ppb of methyl mercuric chloride.

DISCUSSION AND CONCLUSION

The micronucleus test applied to amphibian larvae provides an indication of the real impact of aquatic genotoxins on the whole organism. It takes account of a variety of possible interactions, especially those between the environment and the animals, as well as synergistic or antagonistic effects of chemical mixtures in water.

Among the three species used for the test, it is not possible to state which is best suited for detection of genotoxic substances in water. We can, however, draw up a list of the technical constraints for each system and the results obtained.

For the three species the optimal treatment duration is quite similar, and rearing the breeders does not present any particular problem.

Urodele larvae eat exclusively live prey, whereas xenopus tadpoles can be fed on dehydrated aquarium fish food. The interval between egg laying and testing is 6 weeks for urodele and only 2 weeks for xenopus, giving a technical advantage to the latter species. However, recording micro-nucleated cells is somewhat easier in urodeles than in xenopus, since RBCs and micronuclei are larger in the former. Nevertheless, recording micro-nuclei for all three species is more straightforward than in the rodent micronucleus test.

When a sample induces micronuclei it is of interest to find out whether the genotoxic agent causes chromosome breaks or whether it disrupts mito-sis. This is easy to verify in urodele (Zoll et al., 1988) whose chromo-somes are long, but more difficult in xenopus whose chromosomes are small and particularly difficult to scatter.

Pleurodele and xenopus tests differ in sensitivity depending on the compound considered. For example, for cyclophosphamide the detection threshold is 0.5 ppm in pleurodele (Fernandez et al., 1988), and 5 ppm in xenopus (unpublished results). Similarly for BaP, the test is positive with 0.025 ppm in pleurodele (Fernandez et al., 1988) and 0.06 ppm for xenopus (van Hummelen et al., 1989). Conversely the detection threshold for for methyl mercury is 2.5 ppb in xenopus (unpublished results) and 12 ppb in pleurodele (Zoll et al., 1988).

All three species can demonstrate the genotoxicity of both direct and indirect mutagens (i.e. those requiring metabolic activation).

Since EMS, which has a short half-life and low clastogenicity (Sega, 1984) induces a significant increase of RBCs with micronuclei in all three species, the test could be used to detect short-lived genotoxic molecules in polluted water.

Various factors contribute to the high sensitivity of our biological models. The larvae strongly accumulate pollutants from the surrounding medium. This has been well demonstrated for compounds such as methyl mercury (Zoll et al., 1988 and unpublished results) and BaP (Grinfeld et al., 1986, Marty et al., 1988 and unpublished results).

The duration of the assay (between a week to a week and a half) may represent a limitation for these tests. However, it may be long enough to enable detection of the effects of chronic genotoxic intoxication.

These three test systems thus seem suitable for monitoring genotoxic aquatic pollution and/or for the quality control of drinking water. They may also be applied to the detection of some waterborne carcinogens.

ACKNOWLEDGMENTS

The authors wish to thank all their colleagues for their contributions in setting up the amphibian micronucleus test. They are indebted to Dr. P. Cochard for his invaluable help in the preparation of this manuscript, and to C. Daguzan for her excellent technical assistance over the years. This work was supported by the CNRS and by grants form the Ministere de l'Environnement, France, and the Universite Toulouse le Mirail.

REFERENCES

AFNOR, 1987, Essais des eaux: Detection en milieu aquatique de la geno-toxicite d'une substance vis-a-vis de larves de batraciens (Pleurodeles waltl et Ambystoma mexicanum). Essai des micronoyaux. Association Francaise de Normalisation (AFNOR), T 90:325.

Chouroulinkov, Y., and Jaylet, A., 1989, Contamination of aquatic system and its genetic effects, in: "Aquatic Ecotoxicology: Fundamental concepts and methodologies," A. Boudou, and F. Ribeyre, eds., CRC Press, Boca Raton, FL, (in press).

Evans, H.J., Neary, G.J., and Willamson, F.S., 1959, The relative biological efficiency of single doses of fast neutrons and rays on Vicia faba roots and the effects of oxygen. II. Chromosome damage, the production of micronuclei., Int. J. Radiat. Biol., 1:216.

Fernandez, M., Gauthier, L., and Jaylet, A., 1988, Use of newt larvae for in vivo genotoxicity testing of water: results on 19 compounds evaluated by the micronucleus test., Mutagenesis, (in press).

Fernandez, M., and Jaylet, A., 1987, An antioxidant protects against the clastogenic effects of benzo(a)pyrene in the newt in vivo., Mutagenesis, 2:293.

Giraud, M., and Guillet, H., 1972, Teneur en mercure des milieux naturels, in: "La pollution par le mercure et ses derives," Rapport Ministere de l'Environnement, France.

Grinfeld, S., Jaylet, A., Siboulet, R., Deparis, P., and Chouroulinkov, I., 1986, Micronuclei in red blood cells of the newt Pleurodeles waltl after treatment with benzo(a)pyrene, dependence on dose, length of exposure, post-treatment time and uptake of the drug, Environ. Mutagenesis, 8:41.

Heddle, J.A., Hite, M., Kirkhart, B., Mavournin, K., MacGregor, J.T., Newell, G.W., and Salamone, M., 1983, The induction of micronuclei as a measure of genotoxicity, A report of the U.S. Environmental Protection Agency Gene-Tox Program, Mutat. Res. 123:61.

Hooftman, R.N., and de Raat, W. K., 1982, Induction of nuclear anomalies (micronuclei) in the peripheral blood erythrocytes in the eastern mudminnow Umbra pigmaea by ethyl methansulfonate, Mutat. Res., 104:147.

Jaylet, A., Deparis, P., Ferrier, V., Grinfeld, S., and Siboulet, R., 1966a, A new micronucleus test using peripheral blood erythrocytes of the newt Pleurodeles waltl to detect mutagens in fresh water, Mutat. Res., 164:245.

Jaylet, A., Deparis, P., and Gaschignard, D., 1986b, Induction of micronuclei in peripheral erythrocytes of axolotl larvae following in vivo exposure to mutagenic agents, Mutagenesis, 1:211.

Jaylet, A., Gauthier, L., and Fernandez, M., 1987, Detection of mutagenicity in drinking water using a micronucleus test in newt larvae (Pleurodeles waltl), Mutagenesis, 2:211.

Jaylet, A., and Zoll, C., 1989, Detection of pollution in fresh water using a mutagenicity test, CRC Critical Reviews in Aquatic Toxicology, Boca Raton, FL, (in prep).

Krauter, P.W., Anderson, S.L., and Harrisson, F.L., 1987, Radiation-induced micronuclei in peripheral erythrocytes of Rana catesbeiana: an aquatic animal model for in vivo genotoxicity studies, Environ. and Molec. Mutagen. 10:285.

Marty, J., Lesca, P., Jaylet, A., Ardourel, C. and Riviere, J.L., 1988, In vivo and in vitro metabolism of benzo(a)pyrene by the larva of the newt, Pleurodeles waltl, Comp. Biochem. Physiol., (in press).

Nieuwkoop, P.D., and Faber, J., 1956, Normal table of Xenopus laevis, P.D. Nieuwkoop, and J. Faber, eds., North Holland Publishing Company, Amsterdam, p.1.

Sandhu, S.S., and Lower, W.R., 1987, "In situ monitoring of environmental genotoxins," in: Short-Term Bioassays in the Analysis of Complex Environmental Mixtures V. Shahbeg Sandhu, David DeMarini, Marc Mass, Martha Moore and Judy Mumford, eds. Plenum Press, New York, p.145.

Schmid, W., 1976, "The micronucleus test for cytogenetic analysis," in: "Chemical Mutagens," A. Hollaender, ed., Plenum Press, New York, 4:31.

Sega, G.A., 1984, A review of the genetic effects of ethyl methanesulfonate, Mutat. Res., 134:113.

Siboulet, R., Grinfeld, S., Deparis, P., and Jaylet, A., 1984, Micro-
    nuclei in red blood cells of the newt _Pleurodeles waltl Michah_:
    induction with X-rays and chemicals, _Mutat. Res_. 125:275.
Van Hummelen, P., Zoll, C., Paulussen, J., Kirsh-Volders, M., and Jaylet,
    A., 1989, The micronucleus test in xenopus: a new and simple _in vivo_
    technique for detection of mutagens in fresh water, _Mutagenesis_
    (accepted for publication).
Zoll, C., Saouter, E., Boudou, A., Ribeyre, F., and Jaylet, A., 1988,
    Genotoxicity and bioaccumulation of methyl mercury and mercuric
    chloride _in vivo_ in the newt _Pleurodeles waltl_, _Mutagenesis_, p.337.

PREDICTING THE ECOLOGICAL SIGNIFICANCE OF EXPOSURE TO GENOTOXIC

SUBSTANCES IN AQUATIC ORGANISMS

Susan L. Anderson

Applied Sciences Division
Lawrence Berkeley Laboratory
1 Cyclotron Road, Building 70-110A
Berkeley, California 94720

Florence L. Harrison

Environmental Sciences Division
Lawrence Livermore National Laboratory
Livermore, California 94550

INTRODUCTION

The effects of genotoxic substances are not adequately considered in aquatic ecological hazard assessment, and comparatively little has been done to develop methods for predicting these effects. It would be short sighted to ignore the potential for detrimental ecological impacts resulting from the disposal of genotoxic substances in aquatic environments. Advancements in this area must use our extensive knowledge of the mechanisms of effect of genotoxic substances in mammals, recognizing the factors that are unique to aquatic ecotoxicology. Because so much is known about basic mechanisms of effect and methods for evaluating the effects of genotoxic substances in mammals, this information must be applied to other species. However, as the methods for evaluating genotoxic effects in aquatic organisms are developed, we must go beyond method development and attempt to predict ecologically significant impacts from data on genotoxic responses. To accomplish this goal, fundamental processes that elucidate both similarities and differences in responses between or among species must be identified, and unifying concepts that may aid in predicting genotoxic effects in varied species must be formulated.

Three potential effects of exposure to genotoxic substances probably occur in both mammals and aquatic species. First, reduced fertility may occur if genetic damage induces cell death in dividing gametes. Second, reproductive success may be impaired if dominant- and recessive-lethal mutations are induced, causing embryo mortality or abnormality. Third, exposure to genotoxic contaminants may cause cancer.

Existing studies of genotoxic effects on aquatic organisms may not have focused on the most ecologically significant endpoints. In general, the effects of genotoxic substances on reproductive processes in aquatic organisms have been ignored, while the study of cancer epidemiology in fish has blossomed (Mix, 1986). Studies of cancer in fish are probably more common

In Situ Evaluations of Biological Hazards of Environmental Pollutants
Edited by S. S. Sandhu *et al.*
Plenum Press, New York, 1990

for three reasons. First, the lesions are often aesthetically unappealing, and this visual component makes "cancer in fish" a topic that can be communicated to the public. Second, the lesions are relatively easy to quantify, and third, cancer epidemiology is the cornerstone of research on effects of genotoxic substances in human health. It is possible that detrimental reproductive effects occur in many more species than does cancer and that these effects have greater impact on the integrity of populations. This may be a case in which the unique perspectives of aquatic ecotoxicology have been overshadowed by the human health approaches.

The potential significance of reproductive effects of exposure to genotoxic substances in aquatic organisms clearly deserves further attention. Cytogenetic responses have been studied in several species, but with few exceptions (Liguori and Landolt, 1985), these data have not linked the cytogenetic response to significant detrimental consequences to the organism or its progeny.

In this paper, we propose a simple conceptual model that delineates interrelationships between short-term cytogenetic responses and detrimental reproductive effects in one aquatic species. We incorporate information on the mechanisms of effect of ionizing radiation from research on mammals as well as from knowledge of specific characteristics of aquatic organisms. This model was developed using results we obtained on the effects of ionizing radiation in the polychaete worm Neanthes arenaceodentata. While results of the individual studies have been previously reported (Harrison et al., 1986; Anderson et al., 1989; Harrison and Anderson, 1988a,b), this is the first time that selected findings of all of the studies have been synthesized. Our goal in formulating the model was to identify concepts and common processes that may aid in predicting the reproductive and population-level consequences of exposure to genotoxic substances in varied aquatic species.

EXPERIMENTAL MODEL

Our model describes interrelationships between selected effects of ionizing radiation on N. arenaceodentata at the subcellular, cellular, organismal, and population levels (Figure 1). The model has two components. On the left side of Figure 1, we have listed the key effects (occurring at each level of biological organization) that may ultimately be related to a change in reproductive success or population size in this species. On the right, we have listed some of the factors that can increase or decrease the magnitude of the effects observed at a given level. These factors are significant because, to predict effects on one level of biological organization from effects observed at another level, we must assume that confounding variables will alter the responses. Failure to evaluate the significance of these factors would be overly simplistic.

The key effects that link cytogenetic responses and reproductive effects are relatively simple (Figure 1, left side). First, DNA damage induced by ionizing radiation is realized as either cell death or induction of mutations. If cell death occurs in dividing gametes, reduced fertility may result. If dominant- and recessive-lethal mutations are induced in gametes, then embryo mortality and abnormality occur in subsequent generations. These two genotoxic effects can result in decreased reproductive success at the organismal level. If a sufficient decrease in reproductive success is incurred, then changes in stable population size may result.

Several factors may modify the magnitude of the effects of ionizing radiation in N. arenaceodentata at each level of biological organization (Figure 1, right side). For example, cytogenetic responses at any given

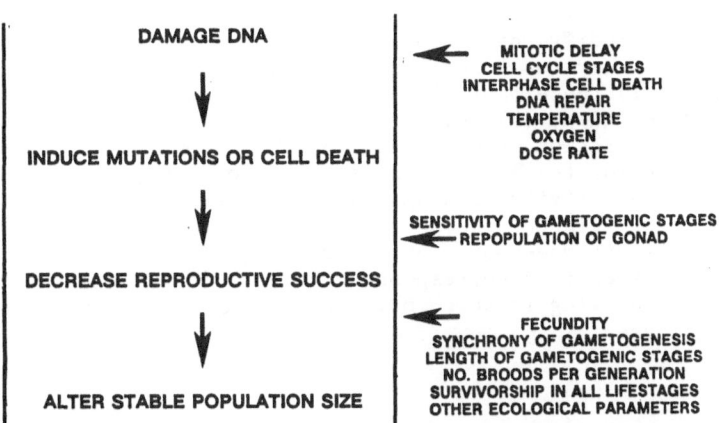

Figure 1. Conceptual model of potential interrelationships between genotoxic responses and alterations in reproductive success in N.
arenace-odentata exposed to ionizing radiation. Effects at four
levels of biological organization that may link genotoxic responses to effects on reproductive success, and potentially population changes are listed on the left. On the right, selected
factors that may either increase or decrease the levels at which
effects are observed at each level are provided. These factors
are listed because because they modify our ability to predict
effects on one level of biological organization from results to
obtained at another level.

dose may vary depending on the stage of the cell cycle at the time of irradiation, DNA repair capacities of each cell type, oxygen tension, dose
rate, and temperature. At high doses, mitotic delay and cell death in
interphase must also be evaluated. These are factors that have been extensively studied in biomedical research. In reproductive assays, effects
may vary in severity depending on the stages of gametogenesis at the time
of irradiation and the potential for damaged oocytes to be replaced by
dividing germ cells. Of course, mammals do not have the capacity to replace damaged oocytes, but this phenomenon is common in many invertebrates
and lower vertebrates. To predict effects at the population level from
data obtained in reproductive assays, we need to know how severity of effects may vary with factors such as fecundity, synchrony of gametogenesis,
duration of gametogenic stages, number of broods per generation, survivorship in individual lifestages, and ecological parameters such as intraspecific competition.

APPROACH

Our selection of important responses and modifying factors for the
model is supported by our experimental data on levels of effect and mechanisms of effect of ionizing radiation in the polychaete worm. We conducted
four studies using assays to quantify effects at three levels of biological
organization. To determine the levels at which DNA damage is observable,
we quantified sister chromatid exchange (SCE) and chromosomal aberration
frequencies in tissues of hatchling and juvenile N. arenaceodentata
(Harrison et al., 1986; Anderson et al., 1989). Studies at the cellular
and organismal levels were conducted to evaluate changes in reproductive
success after acute and lifetime exposure to ionizing radiation and to determine mechanisms of reproductive effect (Anderson et al., 1989; Harrison
and Anderson 1988a,b). The final phase of our program is now underway. We
are modeling potential population-level impacts of decreased reproductive
success in N. arenaceodentata using the results of our previous studies.

Factors that modify the responses we studied are numerous (Figure 1, right side), and they could not all be addressed here. Factors that we evaluated were (1) the effect of cell death and mitotic delay on chromosome aberration frequencies, (2) the effect of cell-cycle stage at irradiation on chromosome aberration frequencies, (3) the potential for repopulation of the gonad to ameliorate the impacts of radiation exposure, (4) potential differences in sensitivity between gametogenic stages, and (5) DNA repair capacities of germ tissue. These factors were studied in varying levels of detail. This effort does not provide us with a definitive knowledge of how each factor modifies a given response. Rather, it highlights the importance of these modifying factors in making extrapolations between different types of responses.

LIFECYCLE OF NEANTHES ARENACEODENTATA

The marine polychaete worm Neanthes arenaceodentata is an ideal species for our evaluations of genotoxic and reproductive effects of ionizing radiation. N. arenaceodentata has been proposed as a model species for marine genetic toxicology (Pesch et al., 1981), and full lifecycle reproductive tests have been described (Reish, 1980). Another advantageous characteristic of this species is that females spawn only once in a lifetime (approximately 300 embryos per brood). Consequently, reproductive success is more easily evaluated in Neanthes arenaceodentata than in a species with repeated spawning, asynchronous oocyte development, or more prolific gamete production. In short, the effects and processes to be studied are simple in nature and relatively well characterized.

The lifecycle of N. arenaceodentata is composed of five stages (Figure 2). In the embryonic stage, the developing brood is tended by the male in a tube of mucous and detrital matter. The embryos hatch in approximately 12 days, spending approximately 20 more days in the tube as hatchlings. Juvenile worms leave the tube and become more freely mobile. The juvenile stage (approximately 60 days) ends when the worms become sexually mature adults and begin to pair up. The worms exist as sexually mature adults for approximately 40 days. The fifth and final stage is the sexually reproducing adult. In this brief but important stage, fertilization and spawning occur; the female dies when her body wall ruptures at spawning. Completion of the entire lifecycle requires approximately 3 to 4 months. Reproductive success is evaluated by either enumerating the number of embryos immediately after spawning or by estimating the number of hatchlings six days after spawning.

CYTOGENETIC EFFECTS

Cytogenetic effects were quantified in two studies to determine the levels at which DNA damage occurred in response to ionizing radiation exposure (Harrison et al., 1986; Anderson et al., 1989). This is the first step in the model (Figure 1, left side). Harrison et al., 1986, determined that both sister chromatid exchanges and chromosomal aberrations were induced in hatchling worms at relatively low doses of ionizing radiation (0.6 Gy and 2.0 Gy, respectively). Subsequently, Anderson et al., 1989, quantified frequencies of chromosomal aberrations in tissues of juvenile N. arenaceodentata and evaluated the levels at which mortality, lifespan, and reproductive success were impaired. In the latter study, cells were sampled at predetermined times after irradiation to provide a sampling of cells that were at different stages of the cell cycle at the time of irradiation.

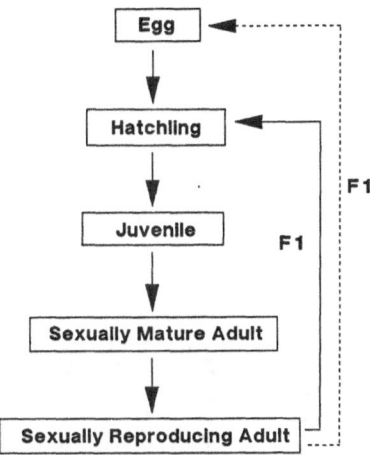

Figure 2.   Lifecycle of the marine polychaete worm <u>Neanthes</u> <u>arenace-</u>
<u>odentata</u>. Reproductive success (F1) is evaluated by either
enumerating the number of embryos immediately after spawning
(dotted line) or by enumerating the percentage of abnormal or
dead embryos 6 days after spawning (solid line) to obtain an
estimate of the humber of hatchlings.

Anderson et al., 1989, observed a significant increase in chromosomal
aberrations at 2.0 Gy, and the response to radiation exposure was dependent
on dose and the time after irradiation (Table 1). A significant increase
in the percent abnormal cells occurred at the lowest dose tested for the
6-h and 51-h sampling times. In addition, a significant difference in sensitivity of cells irradiated at the two sampling times and significant
differences in the types of aberrations were observed. The data also indicated increased variation in mean values between replicates as the dose
increased (Table 1). This variation was attributed to a potentially greater incidence in interphase cell death and mitotic delay at the higher
doses. This hypothesis was supported by an observed decline in the number
of scorable metaphase cells as the radiation dose increased (Table 1).

Results of these studies demonstrate that SCE and chromosomal aberrations are induced at relatively low doses, and three important factors
were found to modify effect-level estimates. They are the stage of the
cell cycle at the time of irradiation and either interphase cell death or
mitotic delay at higher doses (Anderson et al., 1989). These results demonstrate that cytogenetic responses obtained at high dose rates could be mis-
interpreted. If low effect-levels are obtained because cell death is occurring, then one might erroneously conclude that effects at the organismal
level might not be observable. It is also possible to make inaccurate
predictions from cytogenetic data if cell cycling is not well understood.

REPRODUCTIVE EFFECTS

We evaluated reproductive success of <u>N</u>. <u>arenaceodentata</u> by quantifying
decreased brood size and embryo survivorship in response to ionizing
radiation exposure. Three studies were performed. The first study compared the levels at which reproductive effects were observed to the levels
at which cytogenetic responses, lifespan changes, and mortality occurred.
The next two studies evaluated the mechanisms of reproductive effects in
response to acute and chronic exposures. Data from these experiments were
used to link the first and second steps of the model and to formulate the
third and fourth steps (Figure 1, left side).

Table 1. Chromosomal Aberrations Observed in Cells from Irradiated Neanthes arenaceodentata Juveniles.

Sample time (h)	Dose (Gy)	Cells scored	Abnormal cells	Types of aberrations observed[a]		Frequency of aberrations per cell[c]	Percent abnormal cells[c]
				Chromatid	Chromosome		
51	0	147	2	--	2 terminal deletions[b]	0.012 ± 0.01	0.6 ± 1.2
51	2.0	113	14	2 deletions	15 terminal deletions; 2 dicentrics	0.17 ± 0.11	12 ± 7.9*
51	4.0	111	19	2 deletions	14 terminal deletions; 3 dicentrics	0.19 ± 0.089	19 ± 8.9
51	8.0	77	14	4 deletions; 1 asymmetrical chromatid interchange	14 terminal deletions	0.26 ± 0.18	20 ± 12
51	16.0	43	13	3 deletions; 1 isochromatid interchange	12 terminal deletions; 2 dicentrics	0.39 ± 0.28	32 ± 16
6	0	177	4	2 deletions	3 terminal deletions	0.026 ± 0.044	2 ± 3
6	2.0	83	24	19 deletions; 1 asymmetrical chromatid interchange	16 terminal deletions	0.42 ± 0.17	29 ± 14*

[a] Chromosome-type aberrations involve breakage of one or more entire chromosomes, whereas chromatid-type aberrations involve breakage of one or more chromatid arms on one or more chromosomes.

[b] Terminal deletions cannot be distinguished from isochromatid deletions (a lower probability chromatid-type occurrence), and it is therefore possible that some aberrations classified as terminal deletions are isochromatid deletions, especially at the 6-h sampling time.

[c] Mean and standard deviation of four replicate experiments. * Asterisks indicate lowest doses at which responses were significantly different from controls, (x^2 = 11.4, P <0.01). For both the frequency of aberrations per cell and the percentage of abnormal cells, the test for deviation from zero slope was highly significant ($F_{1,15}$ = 10.55, P <0.01 and $F_{1,15}$ = 15.91, P <0.0015, respectively) indicating a significant dose response.

Comparison of Effect-Levels

Anderson et al., 1989, irradiated 17 adult female, adult male and juvenile worms per dose with one of five doses and a control (1, 4, 8, 46, and 102 Gy) and estimated brood size after spawning. Results of this preliminary dominant-lethal study demonstrated a significant decline in brood size at 4 Gy. In the same study (cited above), chromosomal aberrations were induced at 2 Gy and significant mortality was observed at 500 Gy. Lifespan of males surviving over 600 days postirradiation was significantly reduced at 102 Gy.

These data provided initial support for some of the interrelationships proposed in the experimental model (Figure 1, left side). Cytogenetic damage and reproductive impairment occurred in the same dose range, although the estimates of reproductive effect were considered preliminary. These same levels of radiation exposure did not affect survivorship of adult or juvenile male worms maintained in the laboratory for almost 2 years. Therefore, the evidence indicated that reproductive impairment in this organism occurred due to DNA damage and could ultimately determine the sensitivity of this species to ionizing radiation.

Mechanisms of Reproductive Effect

Next, our goal was to determine what the mechanism was for decreased reproductive success as well as to obtain refined estimates of reproductive effects. Impaired reproductive success could be caused by both reductions in fertility due to cell death and decreased embryo survival due to the induction of dominant- and recessive-lethal mutations. Two sets of experiments (Harrison and Anderson, 1988a,b) were conducted to address this issue and to examine differences in effect-level estimates obtained using acute and chronic exposures.

In the acute exposure experiments (Harrison and Anderson, 1988a), we irradiated mated pairs at the time when oocytes were first visible in the female, and they received a single dose of ^{137}Cs of either 0, 0.5, 1.0, 2.0, 5.0, 10.0, or 50.0 Gy. The broods from the mated pairs were sacrificed 6 days after spawning; information was obtained on brood size, number of normal and abnormal embryos, number of embryos that were living, dying, and dead, and estimated hatch size. The number of dying and dead embryos was determined using a Trypan Blue staining technique that we developed to distinguish live and dead embryos (Harrison and Anderson, 1988a,b). The primary effect on mated pairs of acute irradiation exposure was increased mortality of the embryos. Except for those mated pairs that received 10 or 50 Gy, there was no evidence of gamete loss or reduced fertilization success. There was no significant decrease in the number of embryos in the brood at doses below 10 Gy. The results on embryo mortality indicate that lethal mutations were induced in the germ cells (at 0.5 Gy) and these affected survival of early lifestages.

At the same time that the acute exposure experiments were underway, we conducted chronic exposure experiments in a ^{60}Co facility that was dedicated to this work. Brood size and embryo mortality were quantified for over 300 broods exposed to either one of three radiation dose rates (0.19 mGy/h, 2.1 mGy/h, and 17 mGy/h) or a control. Average total doses received by the irradiated worms were 0.55, 6.5, and 54 Gy. Beginning within one day after spawning, worms were cultured in front of the cobalt source at predetermined distances. Exposure was terminated immediately after the subsequent spawning. Brood size analysis was conducted using the same methods used for the acute exposure experiments.

Data from lifecycle exposures documented reduced reproductive success at 0.55 Gy total dose when delivered over 3 to 4 months at a rate of 0.19

mGy/h. Decreased frequency of reproductive success at the low doses (0.55 and 6.5 Gy total dose) was attributed to an increased frequency of dead embryos. Only at 54 Gy was a decrease in brood size also observed.

Implications of Findings on Reproductive Effects

Our observation that embryo mortality and abnormality are induced at lower doses than are brood-size changes may indicate difference and radio-sensitive targets at low doses as compared to high doses. This observation may indicate one of two things. First, it is possible that the mutations causing at least some portion of the embryo mortality are induced at lower doses than are the aberrations that are more likely to cause cell death and reduced fecundity. However, it is also possible that the methods used to quantify embryo survivorship and abnormality are inherently more sensitive than the brood size measurements. Brood size varies with size of the worm and is relatively variable even between worms in control groups. However, there is one highly speculative line of reasoning that would support the hypothesis that mutations are detectable at lower doses than are aberrations. In the cytogenetic studies described above, Harrison et al., 1986, found significant induction of SCE at 0.6 Gy whereas, chromosomal aberrations were induced at 2.0 Gy. Because SCE have been correlated with mutagenesis, it is possible that they are predictive of the levels at which lethal mutations are induced in germcells. In contrast, assays of chromosomal breakage and rearrangement would be more predictive of decreased brood size due to cell death.

Comparison of embryo survivorship data obtained in the acute and chronic exposure experiments (Figure 3) reveals that DNA repair may not be very active in gametes of this species. For total doses at or below 10 Gy, the severity of effects observed in chronic exposures was the same as that observed after acute exposures. These data indicate that DNA damage in gametes or selected gametogenic stages is cumulative and that DNA repair activity in these cells may be relatively limited.

At 54 Gy, the results of the chronic experiment demonstrated more severe effects than were obtained in the acute experiment. The most plausible explanation for this difference is that more sensitive gametogenic stages may have been irradiated in the chronic exposure experiment. The most sensitive stages of gametogenesis are unknown, and they may not have been present in the worms at the time of irradiation in the acute experiment.

The results of the reproductive studies provide further support for the interrelationships described in the model (Figure 1, left Side). Both gamete death and mutagenesis appear to be mechanisms causing reduced reproductive success. However, the relative importance of these two processes may vary with dose. We acknowledge that alternative mechanisms such as membrane damage in oocytes may explain some of the reproductive effects observed. However, the data obtained to date indicate that the mechanisms proposed are highly plausible. Efforts must continue to identify additional alternative hypotheses and to explore potential targets.

Three important factors that modify the levels at which effects are observed in the reproductive studies were examined (Figure 1, right side). Data obtained indicate that there may be little DNA repair in some gametogenic stages because the doses received in the chronic exposure experiments appeared to have either a cumulative impact on the gametes or an impact on a highly sensitive gametogenic stage. Second, the sensitivity of different gametogenic stages may vary because differences were observed between results obtained in the acute experiments and results obtained in the chronic experiments at high doses. This result is not surprising because it is

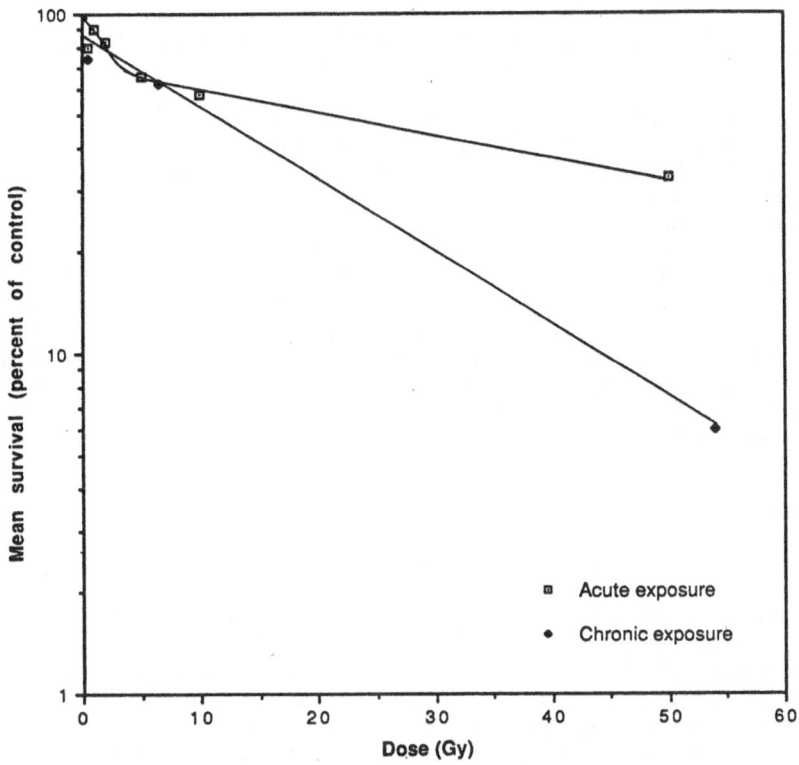

Figure 3.   Comparison of embryo survival in acute and chronic exposure ex-
            periments. Mean percent survival of embryos is expressed as a
            percentage of control survival and plotted on a log scale
            against total dose.

is well known that the radiosensitivity of gametogenic stages in fish
varies widely (Anderson and Harrison, 1986). Third, repopulation of the
gonad by replacement of damaged gametes did not appear to be occurring to
an appreciable extent in the chronic exposure experiments.

POPULATION-LEVEL EFFECTS

     Our experiments have provided convincing evidence that, for this one
species, the gametes are the critical cells that limit the sensitivity of
the species to ionizing radiation exposure. Of course, gametes are also
the cells that transfer genetic information from one generation to the
next. This brings us to the fourth and final step of our conceptual model.

     We are now developing a population model that not only evaluates the
impact of embryo mortality and brood-size changes at the population level
but also evaluates the impact of heritable mutations on subsequent generations. We have, therefore, begun to combine single cohort population-
projection models with population genetics in ecological risk assessment.

     The life history traits of N. arenaceodentata are ideal for this type
of modeling because they are extremely simple. Because the females die at
spawning, we need only evaluate reproductive contributions at the end of
the lifecycle. If this worm spawned several times at different ages, the
problem would be far more complex. Moreover, if asynchronous spawning
occurred, it would have been difficult to observe such clear-cut alterations in reproductive success. Information on life history characteristics of N. arenaceodentata such as survival at each lifestage, time to

spawning, and the number of embryos spawned will be derived from our own laboratory studies and from those of Pesch et al., 1987.

In evaluating effects of genotoxic substances on aquatic organisms at the population level, the critical link between the levels of biological organization is the gametes. With one exception (Marshall, 1986), eco-toxicological studies that model impacts of contaminant exposure at the population level have not specifically evaluated the effects of genotoxic substances. Therefore, we know of few studies on aquatic organisms that have attempted to model the population-level impacts of transmitting DNA damage in gametes from one generation to the next. Basic principles of population genetics can be applied to this problem, and they can be linked to basic life history studies in aquatic ecology. An outgrowth of modeling efforts of this type would be to determine life history traits that make organisms susceptible to population-levels impacts from exposure to geno-toxic contaminants.

DISCUSSION

In this paper, we link results of four previously reported studies in a conceptual model. The purpose of developing a model is to formulate unify-ing concepts that may aid in predicting reproductive and population-level effects from data on genotoxic responses. While the model was developed using the responses of a single species exposed to ionizing radiation, we anticipate that our approach may be valuable in evaluating effects on other species and in evaluating the effects of different genotoxic substances. Formulation of the model has also aided us in prioritizing suggestions for future research on the effects of genotoxic substances on aquatic organisms and the factors that alter those effects.

Value of the Experimental Model

Some of the most fundamental concepts that were identified in the model are summarized below. First, DNA damage appears to cause alterations in reproductive success by inducing in gametes lethal mutations that result in embryo mortality. Cell death in gametes is also induced at high doses. We have found, therefore, that genotoxic effects induced in gametes can cause detrimental effects on the reproductive success of individual organisms. In addition, a proportion of the DNA damage in gametes causes heritable effects that can be quantified in subsequent generations. The extent to which these heritable effects have an impact on population size depends largely on the life history traits of individual organisms.

Because the mechanism of effect of ionizing radiation is probably similar in cells of different organisms, and because the gametes were found to be the critical link in these studies, it is likely that aspects of this model can be applied, at least in part, to other invertebrates and perhaps fish. Other genotoxic substances could also be considered if studies were conducted in parallel to ensure that the active form of the substance had a direct impact on DNA in germ cells; new modifying factors would have to be considered. This model, then, may provide others with a framework for eval-uating the magnitude and significance of reproductive effects of exposure to genotoxic substances in varied aquatic organisms.

Numerous studies support the hypothesis that mechanisms of DNA damage due to radiation exposure are common throughout the animal kingdom. This finding implies that portions of the model may be broadly applicable. Anderson and Harrison (1986) reviewed data on effects of ionizing radiation in aquatic organisms and compared these data to observations of similar low effect-levels in mammalian species. For aquatic species, lowest effect-levels were observed in studies evaluating fecundity or early development of fish. Similarly, the lowest effects observed using terrestrial species (including mammals) involved studies of impaired fecundity or mutagenesis in early development. In addition, Woodhead (1976) has demonstrated that cytogenetic responses to ionizing radiation in cells of a fish species, a toad species, and man have similar dose effects. These observations further indicate that there is little variation in the cellular and sub-cellular levels of sensitivity to ionizing radiation throughout the animal kingdom. Clearly, comparative studies on the effects of genotoxic chemicals are needed.

In our studies, genotoxic responses were predictive of the levels at which detrimental reproductive effects occurred in individual worms. This was true despite the many factors that may modify the levels at which these responses are observed. Further, we are cautiously optimistic that geno-toxic responses can be used to predict detrimental impacts at higher levels of biological organization in aquatic organisms. However, we also suggest that erroneous conclusions can be made if the factors that modify the levels at which cytogenetic responses are observed are not clearly delin-eated in an experimental model. For example, comparisons between results obtained at varied field sites must consider confounding variables, such as changes in temperature and lifestages of the organisms studied.

Future Research Needs

Future studies of the effects of genotoxic substances on aquatic organ-isms must use principles obtained from research on mammals, and they must also focus on unique differences between mammals and aquatic organisms. Results of studies on mammals can be used to identify probable mechanisms of effect at the subcellular level. In addition, we should continue to apply the general techniques of investigation on mammals for use with aquat-ic animals. However, additional studies must evaluate the important dif-ferences between research directed toward human health and ecotoxicology.

There are three particularly striking differences between these two fields. First, the tools of investigation could eventually be very dif-ferent. Dominant- and recessive-lethal assays are much more easily conduct-ed in aquatic species than in mammals. Assays of this nature should be fur-ther developed, and they can be designed to estimate effects of cumulative doses of genotoxic substances on gametes. In fact, the extremely varied life history traits of aquatic organisms make it possible to tailor the design of assays to a range of possible questions.

Second, the ecological significance of genotoxic effects is different in aquatic organisms than it is in mammals, because effects at the indi-vidual level are generally considered to be less significant. Research is needed to elucidate patterns of susceptibility to genotoxic substances for aquatic species with different life history traits. This would provide us with a better understanding of potential magnitude of population-level re-sponses for groups of organisms having common characteristics. In addi-tion, it would be valuable to determine whether reproductive effects of genotoxic exposure are more significant at the population level than are carcinogenic effects. This question could be evaluated using a modeling approach. It is possible to simulate removal of selected numbers of fish from older age groups due to cancer mortality. These data would then be compared to a simulation of the effects of removing selected numbers of embryos due to mutagenic effects expressed in early development.

Finally, some of the factors that modify genotoxic responses differ in significance between mammals and aquatic species. These factors should be studied more extensively. For example, temperature changes clearly have a more significant effect on the cellular metabolism of poikilotherms than homeotherms. It is possible that cold-adapted species may be particularly vulnerable to cumulative doses of genotoxic substances. In addition, effects induced at low temperatures may be be difficult to predict. Decades ago, researchers discovered that oocyte damage in frogs irradiated at cold temperatures was realized only after the frogs were transferred to warmer chambers. In addition, irradiated frogs survived up to 4 months at cold temperatures, but they died within 3 days after being transferred to warm temperatures (reviewed in Bacq and Alexander, 1961). These findings suggest important research needs into the unique effects of genotoxic substances in the environment. These problems, and others of this magnitude, have remained unexplored for over 30 years.

ACKNOWLEDGMENTS

We extend our gratitude to our project officer Marilyn Varela of the U.S. Environmental Protection Agency for her commitment to this program over the course of several years. This research was supported by the U.S. Environmental Protection Agency, Office of Radiation Programs (DOE-EPA Interagency Agreement DW 89930414-01-1) and was performed at the Lawrence Livermore National Laboratory under the auspices of the U.S. Department of Energy Contract W-7405-ENG-48. Participation of the senior author is this conference and the final development of the manuscript was supported by the University of California Berkeley Superfund Program Project (NIH#P42-ES04705-3). We thank Dr. William Suk the NIEHS Superfund Program Director and Dr. Martyn Smith, Berkeley Program Project Manager for their support. Dr. John Harte provided valuable insights into the population modelling approaches.

REFERENCES

Anderson, S.L., and Harrison, F.L., 1986, Effects of Radiation on Aquatic Organisms and Radiobiological Methodologies for Effects Assessments, Office of Radiation Programs, U.S. EPA, Washington, D.C., EPA 520/1-85-016, p.128.

Anderson, S.L., Harrison, F.L., Chan, G., and Moore, D.H.,II, 1989, Comparison of cellular and whole-animal bioassays for estimation of radiation effects in the polychaete worm Neanthes arenaceodentata., Arch. Environ. Contam. Toxicol., 19:in press.

Bacq, Z.M., and Alexander, P., 1961, Fundamentals of Radiobiology, Pergammon Press, Oxford, England, p.562.

Harrison, F.L., and Anderson, S.L., 1988a, Effects of Acute Radiation on Reproductive Success of Polychaete Worm Neanthes arenaceodentata, Office of Radiation Programs, U.S. EPA, EPA 520-1-88-003, p.28.

Harrison, F.L., and Anderson, S.L., 1988b, Effects of Chronic Radiation on Reproductive Success of Polychaete Worm Neanthes arenaceodentata, Office of Radiation Programs, U.S. EPA, EPA 520/1-88-004, p.35.

Harrison, F.L., Rice, D.W., Jr.,, Moore, D.H., II, and Varela, M., 1986, Effects of radiation on frequency of chromosomal aberrations and sister chromatid exchange in the benthic worm Neanthes arenaceodentata, in: "Oceanic Processes in Marine Pollution," Vol. 1, J.M. Capuzzo and D.R. Kester, eds., Krieger, Malibar, Florida.

Liguori, V.M., and Landolt, M.L., 1985, Anaphase aberrations: An in vivo measure of genotoxicity, in: "Short Term Bioassay in Analysis of Complex Environmental Mixtures IV," M.D. Waters, S.S. Sandhu, J. Lewtas, L. Claxton, G. Strauss, and S. Nesnow, eds., Plenum Press, New York.

Marshall, J.S., 1986, Populations dynamics of Daphnia pulex as modified by chronic radiation stress, Ecology, 47:561.

Mix, M.C., 1986, Cancerous diseases in aquatic animals and their associa-
    tion with environmental pollutants: A critical literature review, Mar.
    Environ. Res., 20:1.
Pesch, C.E., Zajac, R.N., Whitlach, R.B., and Balboni, M.A., 1987, Effect
    of intraspecific density on life history traits and population growth
    rate of Neanthes arenaceodentata (Polychaeta: Nereidae) in the labora-
    tory, Mar. Biol. 96:545.
Pesch, G.G., Pesch, C.E., and Malcolm, A.R., 1981, Neanthes arenaceo-
    dentata, a cytogenetic model for marine genetic toxicology, Aquat.
    Toxicol. 1:301.
Reish, D.J., 1980, The effect of different pollutants on ecologically
    important polychaete worms, U.S. EPA, EPA 600/3-80-053, p.138.
Woodhead, D.S., 1976, Influence of acute irradiation on induction of
    chromosome aberrations in cultured cells of the fish Ameca splendens.
    in: "Proc. Symp. Biol. Environ. Eff. Low-Level Radiat., Vol. I.,"
    International Atomic Energy Agency, Vienna, p.67.

FIELD STUDIES: TERRESTRIAL SYSTEMS

FLOW CYTOMETRY AS A TECHNIQUE TO MONITOR THE EFFECTS OF ENVIRONMENTAL

GENOTOXINS ON WILDLIFE POPULATIONS

John W. Bickham

Department of Wildlife and Fisheries Sciences
Texas A & M University
College Station, Texas 77843

INTRODUCTION

A critical environmental issue facing society is the problem of toxic
waste disposal. The escape of toxins into the environment has resulted in
local (e.g., Love Canal) or regional (e.g., Chernobyl) human health problems. The threat to human health posed by such pollution is profound,
including both short-term and long-term effects. Environmental mutagens
can cause somatic mutations that result in cancer or germ cell mutations
that lead to impaired fertility or birth defects. Heritable mutations can
affect levels of genetic polymorphism and evolutionary processes. Thus,
the effects of exposure to such contaminants might be expressed years after
the immediate problem has been resolved. Indeed, future generations as
well as ourselves are at risk. Therefore, the detection of environmental
mutagens, and the development of techniques to accomplish this, should be a
high research priority for both health-related and environmental-related
agencies. The primary focus of the author's laboratory has been the comparative cytogenetics and population genetics of various wildlife species.
Investigations into vertebrate chromosomal evolution have led to an
interest in mutagenesis and the role that environmental mutagens play in
ecological and evolutionary processes.

Chromosomal evolution involves two distinct events: the production of
variation through mutation and the subsequent fixation of new rearrange-
ments in populations. Nearly all evolutionary studies focus on the process
of fixation. Mutagenesis largely has been ignored as an evolutionary
phenomenon. As a result, most models of chromosomal evolution consider
mutation rates to be constant, yet we know they probably are not (Baker and
Bickham, 1980).

Remarkable patterns of chromosomal variation exist among vertebrates,
the results of differing processes of rearrangement fixation and possibly
mutation. For example, the karyotypes of some taxa are highly stable through evolutionary time. The chromosomes of turtles have been conserved without change in numerous species for as long as 200 million years (Bickham,
1981). Some groups of mammals are also highly conservative. The bat genus
Myotis is among the most speciose genera of mammals, but of the 41 species
studied, only a single species differs from the predominant 2n=44 karyotype
in any significant way (Bickham et al., 1986). Many groups of mammals exhibit a pattern of chromosomal variation in which only a single type of

In Situ Evaluations of Biological Hazards of Environmental Pollutants
Edited by S. S. Sandhu *et al.*
Plenum Press, New.York, 1990

rearrangement, such as pericentric inversion or heterochromatic addition, predominates during evolution. Vespertilionid bats related to <u>Myotis</u>, for example, show an overwhelming propensity for centric fusions (Bickham, 1979), and rodents of the genus <u>Peromyscus</u> incorporate only pericentric inversions and heterochromatic additions. This process is called karyo-typic orthoselection and for <u>Peromyscus</u> has been explained as a result of an unusual method of chromosomal synapsis during meiosis (Greenbaum et al., 1986).

The most radical pattern of chromosomal evolution is that of karyotypic megaèvolution (Baker and Bickham, 1980) in which closely related species have rearranged their karyotype so completely that banding analyses fail to identify any, or very little, homology between the two. This indicates a very rapid evolutionary rate involving many different types of rearrange-ments. Thus, various vertebrate groups differ greatly in the ways and rates at which their chromosomes change. It is possible that the process of mutagenesis, including differential susceptibility to mutation among species, plays an important role in these processes. It has been shown that there is significant variation among humans in regard to susceptibil-ity to mutagens (Hsu, 1983, 1986; Hsu et al., 1985) and it is reasonable that species level variation might exist as well. Moreover, it has been suggested that populations occurring in geographic localities contaminated with environmental mutagens might diverge genetically from other popula-tions (Yosida and Parida, 1980), although this has never been adequately tested.

To test whether mammalian species differ in their susceptibilities to environmental mutagens, two species of rodents from a toxic waste disposal site adjacent to the Firemen's Training School (College Station, Texas) were analyzed using cytogenetic and flow cytometric techniques (McBee, 1985). The two species indeed were found to differ in their response to environmental pollution (McBee et al., 1987). Although this does not prove a species-specific difference in mutation rates, it suggests the possibil-ity that species with high susceptibility to environmental mutagens might serve as sentinels. These could be highly sensitive indicators of muta-genic exposure that occur naturally at polluted sites. Such populations could be monitored through time to determine the magnitude of mutagenic exposure and possibly the success of clean-up efforts.

This paper summarizes studies that have centered on the identification of the effects of environmental mutagens at chemical dump sites upon nat-urally occurring wildlife populations. These studies have shown that cyto-genetic and flow cytometric procedures are sensitive and that wildlife are demonstrably affected by environmental mutagens at two well-characterized sites differing markedly in both habitat (terrestrial and aquatic) and origin of contaminating wastes (nuclear reactors and petrochemical wastes).

The cytogenetic and flow cytometric procedures used in this laboratory provide similar results in laboratory dosing experiments. Studies with a known chemical mutagen (triethylenemelamine or TEM) that causes chromosome breaks including heritable translocations (Rutledge et al., 1986) further documents the sensitivity of the procedures (see below) and validates this technical approach.

METHODS

Flow cytometry

Flow cytometry (FCM) is a rapid and simple means of quantifying cel-lular characteristics from populations of suspended cells. Any cellular or subcellular entity that can be labeled with a fluorescent tag can be quantified in large numbers of cells--hundreds of thousands or even millions.

Flow cytometers work on the principle that fluorescence emissions from cells stained with DNA-binding fluorochromes can be measured and used to differentiate cells with different DNA content. Cells in suspension are stained with an appropriate fluorochrome, such as DAPI (4,6-diamidino-2-phenylindole), the binding of which to DNA is stoichiometric; fluorescence intensity is linear and positively associated with DNA content. The stained cells are passed through the flow system that carries them in a single cell stream through the excitation beam. This beam from a laser, mercury vapor bulb, or other appropriate source excites the dye and causes an emission of visible light from the cell. This is measured by a photometer and processed by a pulse height analyzer that produces a pulse height distribution or DNA flow histogram. A cell is counted at a particular channel depending upon DNA content.

Figure 1 shows an example of a flow histogram in which the cells from a dividing population are apportioned into three major groups based upon DNA content. The large peak at channel 100 represents the proportion of the population in G_1 of the cell cycle which is prior to the synthesis stage. The small peak at channel 200 represents cells in G_2 and mitosis. The channels between these peaks represent the cells undergoing DNA synthesis (S phase). Therefore, information as to cell cycle dynamics can be obtained directly from the DNA histograms.

A recent advance in flow cytometry is the development of a microscope based flow cytometer (FCM) by Steen and Lindmo (1979) that is similar to the Leitz MPV flow cytometer used in the author's laboratory. The utility of these instruments results from the speed, precision, and ease with which large numbers of cells can be analyzed for DNA content. For example, the Leitz instrument can analyze up to 5,000 cells per second with a coefficient of variation (CV) of 1% or less. The technique is sensitive enough to differentiate, at a statistically significant level, cells that differ by as little as 1% in DNA content. Otto et al. (1981) demonstrated that mice exposed to a mutagen (cyclophosphamide) showed higher CVs than did controls. The magnitude and duration of the increase was dose dependent. The application of FCM for environmental mutagen testing has been reviewed by Deaven (1982).

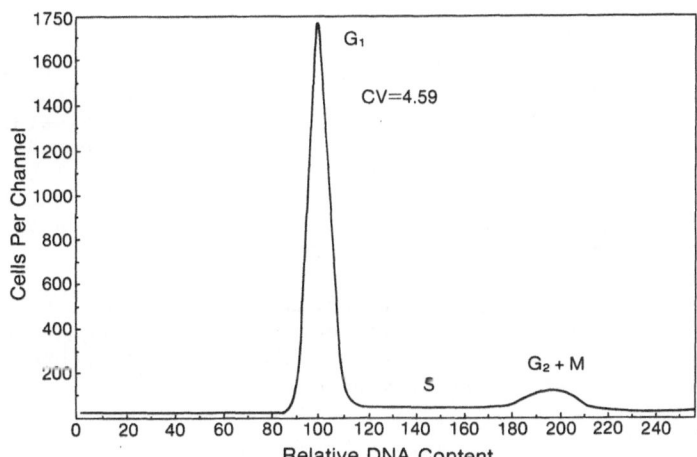

Figure 1. Representative DNA histogram showing the differentiation of a proliferating population of cells into the phases of the cell cycle.

Cell suspensions are made from mammalian spleen or bone marrow or reptilian nucleated erythrocytes, frozen in Ham's F-10 growth medium with 18% fetal calf serum and 10% glycerine, and stored at -80°C. After thawing in warm water, the cells are filtered through 37 um mesh nylon screen, washed once with fresh Hanks' BSS, and stained overnight (or longer) in 2 mL of 0.5 mg DAPI in 85 mL Tris HCl buffer. This staining procedure gives stable and DNA proportional dye binding.

Nuclear fluorescence is measured on a Leitz MPV microscope based flow cytometer. The machine is set for alignment and focus by running 0.97 um fluorescent microspheres (Polysciences Inc.) at the beginning of each day or whenever instrumental drift becomes a problem. Epi-illumination is performed with a 100 W mercury vapor (HBO-100-W2) bulb. DAPI is illuminated using an A2 filter cube with a wavelength range of 340-380 nm. All samples are coded so that dosage or experimental group is unknown when analyzed. Cells are run in stain and the G_1 peak is adjusted to channel 100 using the gain controls. A total of 10,000-20,000 cells is counted in the G_1 (2C) peak for five replicates of each run for all samples. Flow rates are from 100-200 cells per second and the system is flushed with distilled water between each sample. Pulse height distributions are analyzed on a Nuclear Data pulse height analyzer. Data are stored on diskettes using an Apple II microcomputer with a program written by Michael J. Smolen that also computes CVs.

Standard Karyotype Analysis

The procedure used in the author's laboratory is that of Baker et al. (1982), which circumvents the problems with preparation quality, consistency and mitotic indices which have been noted with the classical method (Schmid, 1975; Heddle, 1973). The technique has been modified for field use so that the necessary bone marrow cells may be obtained and fixed in the field, stored in liquid nitrogen, and returned to the laboratory for karyotype analysis.

Animals captured in the field are injected with a yeast solution to increase bone marrow mitoses (Lee and Elder, 1980) and held for 24 hours. The animals are sacrificed and bone marrow samples are removed from the femurs and tibias. These can then be field or laboratory processed to produce cell suspensions suitable for staining. Five slides are prepared for immediate examination of standard metaphase spreads, and the rest of the cell suspension is stored in liquid nitrogen for return to the laboratory. All slides are number coded and scored blind.

Fifty standard metaphase spreads are examined for each individual, and a minimum of ten individuals of each species are examined from each site. This is more animals than is customarily used in laboratory studies (Preston et al., 1981), but is reasonable for field studies where non-uniform dosing with clastogens is expected. All chromosomal breaks are scored (Figure 2). Chromatid and chromosome breaks are scored as a single lesion and rearrangements such as translocations, dicentrics and rings are scored as two lesions (cf. Hsu et al., 1981).

The described karyotype procedure works well for mammalian bone marrow. However, it is much more difficult to obtain adequate preparations from lower vertebrates, such as turtles, which is why the development of FCM as an alternative procedure is desirable.

PREVIOUS STUDIES

Three conclusions can be drawn from preliminary studies. First, cytogenetic procedures including chromosomal analysis and flow cytometry are

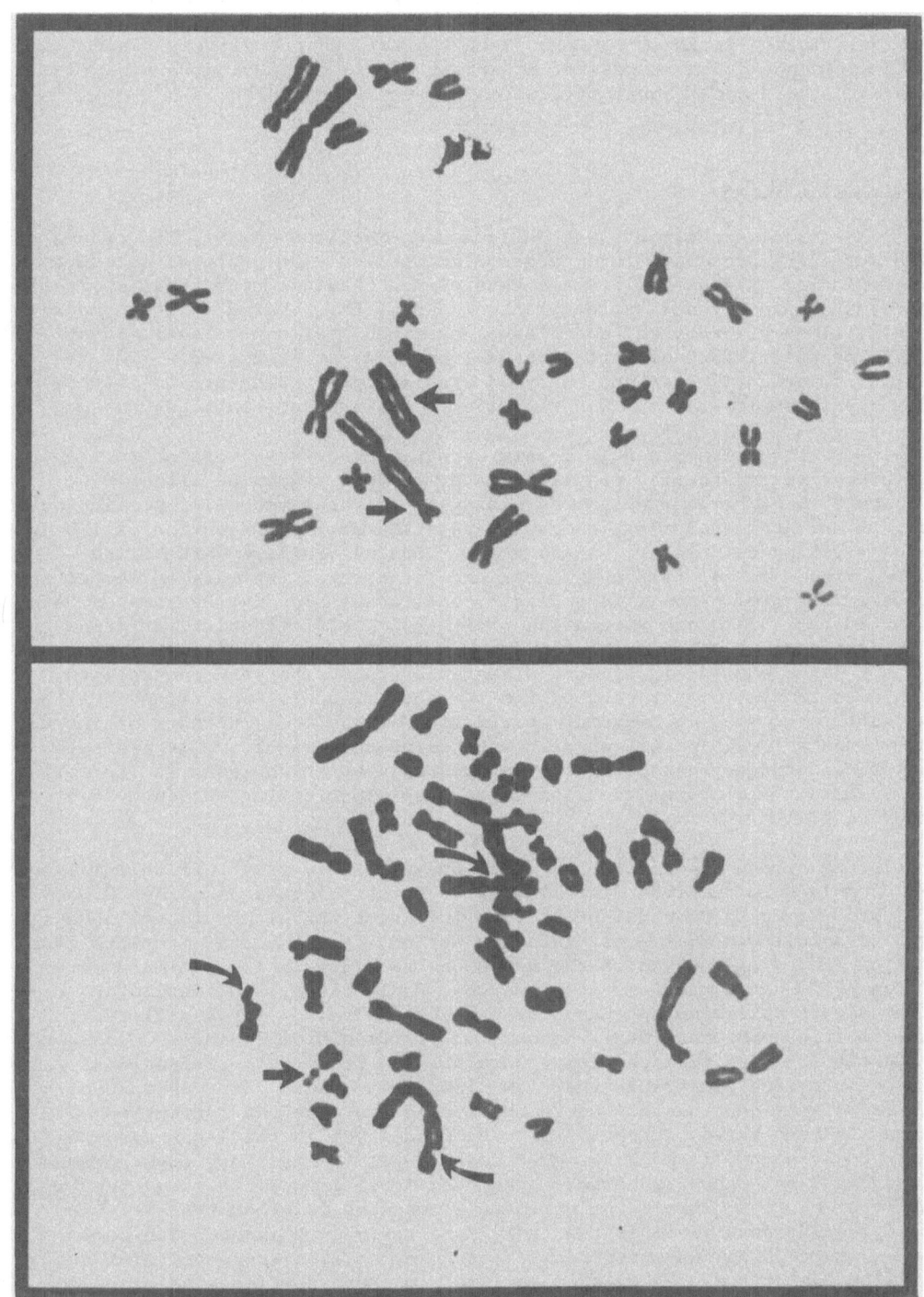

Figure 2. Mutagen-induced chromosomal aberrations. The top cell possesses a chromosomal rearrangement which has resulted in a dicentric chromosome and an acentric fragment (arrows). The bottom cell shows numerous chromatic breaks (arrows).

sensitive indicators of clastogenicity. Second, these same procedures are useful in the detection of the clastogenic effects of environmental mutagens in exposed wildlife species that in turn are potential sentinel species. Third, variation exists among the response of different species to the presence of environmental mutagens. These conclusions are based upon both field and laboratory studies as summarized below.

Laboratory Studies

A series of experiments was performed to test the use of flow cytometry as a screening procedure for mutagenic chemicals. The protocol included the dosing of Sprague-Dawley male rats with a chemical with known mutagenic and clastogenic effects--triethylenemelamine (TEM). Using various dosage levels, times of exposure, and tissues examined, cells were analyzed by FCM. The coefficient of variation was computed for each sample from the various treatment groups and analyzed statistically to determine if chemical exposure elevated the CVs relative to non-exposed controls.

In the first experiment, 15 male Sprague Dawley rats were administered TEM dissolved in sterile, distilled water at dose levels of either 0.0, 0.5, or 0.75 mg/kg body weight by a single, intraperitoneal (i.p.) injection. At 22 h after dosing, animals were given an i.p. injection of 1.0 mg/kg body weight colchicine. At 24 h after dosing, animals were killed. Bone marrow, testis, spleen, kidney, and large intestine were collected and prepared for flow cytometric analysis; additional bone marrow samples were prepared for chromosome aberration analysis. Coefficients of variation were computed for DNA flow histograms generated from all tissues. Student's t Tests showed significant differences in CVs between control and both dose groups but not between the two dose groups. Comparison of flow cytometry data with chromosomal aberration data showed a pattern of increasing values for each parameter with increasing dose level. This indicates that flow cytometry parallels the chromosome aberration assay in its ability to detect a clastogenic response in rats experiencing an acute 24 hr. exposure to TEM.

In the second experiment, 45 male Sprague Dawley rats were administered TEM dissolved in sterile, distilled water at dose levels of either 0.0, 0.1, or 0.25 mg/kg body weight by i.p. injection for either 1 day, 5 days, or 5 days followed by a 7-day recovery period. At 24 hours after the last dose of TEM or at the end of the seven day recovery period, animals were killed by CO_2 asphyxiation. Bone marrow, whole blood, and testicular tissues were collected and prepared for FCM analysis. Three different DNA-specific fluorochromes (DAPI, Hoechst 33258, Propidium Iodide), which have different DNA binding mechanisms, were used to stain cell suspensions. All tissues provided analyzable DNA flow histograms. When compared to control values, significant elevations in CVs around the mean DNA content were detected in bone marrow at the 0.25 mg/kg dose level in the 1-day dose regimen, and at both 0.1 and 0.25 mg/kg dose levels in the 5-day dose regimen. After the 7-day recovery period, both dose level groups had recovered to the level of the control. Significant elevations in blood CVs were seen in the 0.1 mg/kg dose level in the 1-day dose regime and in both TEM dose levels in the 5-day dose regimen. Testicular samples showed significantly elevated CVs at the 0.25 mg/kg dose level in the 5-day dose regimen. The only significant elevation in CV from the 5-day dosing plus 7 days recovery regimen was seen in testicular samples at the 0.25 mg/kg dose level. All three fluorochromes yielded similar results in bone marrow indicating that increases in CV probably are due to perturbations in the DNA resulting from interactions with the clastogen, TEM, rather than differences in how the dyes bind to DNA relative to mutagen binding.

The above experiments indicate that FCM analyses parallel chromosome analyses and both are sensitive indicators of chemical clastogenicity. Moreover, the data suggest FCM is useful to detect an acute response after 24 h exposure or 5 days of exposure in a variety of tissues. In rats exposed for five days and allowed to recover for seven days, the somatic tissues recovered but germinal tissue (testis) did not. The above described studies are being prepared for publication and were presented at the 1987 Environmental Mutagen Society annual meeting (Sawin et al., 1987a, b).

Environmental Studies

Studies of the effects of environmental mutagens upon wildlife populations have been carried out at the Firemen's Training School (FS) and the Savannah River Plant (SRP). The FS study included both chromosomal analysis and FCM analysis and was reported in a doctoral dissertation (McBee, 1985) part of which is published (McBee et al., 1987; McBee and Bickham, 1988). Studies of the SRP site are ongoing (Bickham et al., 1988; Lamb et al., submitted). The following is a summary of the findings of these studies.

Firemen's Training School. Small mammals of two species (Peromyscus leucopus and Sigmodon hispidus) were trapped at a locality polluted with a complex mixture of petrochemical waste products, heavy metals, and PCBs, and from two matched, uncontaminated localities. Contamination and mutagenic potential of the FS site has been well documented by chemical and biological assays (Atlas et al., 1985; Brown, 1980; Brown and Donnelly, 1982). Cytogenetic and flow cytometric techniques were employed to evaluate the use of these resident small mammals as indicators of environmental mutagenesis. Both techniques also were assessed for power of resolution in characterizing the action of environmental mutagens.

Standard karyological analysis and flow cytometric analysis both clearly indicated significant differences in chromosomal aberrancy between animals collected at the polluted site and the uncontaminated site. Standard karyology showed increases in lesions per cell and aberrant cells per individual for both species at the polluted site (McBee et al., 1987). Peromyscus apparently was more susceptible to chromosomal aberration than Sigmodon at both the polluted site and the control site.

Examination of flow DNA histograms of Peromyscus from the polluted site (McBee and Bickham, 1988) revealed broadened and flattened G_1 peaks and increases in CVs for DNA content. CVs in animals from the polluted site consistently fell outside the 95% confidence limits set around values from animals collected at the control site. These patterns are characteristic of laboratory animals challenged with powerful clastogens (see above) and suggest that individuals at the polluted site may be experiencing similar clastogenic events.

This study demonstrated that small mammals are a feasible test model for evaluating environmental mutagenesis. Moreover, FCM analysis was found to be as effective, and much less time consuming and costly, than the chromosome assay. In these respects, the field study paralleled the laboratory studies and validated the use of FCM alone, particularly for animals for which chromosome analysis is impractical.

Savannah River Plant. The genetic effects of low-level radiation on natural populations of vertebrates are poorly understood. No widely used assays for these effects exist. In this study FCM was used to document the levels of DNA content variation in blood cells of slider turtles (Pseudemys scripta) living in catchment basins contaminated with low levels of radioactivity at the Savannah River Plant, a nuclear fuels production facility of the U.S. Department of Energy. This species was selected for two reasons. First, Pseudemys scripta is a long-lived species, with broad ecological tolerances, and widely distributed in the eastern United States. Individuals live up to 20 years (Gibbons and Semlitsch, 1982); thus, the effects of long-term exposure to radiation might be similar to other long-lived species such as man. Second, the ecology and population demography of this species has been studied intensively at the SRP, which offers the possibility that correlations between the genetic effects of environmental mutagens and effects upon life history parameters might be discovered.

In this study (Bickham et al., 1988), turtles were collected from an uncontaminated control site (N=6) and from radioactively and chemically contaminated seepage basins (N=16). FCM analysis was performed on blood cells taken from live turtles, each of which was measured for radioactivity. Coefficients of variation of the G_1 peaks from the flow histograms were computed for all individuals. The mean CV for the control population (X=2.983%) differed significantly from that of the experimental population (X=3.838%) using the Mann-Whitney test (P<.001). Only individuals with a normal (nonmosaic) DNA histogram were used in the CV analysis.

Ninety-five percent confidence limits were calculated for the CVs of the control population (2.629% < u < 3.337%). The upper confidence limit of the controls was exceeded by 12 individuals (75%) of the experimental population (N=16). None of the experimental population animals fell below the lower confidence limits, and four fell within the confidence limits of the control population. A single individual (16.6%) of the control group (N=6) exceeded the confidence limits.

Evidence of aneuploidy mosaicism was found in four of twenty experimental site turtles. Those individuals possessed histograms in which the G_1 peak had a distinct shoulder or adjacent peak. No mosaic individuals were found among the six control site animals. Statistical analysis using Fisher's Exact Test indicated the observed difference in frequency of mosaicism between the two populations was nonsignificant (P>.05).

Spearman's Rank Correlation Test and the Hotelling-Pabst Test were used to test for a significant correlation between CV and plastron length for nonmosaic male (N=10) and female (N=6) experimental turtles. Both tests revealed a significant positive correlation (P<.10) indicative of a trend of increasing CV with size (and age) of the animal. Limitations of sample size did not permit this analysis to be undertaken with control females, but control males did not possess a significant correlation (P>.10).

The turtles inhabiting the seepage basins (the experimental populations) had levels of radioactivity ranging from 500-35,000 CPM. These populations possessed a significantly increased mean CV of the G_1 peaks compared to the control population. Similar results have been obtained for both chemical- and radiation-induced mutagen studies, and we conclude that our results indicate an increased level of radiation-induced chromosomal damage in the experimental populations. Moreover, some specimens possessed abnormal DNA histograms indicative of aneuploid cell populations. Sample sizes were too small for this observation to be statistically significant. Nonetheless, it can be concluded that FCM is a sensitive assay for the detection of aneuploidy mosaicism (McBee, 1985) and that such individuals were present in the experimental site population.

Our data also revealed a significant positive correlation between CV and plastron length in the experimental turtles. This correlation was not observed among control site males. Thus, it appears that increased CV might be the result of the gradual accumulation of mutations in the cell population, with older, larger turtles having greater numbers of aberrant

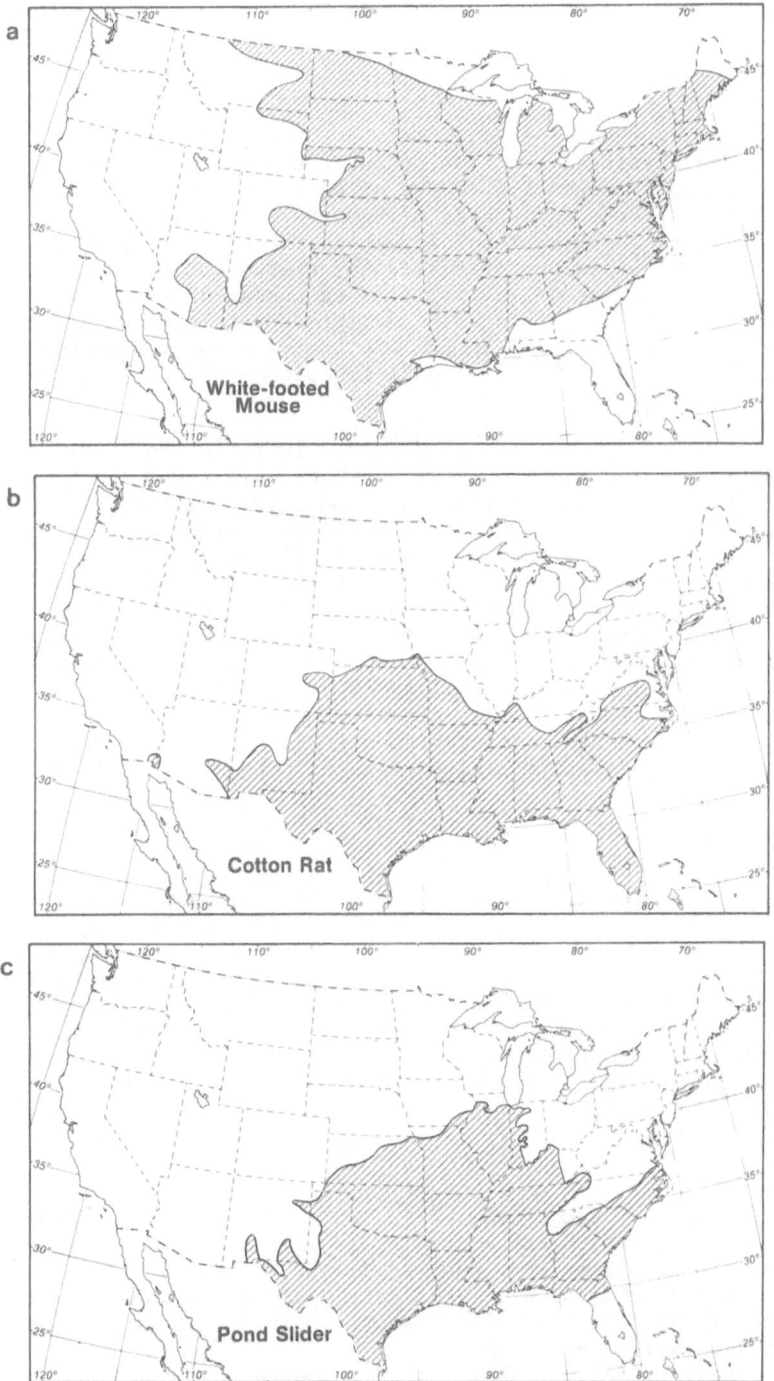

Figure 3.   Geographic distributions of the (a) white-footed mouse
(<u>Peromyscus</u> <u>leucopus</u>), (b) cotton rat (<u>Sigmodon</u> <u>hispidus</u>), and
(c) the pond slider (<u>Pseudemys</u> <u>scripta</u>) in the United States.

cells. If this is true, then it suggests that long-lived species, such as turtles, might be better indicators of the mutagenic effects of low level radiation than short-lived species such as rodents.

DISCUSSION AND CONCLUSIONS

The above environmental studies indicate that the procedures used are sensitive to the clastogenic action of environmental mutagens. FCM, in particular, is highly sensitive to the effects of both chemical mutagens and ionizing radiation. Advantages of FCM over chromosomal assays are that it is of lower cost and less labor intensive, much more rapid, and possibly more sensitive. In addition, virtually any tissue can be used and organisms that do not lend themselves well to chromosome analysis can easily be assayed with FCM. Laboratory studies using a potent mutagen indicate FCM analysis detects perturbations in the DNA that are correlated with visible chromosomal aberrations that are a well accepted endpoint in genotoxicity assays.

The selection of species to be monitored in field studies is an important decision (McBee and Bickham, 1989). The species used in the previous studies, white-footed mouse, cotton rat, and slider turtle, are all common in their respective environments, have relatively broad ecological tolerances, and have broad distributional ranges (Figure 3). All of these characteristics contribute to the capability of these species as being excellent monitors (McBee and Bickham, 1989). An additional consideration is the sensitivity of a species to the effects of mutagens. The white-footed mouse was found to have higher levels of chromosomal aberrations than the cotton rat, thus making the possibility of a hierarchy of species, from relatively resistant to relatively sensitive species, a possibility. The determination of where humans fit in such a hierarchy would be important in the establishment of relative health risks of contaminated sites.

ACKNOWLEDGMENTS

The genetic toxicology studies discussed in this paper were funded by Shell Development Co., Shell Oil Co., the U.S. Department of Energy, the University of Georgia's Institute of Ecology, and the Texas Agricultural Experiment Station. The flow cytometer used in these analyses was purchased by a grant from the Caesar Kleberg Foundation for Wildlife Conservation. Sincere thanks are expressed to the numerous students and colleagues who have participated in these studies, especially K. McBee, T. Lamb, J. W. Gibbons, and V. Sawin. L. Ruedas critically reviewed an early draft of the manuscript.

REFERENCES

Atlas, E.L., Donnelly, K.C., Giam, C.S., and McFarland, A.R., 1985, Chemical and biological characterization of emissions from a fireperson training facility, Am. Ind. Hyg. Assoc. J., 46:532-540.
Baker, R.J., and Bickham, J.W., 1980, Karyotypic evolution in bats: evidence of extensive and conservative evolution in closely related taxa, Syst. Zool., 29:239-253.
Baker, R.J., Haiduk, M.W., Robbins, L.W., Cadena, A., and Koop, B.F., 1982, Chromosomal studies of South American bats and their systematic implications, in: "Mammalian Biology in South America," M.A. Mares, and H.H. Genoways, eds., Spec. Publ. Ser., Pymatuning Lab. Ecol. VI:303-327.
Bickham, J.W., 1979, Chromosomal variation and evolutionary relationships of vespertilionid bats (Mammalia: Chiroptera), J. Mammal., 60:350-363.
Bickham, J.W., 1981, 200 million year old chromosomes: Deceleration of the rates of karyotypic evolution in turtles, Science, 212:1291- 1293.

Bickham, J.W., Hanks, B.G., Smolen, M.J., Lamb, T., and Gibbons, J.W., 1988, Flow cytometric analysis of low level radiation on a natural population of turtles (Pseudemys scripta), Arch. Environ. Contam. Toxicol., 17:837-841.

Bickham, J.W., McBee, K., and Schlitter, D.A., 1986, Chromosomal variation among seven species of Myotis (Chiroptera: Vespertilionidae), J. Mammal., 67:746-750.

Brown, K.W., 1980, Final report to the Firemen's Training School of the Texas Engineering Extension Service. A plan to minimize the volume of runoff water which must be treated and disposed of and an assessment of the feasibility of land disposal of the runoff water and sludge. Soil and Crop Sciences Dept., Texas A & M University, College Station, Texas.

Brown, K.W., and Donnelly, K.C., 1982, Mutagenic potential of water concentrates from the effluent of a waste oil storage pond, Bull. Environ. Contam. Toxicol., 28:424-429.

Deaven, L.L., 1982, Application of flow cytometry to cytogenetic testing of environmental mutagens, in: "Cytogenetic Assays of Environmental Mutagens," T.C. Hsu, ed., p.325-351, Allanheld, Osmun and Company, Publishers, Totowa, New Jersey.

Gibbons, J.W., and Semlitsch, R.D., 1982, Survivorship and longevity of a long-lived vertebrate species: how long do turtles live? J. Anim. Ecol., 51:523-527.

Greenbaum, I.F., Hale, D.W., and Fuxa, K.P., 1986, Synaptic adaptation in deer mice: a cellular mechanism for karyotypic orthoselection, Evolution, 40:208-213.

Heddle, J.A., 1973, A rapid in vivo test for chromosomal damage, Mutat. Res., 18:187-190.

Hsu, T.C., 1983, Genetic instability in the human population: A working hypothesis, Hereditas, 98:1-9.

Hsu, T.C., 1986, Genetic susceptibility to carcinogenesis, Cancer Bull., 38:125-128.

Hsu, T.C., Au, W.W., Strong, L.C., and Johnson, D.A., 1981, A short-term cytogenetic test for genetic instability in humans, p.217-235, in: "Short-term Tests for Chemical Mutagens," H. Stich, and R.H.C. Sans, eds., Springer-Verlag, New York.

Hsu, T.C., Cherry, L.M., and Samaan, N.A., 1985, Differential mutagen susceptibility in cultured lymphocytes of normal individuals and cancer patients, Cancer Genet. Cytogenet., 17:307-313.

Lamb, T., Bickham, J.W., Gibbons, J.W., Smolen, M.J., and McDowell, S., 1989, Genetic damage in a population of slider turtles (Trachemys scripta) inhabiting a radioactive reservoir (submitted).

Lee, M.R., and Elder, F.F.B., 1980, Yeast stimulation of bone marrow mitosis for cytogenetic investigations, Cytogenet. Cell Genet., 26:36-40.

McBee, K., 1985, Chromosomal aberrations in resident small mammals at a petrochemical waste dump site: a natural model for analysis of environmental mutagenesis, Unpublished Ph.D. diss., Texas A&M University, College Station.

McBee, K., and Bickham, J.W., 1988, Petrochemical related DNA damage in wild rodents detected by flow cytometry, Bull. Environ. Contam. Toxicol., 40:343-349.

McBee, K., and Bickham, J.W., 1989, Mammals as bioindicators of environmental toxicity, in: "Current Mammalogy," Vol. 2., H.H. Genoways, ed., Plenum, New York, pp.37-88.

McBee, K., Bickham, J.W., Donnelly, K.C., and Brown, K.W., 1987, Chromosomal aberrations in native small mammals (Peromyscus leucopus and Sigmodon hispidus) at a petrochemical waste disposal site, I. Standard karyology, Arch. Environ. Contam. Toxicol., 16:681-688.

Otto, F.J., Oldiges, H., Gohde, W., and Jain, V.K., 1981, Flow cytometric measurement of nuclear DNA content variations as a potential in vivo mutagenicity test, Cytometry, 2:189-191.

Preston, R.J., Au, W., Bender, M.A., Brewen, J.G., Carrano, A.V., Heddle, J.A., McFee, A.F., Wolff, S., and Wassom, J.S., 1981, Mammalian in vivo and in vitro cytogenetic assays: A report of the U.S. EPA's Gene-Tox program, Mutat. Res., 87:143-188.

Rutledge, J.C., Cain, K.T., Cacheiro, N.L.A., Cornett, C.V., Wright, C.G., and Generoso, W.M., 1986, A balanced translocation in mice with a neurological defect, Science, 231:395-397.

Sawin, V.L., McBee, K., and Bickham, J.W., 1987a, Flow cytometry as a screen for in vivo clastogenicity: effects in various tissues, Environmental Mutagen Society Annual Meeting, (abstract).

Sawin, V.L., McBee, K., and Bickham, J.W., 1987b, Flow cytometry as a screen for in vivo clastogenicity: response with time. Environmental Mutagen Society Annual Meeting, (abstract).

Schmid, W., 1975, The micronucleus test. Mutat. Res., 31:9-15.

Steen, H.D., and Lindmo, T., 1979, Flow cytometry: a high-resolution instrument for everyone, Science, 204:403-404.

Yosida, T.H., and Parida, B.B., 1980, Karyotype evolution, species differentiation and environmental mutagen, Proc. Jpn Acad., 56:79-84.

DEVELOPING COMPREHENSIVE FIELD STUDIES TO IDENTIFY SUBCHRONIC AND CHRONIC EFFECTS OF CHEMICALS ON TERRESTRIAL ECOSYSTEMS: ECOSYSTEM HEALTH - VI

Edward W. Novak[1] and David J. Schaeffer[2,3]

[1]USA-CERL (EN)
P.O. Box 4005
Champaign, Illinois 61820

[2]Department of Veterinary Biosciences
University of Illinois
2001 South Lincoln Avenue
Urbana, Illinois 61801

[3]Corresponding author

INTRODUCTION

The classical definition of an ecosystem couples interacting living organisms and non-living components of the environment to form one physical system (Tansley, 1935) and grew from the recognition that definable and describable units existed in nature. Ecosystem analysis has been advanced by an improved understanding of how ecosystems are structured and how they function. Ecology has advanced from an emphasis on natural history to consideration of energetics, the relationships and connections between species, hierarchies, and systems theory. Still, we consider ecosystems as entities with a distinctive character and individual characteristics. Measures of human or nonhuman animal health, and the clinical analysis of factors that contribute to a definition of a state of health, provide useful analogs to the problems faced by environmental managers attempting to maintain the integrity of ecosystems (Schaeffer et al., 1988). As found in human/nonhuman animal health studies, disease states of ecosystems must be recognized before disease is of clinical magnitude. This paper discusses a comprehensive, systematic approach to ecosystem analysis based on identification and quantification of factors that define the condition or state of an ecosystem in terms of health criteria.

Toxicity is defined as any harmful effect of a chemical (or drug) on a target organisms. As defined by the Organization for Economic Cooperation and Development (OECD) panel of experts (1981), acute toxicity is the "adverse effects occurring within a short time of (oral) administration of a single dose of a substance or multiple doses given with 24 hours." Subchronic toxicity is "the adverse effects occurring as a result of the repeated daily (oral) dosing of a chemical to experimental animals for part (not exceeding 10%) of the life span." Subchronic exposure can last for different periods of time, but 90 days is the most common test duration. Chronic exposure studies are similar to subchronic studies except the period of exposure is longer, and ranges from 6 months to a lifetime. When

In Situ Evaluations of Biological Hazards of Environmental Pollutants
Edited by S. S. Sandhu *et al.*
Plenum Press, New York, 1990

109

used in ecosystems, these terms are used to define the nature of the "effect" rather than to delineate an exposure regime. Because there are no standard definitions for these terms when they are used as modifiers of "effect" in ecosystems, we begin by developing these definitions.

An ecosystem is commonly defined as an interacting system of living and non-living components in the environment. However, because analysis and testing can seldom encompass all elements of an ecosystem, science generally deals with a subset of an ecosystem. This subset, termed an ecological system, has limits imposed by definition, sampling procedures, or a range of other factors, yet realistically deals with the interaction between living and non-living components of the environment. Like humans, ecological systems have a large number of characteristics that must operate within particular boundaries for that ecosystem to function normally. Some of the characteristics critical to ecosystem maintenance are more likely to be affected by, and lend themselves to measurements of effects by chemicals. These are given in Table 1 (Herricks and Schaeffer, 1987; Schaeffer et al., 1988). A subchronic effect is a small, but measurable change in either the mean or variance of a critical characteristic that persists for one seasonal cycle. A chronic effect is of similar or somewhat greater magnitude, but persists for several seasonal cycles. An example of a subchronic effect is a small reduction in corn tasseling only in 1988 due to pesticide applied that year. When the 1988 application continues to reduce tasseling beyond 1988, the effect is chronic.

Several researchers have independently proposed analogies between the known sciences of human/animal health and the undefined science of ecological system health (Gasto, 1980; Rapport et al., 1985; Slobodkin, 1988; Schaeffer et al., 1988). Our development of the comparison concluded that even the most basic tools used by the physician/toxicologist to ascertain the effects of chemicals on humans/animals are unavailable to the ecological system toxicologist (Table 2). One consequence of these large gaps in ecotoxicologists' systematic knowledge of ecological system diseases is the need for each study to devise a new "standard ecological system" reference.

The ecological reference, a non-dosed area or ecological system similar in other respects to the exposed system, is the ecotoxicologist's equivalent of the laboratory control. In contradistinction to laboratory controls that are maintained under the same constant environmental conditions

Table 1.   System Characteristics Critical for Ecological System Maintenance

---

1. Habitat is suitable for desired diversity and reproduction of organisms.

2. Phenotypic and genotypic diversity among the organisms.

3. Robust food chain to support the desired biota.

4. Nutrient pool adequate for desired organisms.

5. Nutrient cycling adequate to perpetuate the ecosystem.

6. Energy flux adequate for maintaining the trophic structure.

7. Feedback mechanisms for damping undesirable oscillations.

8. Temper toxic effects, including the capacity to chelate, bind, or transfer anthropogenic inputs to a degree that they are no longer toxic within the system.

Table 2. Toxicology of Environmental Systems

Physician/Toxicologist	Ecological System Toxicologist
Compendia are used to identify chemically caused diseases; standard terminology is used to describe patient's condition status, prognosis.	Diseases of ecological systems largely undefined; terminology to describe health status and diseases of ecological systems absent.
Wide body of reference data available for "Standard Man."	Body of data for ecological systems, although large, contains little information on normal patterns of change and variability in ecological systems. Virtually no data available from long-term, whole ecological system monitoring. "Standard ecological system" cannot be defined.
	Data from long-term ecological research sites may eventually allow characterization of "standard ecological systems." Additional analysis and validation of test system selection will provide tools for extending some of these characteristics to other systems.
Several types of diagnostic tools available. Their normal ranges of use and the interpretation of data are well defined.	Virtually no proven diagnostic tools. Interpretation of data is difficult and not standardized.
Toxicological concerns are:	Toxicological concerns are:
(1) Effects of poisons on organisms considered as individuals.	(1) Effects of poisons on structure and function of ecological systems.
(2) Design of therapy.	(2) Identification of alternative futures of stressed ecological systems and factors affecting probability that a particular future will be realized.
	(3) Design of remedial actions, recovery and restoration.

as the test subjects but for exposure, the ecological reference is naturally subjected to a myriad of uncontrolled environmental conditions that may differ from those in the exposed ecological system. Also, whereas a large number of replicate control units (e.g., animals) is required in laboratory experiments, practical requirements generally restrict ecotoxicological experiments to the use of one, or a few, pseudoreplicates (Hurlburt, 1984). An exception to low numbers of pseudoreplicates is offered by the use of open top chambers to isolate small, selected sections of an ecological system (Duchelle et al., 1982).

STUDY DESIGN ASSURANCE (SDA): A SYSTEMATIC APPROACH TO DIAGNOSIS

Diagnosis in ecotoxicology, as in other health sciences, requires use of systematic procedures for discovery of causes and analysis of effects. The uniqueness of each ecological system (i.e., the "patient") requires tailoring a new diagnostic process for each study. This is accomplished using "Study Design Assurance" (SDA) procedures (Schaeffer et al., 1985; 1990; Herricks and Schaeffer, 1987). SDA visualizes the diagnostic process in terms of the conceptual data base that would be obtained if the diagnostic tests were carried out. The conceptual data base is devised in a series of steps which include: (1) specifying the management objectives; (2) specific environmental objectives that are couched as clear, testable statements similar to hypotheses generated from the statement of a general theory; (3) identification and selection of one or more traits (i.e., any characteristic of the organism, population, community, or ecosystem that can be expressed quantitatively), measurements that are used to test the environmental objectives; and (4) appropriate feedback loops (i.e., iterations through the SDA process) to ensure that all relevant goals are identified and are attainable if the proposed diagnostic analysis is carried out. In our use of SDA to design ecotoxicological studies, we use Table 3 (Schaeffer and Beasley, 1989) as a guide to the types of information available from different levels of investigative effort.

RESULTS OF USING SDA TO DESIGN A REAL MONITORING STUDY

Discussion of the ecological relevance of toxicity testing is best done with a specific example. This example is from a proposed study to determine the chronic and ecological effects on a terrestrial/aquatic (forest and field/stream) ecosystem exposed chronically (but not continuously) to hydrocarbons from aerosolized tank diesel fuel. Other concurrently used chemicals contribute additional, but shorter, exposures to heavy metals (e.g., Cd, Zn) and chlorinated hydrocarbons (e.g., hexachloroethane, hexachlorobenzene, traces of phosgene).

Organisms exposed to high concentrations of these complex mixtures (CM) for short periods (0.5 - 4 h) may exhibit acute effects (decreased photosynthesis, death) and chronic effects (germ cell mutations) (Schaeffer et al., 1987). In contrast to readily identifiable effects in individuals, ecosystems will most probably show the effects of CM exposures as chronic effects. These effects would include changes in ecosystem structure such as species presence and abundance, changes in ecosystem function such as decreased nutrient cycling (Schaeffer et al., 1988), and genetic damage (e.g., fitness, birth and reproductive defects) in exposed plants and animals. Examples of chronic effects in ecosystems exposed to pollutants similar to those found in CM are: heavy metal contamination from leaded fuels, disruption of aquatic ecosystems by petroleum hydrocarbons similar to those in tank diesel (Moore et al., 1987), and species declines resulting from uptake of chlorinated organics (birds from DDT, and fish, cattle and humans contaminated by PCBs). Because toxic effects of CM could be produced at different rates and to different extents in each species, important changes in an exposed ecosystem, such as the decline or loss of a species, may occur at rates that are not detectable through casual observation, and changes may be evident only when an appreciable number of key species are examined over numerous seasons and years. Effects of climatic shifts would also have to be taken into account.

Some objectives of a monitoring program to determine the nature and magnitude of ecological effects in this ecosystem can be identified. These objectives imply development of ecological baseline data that can be used to predict the effects of physical and chemical damage on the affected ecosystem and on an unaffected reference ecosystem. These objectives are:

(1) Devise and initiate a monitoring program to establish abrupt changes and trends in ecosystem condition.

Table 3

ECOSYSTEM PROPERTY	I - BASELINE MEASUREMENTS	II - ORGANISM MEASUREMENTS	III - COMMUNITY MEASUREMENTS
A. Energetics (Food web simplification e.g. changes favoring herbivores over carnivores)	Qualitative Measures at Organism Level: (e.g. characterize food sources and feeding relationships)	Quantitative Measures (Changes relative to unexposed controls in individuals' physiology and behavior)	Quantitative Measures (Predation, Competition) Qualitative Measures at Organism Level
B. Structure (Change in species density/diversity)	Missing guilds	Quantitative Measures (Territoriality; adaptation)	Quantitative Measures (Connectivity; Vegetative diversity; Guild composition) Qualitative Measures at Organism Level
C. Life History (Age class structure alterations; Niche specialization changes; Morbidity/mortality rate increase; Reproduction rate or success decrease)	--	Quantitative Measures (Reproduction; Mortality; Survivorship; Fecundity)	Quantitative Measures (Succession patterns; Composition; Interactions)
D. Chemical Exposure Routes (Dispersal and Sinks)	--	Qualitative Measures (Through Guilds, Laboratory Studies, Models)	Qualitative/Quantitative Measure (Environmental chemistry and residues in abiotic media; Bioaccumulation)
E. Genetics (Selection pressure change; Altered gene pool; Mutagenesis and/or resistance increase)	--	Quantitative Measures (Fitness: mutation frequency and rate)	Qualitative Measures
F. Homeostasis (Qualitative: Observation; Stability/resilience decrease; Health decline; Population die-off)	--	--	--

Measurement Objectives:

I.   Identify and characterize bird, vertebrate, insect and plant guilds in terrestrial and aquatic ecosystems
II. Characterize measures such as population dynamics and organism physiology
III. Quantify measures such as gross productivity and total oxygen release

Table 4. A Minimum Ecosystem Assessment Program

TEST	PURPOSE	RESULT
i. <u>BENCH</u> (years 1-4)		
Species presence	Identify species changes between use and reference areas (missing species; unusual species presence)	Is loss of common species or introduction of uncommon ones due to chemicals?
Disease symptoms	Evidence of effect (not necessarily cause)	Increased rate of insect infestation, loss of canopy, burning of vegetation, increased or decreased rate of litter decomposition could signal toxicity.
General patterns	Patterns of occurrence, not just presence or absence, are important since interspecies relationships provide information on successional stage, overall condition, and a species functional role.	Vegetation maps might reveal patterns of association that could be related related to probable exposure.
ii. <u>GENERAL</u> (years 2-4)		
Species numbers	Use of ecosystem capacity provides information on resource use; especially valuable when compared to reference area resource use.	Expect no net decrease in species numbers; replacement of species possible; population sizes and age classes may change.
Species concordance	Concordance is the relationships among species, both absolute and relative to reference areas. Provides information on relative magnitude of the stress and absorption of the stress by the ecosystem; i.e. stress ecosystem buffer capacity is exceeded the ecosystem responds to additional stress with discernible changes in concordance.	Stress should alter structural and functional relationships. Direction and magnitude of these changes cannot be predicted at this time. controlled exposure studies are used to show that ecologically important differences between exposed and reference areas are due to chemicals (CM).
Contaminants Soils Sediments Water Tissues	Are toxic chemicals identifiable? Are they at levels that are known to cause, or are likely to cause, toxicity in exposed species?	Expect to find higher levels of Pb, Cd, Fe, other metals; chlorinated organics. Needed to establish causal chain.
Tree growth	Evidence for effects over time ascertained by comparing exposed with reference areas.	Tree cores provide logbook of effects of all stresses; expect changes in cores to correlate with levels of change in total site usage.

Table 4 (continued)

TEST	PURPOSE	RESULT
iii. DETAILED (years 3-4)		
Tree core residue analysis. (Similar analysis of bark, lichen and algae.)	Trees pick up metals and chlorinated organic compounds and store these in the core and bark, providing permanent exposure record.	Chemical analysis of cores should reveal exposure levels.
Nutrient cycling	Primary functional property of an ecosystem which will change if toxic stress applied.	In absence of compensating mechanical effects from tracked vehicles, CM should decrease cycling rates.
Spatial distribution of contaminants and tests for residual toxicity in soil, sediment, and water.	To demonstrate that an observed effect could have resulted from toxic stress, presence of toxic residuals must be demonstrated.	Expect distribution to reflect appropriate spatial relations to probable sources.
Changes in Critical System Properties: Animal behavior (foraging, use territory) Reproduction/ life-table: recruitment, breeding rate, breeding success, fitness.	Demonstration of these types of effects shows that CM toxicity is ecologically significant.	Chronic toxicity should affect behavior, reproduction, age distribution, and other "critical" indicators of ecosystem health.
(Based on Schaeffer et al. 1987), Use of Tradescantia (native or intro-duced laboratory strains) as bioindicators of CM effect	Tradescantia show many types of toxicity when exposed to CM, including mutation, change in photo-synthesis, and death.	Exposure to CM and other toxic chemicals will cause mutations and other easily measured changes.
Avian studies.	Extent of territorial use for: breeding, foraging, etc.; bioaccumulation, other damage.	Studies at other installations have demonstrated that birds provide a good measure of extent of physical damage. The additional chemical data should allow separating physical and chemical damage effects.
Controlled exposure	Establish exposure response data; demonstrate quali-tative and quantitative effects of CM; establish test plots for use as ecological references.	These studies are studies needed to demonstrate that differences found in training and reference areas could be CM related.

Table 5. Detailed Budget (M $) for Minimum Field Program[*]

Task	1	2	3	4
		Year		
		($ in 1,000)		
Establishment of transects (tr) (20 x 2 day x $0.5/day)	$ 20	5	5	5
Transect surveys (20 tr x 2 day x 3/yr x $0.5/day)	$ 60	60	60	60
Other baseline surveys (Birds, mammals, special vegetation)	$ 30	30	30	30
Traplines	$ 10	10	10	10
Chemical studies				
Soils, sediments	$ 15	20	25	25
Plant tissues	$ 15	25	30	30
Water	$ 10	15	20	20
Nutrient cycling studies	$ 0	15	25	25
Toxicity tests	$ 30	40	45	45
Data base development	$ 10	20	25	25
Integration/Expansion	$ 25	10	10	10
Travel	$ 10	10	10	15
Supplies (Chemicals, sample storage containers, etc.)	$ 10	20	25	25
Subcontracts	$ 20	20	20	20
Other studies (Avian, fish, and benthos uptake; stream surveys; air monitoring of CM)	$ 40	50	50	50
Totals	$305	$350	$390	$395

Total (4 years): $1,440,000

*Based on January 1, 1988 estimates.

(2) Determine the age structure of the current tree population, and develop growth rate equations for critical tree species. Use these data to infer effects of chemicals on the forest canopy and timber production, and confirm these analyses with appropriate sampling.

(3) Monitor to identify chemical effects (from run-off and direct aerosol exposure) on stream communities (benthos, fish, algae).

(4) Establish a chemical-effects monitoring program using specific flora (e.g., lichens) as monitors of adverse chemical effects.

(5) Perform chemical testing of soils and sediments for heavy metals and chlorinated organic compounds.

(6) Regularly measure concentrations of CM (and other chemicals) at selected sites during periods of maximum exposure.

Meeting the minimum data needs implied in these objectives requires a four year baseline study and additional intensive monitoring. At least 20 study sites, including exposed and reference areas, and aquatic and terrestrial ecosystems, must be identified and sampled. Data from baseline monitoring are used to determine the requirements for the more intensive studies needed to establish site-specific exposure-reduction plans. Proposed studies are summarized in broad categories below. Together, these studies will establish site condition and site variability. Based on these intensive studies, a plan for long-term monitoring (Hinds, 1984) can be developed. This plan will probably require monitoring some sites during each season each year. The minimum program in Table 4 emphasizes the terrestrial ecosystem since this ecosystem is presumed to receive the heaviest exposures. Table 5 places this minimum program in economic perspective. A similar one-season/one-year study in a native tallgrass prairie ecosystem failed to identify any significant ecosystem effects due to exposure alone (Schaeffer et al., 1990). A forest ecosystem could respond differently because individual plants can be exposed for years, not just a single growing season.

CONCLUSION

The science of ecotoxicology is young. The science of ecosystem health assessment is yet to be defined and recognized as a legitimate field of study. Our work is directed toward developing and demonstrating a conceptual basis for the science of ecosystem health. Studies to determine the health of the ecosystem cannot be divorced from studies to determine the health of human beings. On the one hand, changes in the health of ecosystems may foretell changes in the health of humans sharing the same environment (Beasley and Schaeffer, 1989); on the other, environmental factors which cause measurable changes in the health of humans undoubtedly also affect other species (Novak and Schaeffer, 1988).

REFERENCES

Beasley, V.R., and Schaeffer, D.J., 1989, The National Animal Poison Information Network database as a tool for ecological risk assessment. Ecotoxicol. Environ. Saf., 10:63-73.
Duchelle, S.F, Skelly, J.M., and Chevone, B.I., 1982, Oxidant effects on forest tree seedling growth in the Appalachian Mountains, Water, Air, Soil Pollut. 18:363-373.
Gasto, J., 1980, Metodologia clinica de ecosistemas, in: "Ecologia: La

Transfomacion de la Naturaleza por el Hombre," Editiorial Universitaria: Stantiago, Chile.

Herricks, E.E., and Schaeffer, D.J., 1987, Selection of Test Systems to Evaluate the Effects of Contaminants on Ecological Systems, UILU-ENG 87-2010, Department of Civil Engineering, University of Illinois, Urbana, IL.

Hinds, W.T., 1984, Towards monitoring of long-term trends for terrestrial ecosystems, Environ. Conservat. 11:11-18.

Hurlbert, S.H., 1984, Pseudoreplication and the design of ecological field experiments, Ecol. Monogr. 54:187-211.

Moore, M.N., Livingstone, D.R., Widdows, J., Lowe, D.M., and Pipe, R.K., 1987, Molecular, cellular and physiological effects of oil-derived hydrocarbons on molluscs and their use in impact assessment, Phil. Trans. R. Soc. London. Ser. B. 316:603-623.

Novak, E.W., and Schaeffer, D.J., 1988, Integrating epidemiology and epizootiology information in ecotoxicology studies, Ecosystem Health. III. Regul. Toxicol. Pharmacol., In press.

OECD, 1981, OECD Test guidelines, Report from the OECD expert groups on short term and long term toxicity, March 31, 1981, (Quoted by P.K. Chan, G.P. O'Hara, and A. W. Hayes, Principles and methods for acute and subchronic toxicity, in: "Principles and Methods of Toxicology," A.W. Hayes, ed., New York:Raven Press, 1972, p.6.

Rapport, D.J., Reiger, H.A., and Hutchinson, T.C., 1985, Ecosystem behavior under stress, Am. Nat. 125:617-640.

Schaeffer, D.J., and Beasley, V.R., 1989, Ecosystem Health. II., Quantifying and predicting ecosystem effects of toxic chemicals: Can mammalian testing be used for lab-to-field and field-to-lab extrapolations? Regul. Pharmacol. Toxicol., 9:296-311.

Schaeffer, D.J., Kerster, H.W., Perry, J.A., and Cox, D.K., 1985, The Environmental Audit. I., Environ. Manage. 9:191-198.

Schaeffer, D.J., Novak, E.W., Lower, W.R., Yanders, A., Kapila, S., and Wang, R., 1987, Effects of chemical smokes on flora and fauna under field and laboratory exposures, Ecotoxicol. Environ. Saf. 13:310-315.

Schaeffer, D.J., Herricks, E.E., and Kerster, H.W., 1988, Ecosystem Health: I. Measuring ecosystem health, Environ. Manage. 12:445-455.

Schaeffer, D.J., Seastedt, T.R., Gibson, D.J., Hartnett, D.C., Hetrick, B.A., James, S.W., Kaufman, D.W., Schwab, A.P., Herricks, E.E., and Novak, E.W., 1990, Use of field bioassessments to select test systems for relevant impact assessments or hazard evaluations, Environmental Audit. IX., Training lands in tall-grass prairie, Environ. Manage., accepted for publication.

Slobodkin, L.B., 1988, Intellectual problems of applied ecology, BioScience 38:337-342.

Tansley, A.G., 1935, The use and abuse of vegetational concepts and terms, Ecology 16:284-307.

EARTHWORM IMMUNOASSAYS FOR EVALUATING BIOLOGICAL EFFECTS OF EXPOSURE TO HAZARDOUS MATERIALS

Lloyd C. Fitzpatrick[1], Arthur J. Goven[1],
Barney J. Venables[1,2], Jorge Rodriguez-Grau[1], and
Edwin L. Coopey[3]

[1]Environmental Effects Research Group
Department of Biological Sciences
University of North Texas
Denton, Texas 76203

[2]TRAC Laboratories, Inc.
113 Cedar Street
Denton, Texas 76201

[3]Laboratory of Comparative Immunology
Department of Anatomy and Cell Biology
School of Medicine
University of California
Los Angeles, California 90024

INTRODUCTION

A noncontroversial and cost-effective system of laboratory and in situ bioassays capable of integrating variables of environmental concentration, route of exposure and bioavailability with a broadly applicable suite of toxic endpoints is needed to assess biological risks of environmental pollutants from hazardous and Superfund waste sites, both before and after clean-up. The system also would be useful in screening or categorizing wastes, such as industrial and municipal solids, combustion residues from incinerated solids, sewage treatment sludge, and dredged sediments for appropriate landfill disposal i.e., sanitary versus hazardous). An extensive literature on the basic biology and ecology of earthworms (Edwards and Lofty, 1977; Satchell and Martin, 1981; Satchell, 1983; Lee, 1985; Fitzpatrick et al., 1989) and from laboratory and in situ toxicity and/or bioaccumulation studies (Appendix 1, No. 1) supports using several earthworm species to develop standardized protocols (Appendix 1, No. 2) for evaluating biological risks of terrestrial pollutants.

Most toxicity studies and protocols with earthworms are for acute or LC50 tests. Only a few have focused on more subtle chronic or sublethal effects (Attachment 1, No. 3). Of those, the giant axon conduction velocity protocol (Drewes et al., 1984; Callahan et al., 1985) shows promise as a reliable companion to mortality tests, especially for neurotoxins. However, additional tests using earthworms are needed to assess a wide range of sublethal toxic endpoints, specifically those that can be used to predict effects in mammals (McLain et al., 1985; Greene et al., 1989) including humans.

In Situ Evaluations of Biological Hazards of Environmental Pollutants
Edited by S. S. Sandhu *et al.*
Plenum Press, New York, 1990

119

Among potential sublethal toxic endpoints that could be assayed by using earthworms, few are of greater contemporary relevance to public health than those concerning immune function. We believe there is sufficient knowledge of earthworm immunobiology (Appendix 1, No. 4) to support development of a suite of in vivo and in vitro sublethal assays for screening chemicals, complex mixtures and materials for their immunotoxic potential in mammals, including humans. The purpose of this paper is to present a brief overview of ongoing efforts in our laboratories at the University of North Texas and UCLA (Goven, et al. 1988) to develop a surrogate system of immunoassays in *Eisenia* *foetida*, *Lumbricus* *terrestris*, and *L.* *rubellus* for use in the field and, with standardized filter paper contact and artificial media or soil exposure protocols (EEC, 1984; Heimbach, 1984; OECD, 1984; Greene et al., 1989) in evaluating immunotoxic risks to mammals from environmental pollutants, and for understanding mechanisms by which xenobiotics interfere with recognition, processing, and effector phases of immune function.

RESEARCH OVERVIEW

We have four principal objectives in our research. The first is to determine which specific immune parameters (or endpoints) in earthworms are sufficiently sensitive for inclusion in an integrative assay protocol to assess immunotoxicity of xenobiotics. Initially, we will determine filter paper contact exposure-uptake/depuration dynamics and tissue concentrations of selected toxicants, and measure tissue dose and response profiles for each immune parameter in two species of earthworms. For this work we selected *E.* *foetida* and *L.* *terrestris*, and a set of phagocytic, cell-mediated, and humoral immune functions. *E.* *foetida* was chosen because of its current use in standardized acute toxicity protocols, and *L.* *terrestris* because of its sensitivity to environmental pollutants and suitability for extrapolating laboratory results to field conditions (Dean-Ross, 1983). The immune functions (Table 1) of both are well known, are analogous to those in mammals, and assay protocols for most of the immune functions have been worked out for earthworms and/or mammals.

Our second objective is to select appropriate xenobiotics as standard or reference toxicants for each immunoassay and determine their environmental exposure concentration/dose-response profiles, using both standardized filter paper contact and soil exposure protocols. Third, we will validate our model by comparing actual dose-responses for reference immunotoxicants between earthworm species, and laboratory mice and rats. Our objective is to determine "earthworm dose equivalents" of the reference toxicants necessary to produce analogous immunosuppressive events in laboratory rodents. Finally, we will standardize the immunoassays for general use in biological risk assessment.

Herein, we present an overview of the immunoassays to be evaluated and the results from a representative experiment with Aroclor 1254 on erythrocyte (E) and secretory (S) rosette formation, and phagocytosis by *L.* *terrestris* coelomocytes (coelomic leukocytes). Aroclor 1254, a commercial mix of polychlorinated biphenyls (PCBs), was chosen because it and similar mixtures have immunosuppressive effects in mammals (Vos and De Roij, 1972; Vos and Van Driel-Grootenhuis, 1972; Street and Sharma, 1975; Thomas and Hinsdill, 1978), are not easily metabolized, and can be detected in animal tissue and environmental samples using well established analytical techniques.

IMMUNE FUNCTIONS AND ASSAYS

We intend to evaluate immunoassays for each of the immune parameters listed in Table 1 and discussed below. Since most of the parameters are analogous or perhaps homologous, in the case of phagocytosis, to those in mammals, the resultant immunoassays have potential for predicting toxicity of xenobiotics to phagocytosis, and cellular and humoral immunity in mammals.

Phagocytosis

The most primitive immune response occurring throughout the animal kingdom is phagocytosis. Both earthworm coelomic leukocytes, and mammalian macrophages, neutrophils, and eosinophils have phagocytic and inflammatory response functions. These cells engulf small particles, encapsulate large particles, digest organic materials, and have bactericidal or bacterio-

Table 1.   Comparison of immune functions in the earthworm <u>Lumbricus</u> <u>terrestris</u>, to be used for assessing immunotoxic potential of environmental pollutants, with similar functions in mammals.

IMMUNE PARAMETER	EARTHWORM	MAMMALS
PHAGOCYTOSIS		
Cells:	Coelomocytes	Macrophages Neutrophils Eosinophils
Function:	Phagocytosis Inflammatory Response	
Common Characteristics:	Engulfment of Small Particles Encapsulation of Large Particles Digestion of Organic Material Bactericidal/Bacteriostatic Effects	
CELL MEDIATED IMMUNITY		
Cells:	Coelomocytes	T - Lymphocytes
Function:	Recognition of Self Rejection of Non-Self Immune Regulation (earthworm ?)	
Common Characteristic:	Xenograft Rejection Allograft Rejection Alloimmune Memory PHA Mitogen Stimulation	
HUMORAL IMMUNITY		
Cells:	Coelomocytes	B - Lymphocytes

Molecules:	Antisomes:
	  Lytic/Agglutination
	  Factors
		Antibodies

Function:	Agglutinate Particular Antigen Lyse Cells (antibody plus complement) Act as Opsonins	
Common Characteristics:	Proteins Specific Induction by Antigen 1° and 2° Specific Response	

static effects. Therefore, suppression of these activities in earthworm leukocytes by xenobiotics should indicate their immunotoxic potential on the phagocytic response of macrophages, neutrophils, and eosinophils in mammals. Examining the action of phagocytic cells (i.e., spreading, migration, ingestion, and bacterial killing) in earthworms exposed to xenobiotics may elucidate mechanisms of their toxic effects on aspects of recognition, processing, and effector phases of the phagocytic immune response.

Cell-Mediated Immunity (CMI)

Mammalian T-lymphocytes are capable of recognizing self, rejecting non-self, and functioning in immune regulation. Coelomic leukocytes in earthworms have similar capabilities, although their function in regulation of immune response has not been clearly demonstrated. Both cell types are involved in xenograft and allograft rejection, and alloimmune memory, and can be stimulated by mitogens.

To determine if earthworms can be used for predicting toxic effects of xenobiotics on CMI in mammals, we are examining their influence on the ability of L. terrestris to accept/reject tissue grafts, and leukocytes of both L. terrestris and E. foetida to undergo mitogen-stimulated transformation and effect natural cytotoxic responses. As with mammals, including humans, earthworms effect primary and secondary rejection of allografts and xenografts, but accept autografts (Cooper, 1971, 1975). Studying the processes of primary and secondary graft rejection (allo- and xenogeneic memory) and histopathological responses in earthworms exposed to xenobiotics should suggest which phases in the immune response (i.e., memory/ recognition, processing or effector) are likely to be compromised in mammals. Antigen-stimulated mitosis in immune cells is a primary immune response in animals and it blockage results in immune suppression. A commonly used diagnostic test for CMI competence is to stimulate T-lymphocytes to undergo mitosis with phytohemagglutinin (PHA) and concanavalin A (Con A). Suppression of PHA or Con A stimulated blast transformation in earthworm leukocytes by a xenobiotic would suggest its immunotoxic potential on T-lymphocyte proliferation in mammals.

Humoral Immunity

Coelomic leukocytes in earthworms produce antibacterial, lytic and agglutination factors, called antisomes, which may be analogous to mammalian antibodies produced by B-lymphocytes. Both antisomes and antibodies, agglutinate particular antigen, lyse cells (antibody plus complement) and act as opsonins. They are inducible by exposure to antigen (recognition) and respond anamnestically (memory) in sensitized (immunized) mammals, and earthworms (Stein and Copper, 1982, 1988). Thus, it should be possible to predict effects of xenobiotics on recognition/memory, processing and effector phases of humoral immunity in mammals, by studying their influence on antisome production (i.e., protein synthesis), release and activity as antibacterial, lytic and agglutination agents in earthworms. Suppression of antisome production by coelomic leukocytes may indicate similar effect on antibody production by mammalian B-lymphocytes

EFFECTS OF AROCLOR 1254 ON EARTHWORM IMMUNITY

Immunotoxicity of Aroclor 1254 (hereafter PCB) was assayed by comparing the ability of leukocytes from control and exposed L. terrestris to form erythrocyte (E) and secretory (S) rosettes against, and to phagocytize, rabbit red blood cells (RRBCs). E-rosettes are those leukocytes having a single layer of four or more RRBCs affixed to their surface. E-rosetting in mammals represents a binding of RRBCs to surface receptor molecules.

S-rosettes are leukocytes with at least two layers of four or more RRBCs attached to their surface: S-rosetting results from leukocytes actively secreting a synthesized humoral agglutination factor or immunoglobulin in the case of mammals. Phagocytic leukocytes are those ingesting at least one RRBC.

Leukocytes from exposed and control earthworms were obtained by a non-invasive extrusion protocol which we developed for sequential collection of cells, with minimal trauma, from individual earthworms during chronic immunotoxicity studies (Eyambe et al., in review). Leukocytes were incubated with RRBCs following procedures described by Stein and Cooper (1982, 1988). After incubation, the number of E- and S-rosettes, and phagocytic leukocytes were counted microscopically. Statistical comparison between control and exposed earthworms was used to assess effects of PCB on the three immune parameters (listed in Table 1).

Range-Finding Tests

Prior to challenging immune responses of rosette formation and phagocytosis in L. terrestris, we conducted "range-finding" experiments to determine appropriate exposure times and doses, relation between nominal filter paper exposure and tissue level concentrations, and estimates of LC50 and LD50 for PCB. Tissue concentrations of PCB were determined by gas chromatography with electron capture detection, following procedures published by the U.S. Environmental Protection Agency (Plumb, 1984). All work was conducted with 5-day exposures on filter paper discs in 1-pint glass jars maintained without light, at 10°C in an environmental chamber.

Acute Toxicity of PCB

Relation between nominal filter paper exposure (0.0025 - 0.80 mg cm^{-2}) and whole-earthworm tissue concentrations (56 - 1426 ug g^{-1} dry mass) of PCB, and respective mortalities gave LC50 and LD50 estimates of 0.30 mg cm^{-2} and 1140 ug g^{-1} dry mass, respectively.

Uptake - Depuration of PCB

To determine effects of PCB on rosette formation and phagocytosis, and their relations to whole body-burden concentrations during and after exposure, we exposed earthworms for five days to a nominal filter paper concentration of 0.01 mg cm^{-2}, one-fifth that of the lowest nonlethal concentration used for LC50/LD50 estimations. During exposure, earthworms were removed periodically to determine body-burden concentrations and to examine the ability of leukocytes to phagocytize and form rosettes against RRBCs. After exposure, remaining earthworms were placed in plastic containers with moist peat moss, nourished with a commercial dry baby cereal and maintained without light at 10°C for up to 4 months. The peat moss was changed weekly and earthworms were removed periodically to determine PCB concentrations, and status of phagocytosis and rosette formation.

PCB uptake was more rapid than its loss from whole earthworms. Detectable after 1 h exposure (ca. 19 ug g^{-1} dry mass), levels rose from 8 h to a maximum (ca. 185 ug g^{-1} dry mass) at five days, when exposure was terminated. After three days, post exposure levels had decreased (ca. 20%), remaining relatively constant during the following two weeks. By the third week levels had decreased by 49%, then steadily declined to ca. ug g^{-1} dry mass (3% maximum dose) during the next seven weeks. Compartmentalization of PCB into carcass and coelomic material (coelomic leukocytes plus fluid) during uptake and depuration showed virtually identical patterns between the carcass and whole-earthworm body levels. PCB was more slowly absorbed by the coelomic material, detected only after 64 h or exposure. At five days of exposure PCB level reached 183 ug g^{-1} dry mass, all of which occurred in the cellular fraction. Levels were below detection in the fluid portion of the coelom. Depuration from the coelomic material was more rapid than from the carcass, ca. 47% lost by the second week and below detection after nine weeks.

Immunotoxicity of PCB

Exposure to PCB showed no effect on E-rosette formation by L. ter-restris leukocytes during uptake or depuration phases. However, comparison between controls and experimentals showed that exposure to PCB significantly affected the humoral immune function of S-rosette formation during exposure and depuration when whole-earthworm tissue concentrations were between 136 and 185 ug PCB g^{-1} dry mass, respectively.

Figure 1 demonstrates the general relation between S-rosette formation and PCB concentrations in coelomic leukocytes during exposure/uptake and depuration. Though PCBs were below detection at 48 h, S-rosette formation was 88% normal (ca. 14 per 100 coelomic leukocytes). By 64 h when PCB was first detected, S-rosette formation was only 55%. Maximal suppression (45% normal) occurred at five days, coincident with the highest PCB level in the coelomic leukocytes. After one week of depuration, S-rosette formation had begun to recover (ca. 73% normal). Complete recovery paralleled loss of PCB from the coelom.

Phagocytic activity of coelomocytes was also affected by PCB exposure, but only at 120 h exposure when tissue concentrations were maximal.

PCB Dose-Immune Response

To determine relation between nominal filter paper exposure concentrations and tissue levels of PCB, and the three immune functions, earthworms were exposed to 0.0025, 0.005, 0.01 and 0.04 mg PCB cm^{-2} for five days. Corresponding whole-earthworm concentrations were ca. 56, 76, 185 and 221 ug g^{-1} dry mass. Formation of E-rosettes was not influenced by PCB at any of the nominal exposure or tissue concentrations. However, formation of S-rosettes against RRBCs was significantly influenced by PCB tissue levels as low as 76 ug g^{-1} dry mass (nominal exposure of 0.005 mg cm^{-2}), showing a significant dose-response from 66 to 221 ug g^{-1} dry mass. Phagocytosis was not inhibited until tissue levels of 185 ug g^{-1} dry mass (0.01 mg cm^{-2} nominal exposure) were reached.

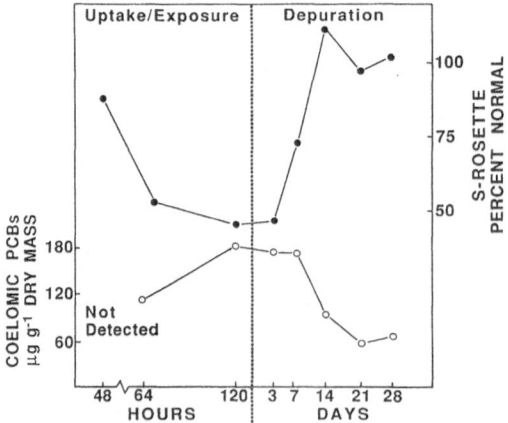

Figure 1. Relation between humoral immune function as demonstrated by S-Rosettes (0) and levels of PCBs (0) in coelomic leukocytes from Lumbricus terrestris.

124

Physiological Effects of PCB

Since suppression of S-rosette formation and phagocytosis could have resulted from an overall physiological effect by PCB and not due to a specific immune effect, we compared the general metabolic "health" of control and experimental earthworms with O_2 respirometry (Fitzpatrick et al., 1987). Mass-specific O_2 consumption rates of earthworms, exposed for five days to nominal filter paper concentrations of 0.01 mg PCB cm^{-2}, were not significantly different from nonexposed controls (31.1 vs. 29.2 uL g^{-1} h^{-1}, respectively).

DISCUSSION

Our first step in developing an earthworm model to screen environmental pollutants for immunotoxic potential in mammals is to determine which immune functions, whether considered as analogous or homologous to those in mammals, are sensitive to xenobiotics known to be mammalian immunotoxins. The experiment with PCB supports the use of one humoral immune function, demonstrated by S-rosette formation, and phagocytosis in L. terrestris to be useful in predicting immunotoxicity in mammals. Absence of an effect on E-rosetting suggests that surface receptors on earthworm leukocytes are not affected by PCB. Aroclor 1254 and similar PCB mixtures, 1248 and 1260, have been shown to suppress humoral immune function (immunoglobulin production) in rabbits (Street and Sharma, 1975), monkeys and mice (Thomas and Hinsdill, 1978) and guinea pigs (Vos and DeRoij, 1972; Vos and Van Driel-Grootenhuis, 1972). Though suppression of an analogous humoral function and a possible homologous phagocytic one was demonstrated in L. terrestris, direct comparisons with published mammalian studies are not yet possible.

Determination of earthworm dose-equivalents with reference toxicants is needed for each of the principal phagocytic, humoral and CMI parameters; the subject of the second and third phases of our research. However, based on published doses of PCBs required for immune suppression in two vastly different mammals, Rhesus monkeys and laboratory mice (Thomas and Hinsdill, 1978), L. terrestris may be more sensitive to PCBs. Suppression of S-rosetting occurred at whole-earthworm tissue concentrations beginning at ca. 76 ug g^{-1} dry mass (15 ug g^{-1} fresh mass) whereas monkeys and mice required levels, as measured in their respective subcutaneous and inguinal fat tissue, of 100 and 3760 ppm Aroclor 1248 to induce humoral suppression. If immune functions in earthworms should prove to be more sensitive than those in mammals, earthworm immunoassays may produce more false positives than false negatives. However, as McLain et al. (1985 unpublished manuscript), pointed out, in using E. foetida to predict acute toxicity in mammals, it is better to err in the conservative direction.

Wildlife tissue levels of PCBs representing a variety of invertebrate, fish and avian species typically fall within the range of 1 - 10 ppm on a fresh mass basis. Thus, the threshold immunosensitivity reported here of 15 ppm (fresh mass) is within the same order of magnitude as actual observed tissue levels in organisms from contaminated sites (Stout, 1986). That sensitivity is considerably higher than the acute of LD50 (76 vs 1140 ug g^{-1} dry mass, respectively).

In addition to being sensitive and having certain immune functions analogous to those in mammals, earthworms possess other advantages for studying immunotoxicity of terrestrial pollutants. Their large surface area to volume ratio and feeding behavior facilitate rapid uptake and tissue distribution of chemicals. Tissues are easily separated, prepared and extracted, allowing for rapid determination of actual xenobiotic doses for the corresponding toxic responses. This is especially so for immunoactive coelomic leukocytes, which we can obtain without harm to individual earthworms (Eyambe et al., in review).

We believe that an earthworm immunoassay model can go well beyond simply providing another "red flag" indicator of environmental hazard. As a surrogate for mammals, earthworms satisfy many of the criteria for developing comprehensive bioassays for use in environmental risk assessment.

ACKNOWLEDGMENTS

This paper derives from ongoing research, partially supported by a grant from the National Institute of Environmental Health Sciences (1 R01 ES34811 -01) to LCF, AJG, BJV and ELC, and Texas Advanced Research Program funds to LCF, AJG and BJV. We thank our colleagues, A.S. Kester and T.L. Beitinger for suggestions and statistical help, respectively.

REFERENCES

Beyer, W.N., and Gish, C.D., 1980, Persistence in earthworms and potential hazards to birds of soil applied DDT, Dieldrin and Heptachlor, J. Applied Ecol., 17:295-307.

Bostrom, U., and Lofs-Holmin, H., 1982, Testing side effects of pesticides on soil fauna - a critical review, Report 12, Swedish University of Agricultural Sciences and Department of Ecology and Environmental Research, Uppsala, Sweden.

Callahan, C.A., 1988, Earthworms as ecotoxicological assessment tools, in: "Proceedings of International Conference on Earthworms in Waste and Environmental Management," July 1984, C.A. Edwards and E.F. Neuhauser, eds., Cambridge, England.

Callahan, C.H., Russell, L.K., and Peterson, S.A., 1985, A comparison of three earthworm bioassay procedures for the assessment of environmental samples containing hazardous wastes, Biol. Fert. Soils, 1:195-200.

Chateaureynaud-Duprat, P., and Izoard, F., 1977, Compared study of immunity between two genera of lumbricians: Eisenia and Lumbricus. in: "Developmental Immunology," J.B. Solomon and J.D. Horton, eds., Elsevier/North Holland, Amsterdam, pp.33-40.

Cooper, E.L., 1968, Transplantation immunity in annelids, I. Rejection of xenografts exchanged between Lumbricus terrestris and Eisenia foetida., Transplantation, 6:322-337.

Cooper, E.L., 1969a, Specific tissue graft rejection in earthworms, Science, 166:1414-1415.

Cooper, E.L., 1969b, Neoplasia and transplantation immunity in annelids, J. Natl. Cancer Inst., 31:655-669.

Cooper, E.L., 1971, Phylogeny of transplantation immunity, Graft rejection in earthworms, Transplant. Proc., 3:214-216.

Cooper, E.L., 1973a, Evolution of cellular immunity, in: "Proceedings of Symposium, Nonspecific Factors Influencing Host Resistance," W. Braun and J. Ungar, eds., S. Karger, Basel, Switzerland, pp.11-23.

Cooper, E.L., 1973b, Earthworm coelomocytes: role in understanding the evolution of cellular immunity, I., Formation of monolayers and cytotoxicity, in: "Proc. III Int. Colloq. Invertebrate Tissue Culture," J. Rehacek, D. Blaskovic, W.F. Hink, eds., Slovak Academy of Sciences, Bratislava, pp.381-404.

Cooper, E.L., 1974a, Phylogeny of leucocytes: earthworm coelomocytes in vitro and in vivo, in: "Lymphocyte Recognition and Effector Mechanisms, Proc. Eighth Leucocyte Culture Conf.," K. Lindahl-Kiessling and D. Osoba, eds., Academic Press, New York, pp.155-162.

Cooper, E.L., 1974b, Transplantation immunity in annelids. IV., Influence of earthworm size on rejection, J. Invertebr. Pathol., 24:260-261.

Cooper, E.L., 1975, Characteristics of CMI and memory in annelids, in: "Immunologic Phylogeny. Adv. Exp. Med. Biol. Vol. 64," A.H. Hildemann and A.A. Benedict, eds., Plenum Press, New York, pp.127-136.

Cooper, E.L., 1976, The earthworm coelomocyte: a mediator of cellular immunity. in: "Phylogeny of Thymus and Bone Marrow-Cursa Cells," R.K. Wright and E.L. Cooper, eds., Elsevier/North Holland, Amsterdam, pp.90.

Cooper, E.L., Stein, E.A., Tochinai, S., Wojdani, A., and Lemmi, C.A.E., 1981, Earthworms, immunology and aging, in: "Workshop on the Role of Earthworms in the Stabilization of Organic Residues," M. Appelhof, Complier, Beech Leaf Press of Kalamazoo Nature Center, Proc. Vol. 1, pp.49-57.

Cooper, E.L., and Roch, P., 1984, Earthworm leukocyte interactions during early stages of graft rejection, J. Exp. Zool., 232:67-72.

Cooper, E.L., and Roch, P., 1986, Second-set allograft responses in the earthworm Lumbricus terrestris: Kinetics and characteristics, Transplantation, 41:514-520.

Cooper, E.L., Roch, P. and Wright, R.K., 1982, Phylogeny of mononuclear phagocytes, in: "Self-Defense Mechanisms: Role of Macrophages," D. Mizuno, Z.A., Cohn, A., Takaya, K., and Ishida, N., eds., University of Tokyo Press, Tokyo, Japan, pp.3-14.

Dean-Ross, D., 1983, Methods for the assessment of the toxicity of environmental chemicals to earthworms, Regul. Tox. Pharm., 3:48-59.

Drewes, C.D., Vining, E.P., 1984, In vivo neurotoxic effects of dieldrin on giant nerve fibers and escape reflex function in the earthworm, Eisenia foetida. Pestic. Biochem. Physiol., 22:93-104.

Drewes, C.D., Vining, E.P., and Callahan, C.A., 1984, Non-invasive electrophysiological monitoring: A sensitive method for detecting sublethal neurotoxicity in earthworms. Environ. Tox. Chem., 3:599-604.

Drewes, C.D., Vining, E.P., and Callahan, C.A., 1988, Electro-physicological detection of sublethal neurotoxic effects in intact earthworms, in: "Proceedings of the International Conference on Earthworms in Waste and Environmental Management," C.A. Edwards, and E.F. Neuhauser, eds., July 1984, Cambridge, England.

Edwards, C.A., 1984, Report of the second stage in development of a standardized laboratory method for assessing the toxicity of chemical substances to earthworms, Report EUR 9360 EN. Commission of the European Communities, Luxembourg.

Edwards, C.A., and J.R. Lofty, 1977, "Biology of Earthworms," Chapman and Hill, London.

EEC (Commission of the European Communities) 1984, Report of the second stage in development of a standardized laboratory method for assessing the toxicity of chemical substances to earthworms, The artificial soil test, DG X1/AL/82/83, Rev. 4, 1984.

Eyambe, G., Goven, A.J., Fitzpatrick, L.C., Venables, B.J., and Cooper, E.L., 1989, Extrusion protocol for use in long-term immunological studies with earthworm Lumbricus terrestris coelomic leukocytes, Laboratory Animal Dev. Comp. Immunol., (submitted).

Fitzpatrick, L.C., Goven, A.J., Earle, B., Rodriguez, J., Briceño, J., and Venables, B.J., 1987, Thermal acclimation, preference and effects on VO2 in the earthworm Lumbricus terrestris. Comp. Biochem. Physiol. 87A:1015-1016.

Goats, G., and Edwards, C.H., 1982, Testing the toxicity of industrial chemicals to earthworms, Rothausted Exp. Stn. Rep., 1982:104-105.

Goven, A.J., Venables, B.J., Fitzpatrick, L.C., and Cooper, E.L., 1988, An invertebrate model for analyzing effects of environmental xenobiotics on immunity, J. Clin. Ecol., 4:150-154.

Greene, J.C., Bartels, C.L., Warren-Hicks, W.J., Parkhurst, B.R., Linder, G.L. Peterson, S.A., Miller, W.E., 1989, Protocols for Short-term Toxicity Screening of Hazardous Waste Sites, U.S. EPA, EPA/600/3-88/029, Corvalis, OR.

Heimbach, F., 1984, Correlations between three methods for determining the toxicity of chemicals to earthworms, Pestic. Sci., 15:605-611.

Karnak, R.E., Hamelink, J.L., 1982, A standardized method for determining acute toxicity of chemicals to earthworms, Ecotoxicol. Environ. Sci., 6:216-222.

Lee, K.E., 1985, "Earthworms: Their Ecology and Relationships with Soils and Land Use," Academic Press, New York.

Lofs-Homlin, A., 1980, Measuring growth of earthworms as a methods of testing sublethal toxicity of pesticides-experiments with benomyl and trichloroacetic acid (TCA). Swed. J. Agric. Res., 10:25-33.

Ma, W., 1984, Sublethal toxic effects of copper on growth, reproduction and litter breakdown activity in the earthworm Lumbricus rubellus, with observations on the influence of temperature and soil pH, Environ. Pollut., (Series A) 33:207-219.

Malecki, M.R., Neuhauser, E.F., Loehr, R.C., 1982, The effects of metals on growth and reproduction of Eisenia foetida (Oligochaeta, Lumbricidae). Pediobiolgia, 24:129-137.

Neuhauser, E.F., Malecki, M.R., Loehr, R.C., 1983, Methods using earthworms for the evaluation of potentially toxic materials in soils, in: "Second Annual ASTM Symposium on testing of hazardous and industrial solid waste," R.A. Conway and W.P. Gulledge, eds., Hazardous and Industrial Solid Waste Testing, ASTM STP 805, ASTM, Philadelphia, pp.313-320.

Neuhauser, E.F., Malecki, M.R., Loehr, R.C., 1984, Growth and reproduction of the earthworm Eisenia foetida after exposure to sublethal concentrations of metals, Pedobioliga, 27:89-97.

Neuhauser, E.F., Loehr, R.C., Milligan, D.L., and Malecki, M.R., 1985, Toxicity of metals to the earthworms, Eisenia foetida, Biol. Fert. Soils, 1:149-152.

Neuhauser, E.F., Durkin, P.R., Milligan, D.L., and Anatra, M., 1986a, Comparative toxicity of ten organic chemicals to four earthworm species, Comp. Biochem. Physiol., 83c:197-200.

Neuhauser, E.F., Loehr, R.C., and Malecki, M.R., 1986b, Contact and artificial soil test using earthworms to evaluate the impact of wastes in soil, in: "Hazardous and Industrial Solid Waste Testing: Fourth Symposium, ASTM STP 886, J.K. Petros, Jr., W.J. Lacy and R.A. Conway, eds., American Society for Testing and Materials, Philadelphia, pp.192-203.

Organization for Economic Growth and Development (OECD) Guidelines for Testing of Chemicals, 1984, Section 2, Effects on Biotic Systems: Earthworm Acute Toxicity Tests, Paris, France:Organization for Economic Cooperation and Development.

Plumb, R.H., Jr., 1984, Characterization of hazardous waste sites: A methods manual, Vol. 3, Available Laboratory Analytical Methods, U.S. Department of Commerce, Environmental Protection Agency, EPA-600/4-84-038.

Roberts, B.L., Dorough, H.W., 1984, Relative toxicities of chemicals to the earthworm Eisenia foetida., Environ. Toxicol. Chem., 3:67-78.

Roberts, B.L., and Dorough, H.W., 1985, Hazards of chemicals to earthworms, Environ. Toxicol. Chem., 4:307-323.

Satchell, J.E., ed., 1983, "Earthworm Ecology: From Darwin to Vermiculture," Chapman and Hall, London.

Satchell, J.E., and Martin, K., 1981, A Bibliography of Earthworm Research, Institute of Terrestrial Ecology, Merlewood Research Station, England: Orange-over-Sands.

Stein, E.A., and Cooper, E.L., 1982, Agglutinins as receptor molecules: A Phylogenetic approach, in: "Developmental Immunology: Clinical Problems and Aging," E.L. Cooper and M.A.B. Brazier, eds., Academic Press, New York, pp. 85-98.

Stein, E.A., and Cooper, E.L., 1988, In vitro agglutinin production by earthworm leukocytes, Dev. Comp. Immunol., 12:531-547.

Stenersen, J., 1979, Action of pesticides on earthworms, Part III:

Inhibition and reactivation of cholinesterases in *Eisenia foetida* (Savigny) after treatment with cholinesterase-inhibiting insecticides, Pestic. Sci., 10:113-122.

Stout, V., 1986, What is happening to PCBs? Elements of environmental monitoring as illustrated by an analysis of PCB trends in terrestrial and aquatic organisms, in: "PCBs and the Environment, Vol. I," J.S. Waid, ed., CRC Press, Inc., Boca Raton, FL, pp. 163-205.

Street, J.C., and Sharma, R.P., 1975, Alternation of induced cellular and humoral immune responses by pesticides and chemicals of environmental concern: Quantitative studies of immunosuppression by DDT, Aroclor 1254, carbaryl, carbofuran, and methylparathion, Toxicol. Appl. Pharmacol., 32:587-602.

Thomas, P.T., and Hinsdill, R.D., 1978, Effects of polychlorinated biphenyls on immune responses of Rhesus monkeys and mice, Toxicol. Appl. Pharmacol., 44:41-51.

Thompson, A.R., 1971, Effects of nine insecticides on the numbers and biomass of earthworms in pasture, Bull. Environ. Contam. Tox., 5:577-586.

Tomlin, A.D. and Gore, F.L., 1974, Effects of six insecticides and a fungicide on the numbers and biomass of earthworms in pasture, Bull. Environ. Contam. Tox., 12:487-492.

Van Gestel, C.A.M., and Ma, W., 1988, Toxicity and bioaccumulation of chlorophenols in earthworms, in relation to bioavailability in soil, Ecotox. Environ. Saf., 15:289-297.

Vos, J.G., and De Roij, T.H., 1972, Immunosuppressive activity of a poly-chlorinated biphenyl preparation on the humoral immune response in guinea pigs, Toxicol. Appl. Pharmacol., 21:549-555.

Vos, J.G., and Van Driel-Grootenhyis, L., 1972, PCB-induced suppression of the humoral and cell-mediated immunity in guinea pigs, Sci. Total Environ., 1:289-302.

SENTINEL SURVEILLANCE SYSTEMS

ANALYTICAL EPIDEMIOLOGY IN PET POPULATIONS FOR ENVIRONMENTAL RISK

ASSESSMENT

Lawrence T. Glickman

Department of Pathobiology
School of Veterinary Medicine
Purdue University
West Lafayette, Indiana 47907

INTRODUCTION

There is a critical need to assess the potential impact on animal and human health of the estimated 70,000 man-made chemicals already in the environment and the approximately 1,000 new chemicals released each year (Hollstein et al., 1979). The process by which the potential adverse health effects of exposures to environmental contaminants are determined is called risk assessment. Risk assessment involves an evaluation of the results of epidemiologic, clinical, toxicologic, and environmental research, and extrapolation of the findings to predict the type, and estimate the extent, of health effects under given conditions of exposure (National Research Council, 1983). Inherent to the process is characterization of the uncertainties of the resulting estimates.

Risk assessment generally contains four basic elements: hazard identification, dose-response assessment, exposure assessment, and risk characterization. The goal of the process is to use factual information from these elements to define the health effects resulting from exposure of individuals or populations to hazardous substances. Once these effects are defined, it is the responsibility of regulatory agencies to weigh policy options and select the most appropriate action, taking into account the risk assessment along with economic, social, political, and technical concerns. This process is referred to as risk management.

TRADITIONAL APPROACHES TO RISK ASSESSMENT

Clinical and epidemiologic information derived from human patients should be used for quantitative risk estimation whenever it is available. When such information is lacking, laboratory animal data usually constitute the primary basis for both qualitative and quantitative risk estimation (Hoel and Hogan, 1988). However, most animal-based human risk estimation is qualitative in nature because of problems in extrapolating from the high dose levels typically employed in laboratory studies to the low dose exposures encountered in occupational and other environmental settings, and in extrapolating results from inbred rodent models to humans.

In Situ Evaluations of Biological Hazards of Environmental Pollutants
Edited by S. S. Sandhu *et al.*
Plenum Press, New York, 1990

Alternative animal models that more closely approximate human exposures to mixtures of chemical contaminants outside the laboratory are needed for both qualitative and quantitative risk assessment. An ideal animal species for risk assessment would be one that is exposed to chemical contaminants in identical habitats to humans and at similar doses. Furthermore, it should respond physiologically to these insults by manifesting a broad spectrum of pathologic conditions including behavioral and reproductive abnormalities, immunologic and biochemical perturbations, and anatomic changes including cancer. Since no animal species used for risk assessment can be expected to respond identically to humans, those whose primary concern is the assessment of chemical hazards to humans must be able to judge whether the animal data are relevant. This necessitates an understanding of the toxic properties of the chemicals in question and the comparative physiologies of the animal species tested and humans (Kendall, 1988).

USE OF NONLABORATORY ANIMALS IN RISK ASSESSMENT: ANIMAL SENTINELS

Animals outside the typical laboratory setting can provide crucial information for hazard identification, dose-response assessment, and exposure assessment. Under appropriate conditions, domestic animals and wildlife can serve to identify unknown chemical contaminants in the environment before they cause human disease, or to clarify the range and degree of risk posed by known chemical contaminants. Regular collection and systematic evaluation of animal data can alert regulatory agencies to the presence of toxic substances and thereby act as a sentinel or early warning signal. This history of intoxications manifesting in animal populations before they affect humans in the same geographic area has been previously reviewed (Buck, 1979). However, rarely have free-living animal populations been used by public health agencies to take advantage of their predictive capacity, except perhaps as indicators of infectious agents in the environment. For example, chickens in cages placed in the field and periodically bled and serologically tested have long been used as sentinels to monitor the level of arbovirus activity, thereby providing early warning for humans and other susceptible animal species. This approach, whereby animals are deliberately placed at high-risk sites in contrast to observations of naturally occurring animal populations, is referred to as an in situ method.

Epidemiologic studies of nonlaboratory animal populations or individual animals with naturally occurring disease provide an alternative approach to human health risk assessment. In contrast to laboratory animals, diseases in wildlife and domestic species result from natural exposures to a wide variety of environmental chemical contaminants including those found in the human diet, infectious agents, or pharmaceuticals. The biological effects of toxic substances in animals can be evaluated with the animal remaining in its natural habitat, i.e., in the field, on the farm, or in the home. Such settings offer an opportunity to assess the intensity of natural exposures, to measure the effects of chemical mixtures, and to determine the biological results of low level exposures over a prolonged period of time. For example, because tumors usually occur in older animals, they probably reflect chronic pathological processes associated with carcinogenesis more closely than do tumors induced in laboratory animals. Yet, while carcinogenesis in most animals results from chronic exposures, because of the animal's relatively short lifespan, there is compression of many pathophysiological responses, which results in relatively short latent periods for animal tumors.

The relationship between assessment studies using laboratory and nonlaboratory animal species is illustrated in Figure 1. Laboratory animal experiments involve administration of high doses of chemicals, usually to rodent species. The strength of this approach is its ability to identify potentially hazardous chemicals in a short time and to determine dose-response relationships under controlled conditions. Its weakness lies in its inability to assess the extent of human exposure or to evaluate health effects resulting from naturally occurring, complex chemical mixtures.

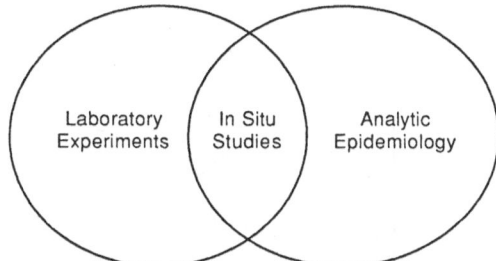

Figure 1.   Use of animals for risk assessment (Modified from: National Research Council, 1989).

Studies of animals outside the laboratory more accurately approximate the relationship between human exposures and toxic effects of chemical contaminants. Epidemiologic studies of animal populations in their natural setting can be conducted either before specific concerns arise regarding toxic effects of environmental pollutants or they can be instituted in response to a particular problem in order to measure exposure patterns and biological effects. In the later situation, it may be desirable to bring animals to a site (in situ) where exposures are suspected rather than conducting epidemiologic investigations on animals already present at the site. The in situ approach is a hybrid comprising desirable elements of controlled laboratory experiments and observational (noninterventional) epidemiologic studies. The strength of the epidemiologic and in situ methods is the capacity to integrate measures of chemical exposures with the controlled monitoring of biological effects under more natural conditions than could be simulated in the laboratory.

OBJECTIVES IN USING ANIMAL SENTINEL SYSTEMS

The term "sentinel" is derived from the French sentinelle, "watch tower." An animal sentinel system is one in which animal data are regularly and systematically collected, summarized, and analyzed in order to identify health hazards to either humans or the animals themselves from chemical or biological contaminants in the environment.

Disease results from highly complex events involving multiple, heterogeneous environmental insults occurring over a broad range of individual susceptibilities. The impact of these interactions can be appreciated only by studying population effects under natural conditions over time. Herein lies the strength of the epidemiologic method which, if vigorously applied, can bring us closer to the truth and provide a clearer picture of what appears as "messy" biology. Animals can often be used more effectively than people as subjects for such investigations.

The risk assessment process can be viewed as a scientific exercise whose goal is to bring us closer and closer to the truth (Figure 2). The measure of success in risk assessment is reflected by decreasing confidence limits surrounding the actual risk estimate. Epidemiologic and in situ animal studies can play an increasingly important role in clarifying some of the uncertainties generated from laboratory animal experiments. Because of their capacity to integrate natural exposures with biological effects, they also provide more relevant data than do fixed station monitors for

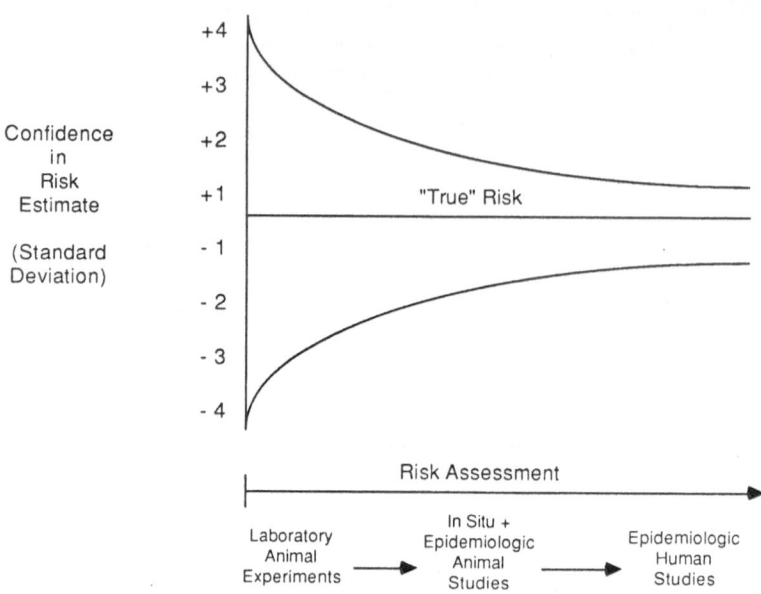

Figure 2.   Moving toward the truth in risk assessment (Modified from: National Research Council, 1989).

many environmental pollutants. The ability to "approximate the truth" in animal epidemiologic studies, as in human studies, depends mainly on scientific application of accepted epidemiologic methods and analytic techniques that control for confounding and reduce the chance of bias.

ANIMAL SENTINELS VERSUS ANIMAL MODELS

Animal models of disease have long been used in biomedical research to better understand disease pathogenesis and therapy. A model in this context is a representation of an object, system, or idea, in some form other than that of the entity itself (Shannon, 1975) whose functions are those of prediction and comparison. Animal models have been classified according to the extent that their genetic composition and their environment are controlled by the investigator (Cypess and Hurvitz, 1974). These include: (1) the laboratory that affords the greatest degree of genetic and environmental manipulation, (2) the domestic habitat, either urban or rural, in which the extent of human interference may vary and in which the animal environment may closely approximate that of humans, and (3) the natural ecological setting in which there is almost no intentional interference.

An animal sentinel is one who watches or guards in this case for the presence of disease-causing organisms or chemical contaminants. A comparison between animal models and animal sentinels is made in Table 1. The key difference between animal sentinels and models, is that for an animal sentinel to provide early warming for an environmental hazard it should be more sensitive to its biological effects than man and it should respond within a shorter latency period. Therefore, not every animal model will necessarily make a good animal sentinel.

PET ANIMALS AS ENVIRONMENTAL SENTINELS

Many animal species have the potential to serve as sentinels for risk assessment. The specific sentinel animal species selected will depend on a

Table 1. Comparison of Animal Sentinels and Animal Models

Feature	Animal Model	Animal Sentinel
Etiology	Same*	Same
Pathogenesis	Similar	Similar
Clinical disease	Similar	Similar
Exposure	Natural or experimental	Natural
Latency period	Similar or shorter	Shorter
Sensitivity to exposure	Similar	Greater

* Compared with humans

number of considerations including the target environment to be monitored (e.g., occupational sites, the home, outdoor pollution), the medium or source of exposure (e.g., air, water, soil), and the desired endpoints to be measured (e.g., exposure, biological effects, or both). In addition, some species may be more appropriate for one type of biological effect than another due to differing sensitivities or the ability to measure that endpoint based on existing biomedical knowledge or access to the species in its natural setting.

The pet population, particularly dogs and cats, is a large and relatively untapped resource for human health risk assessment. In the United States an estimated 52 million dogs and 56 million cats live in 29% to 39% of all households (American Veterinary Medical Association, 1988). Fifty-four percent of dog owners and 38% of cat owners seek regular veterinary care for their animals and many veterinarians are board certified in specialties such as internal medicine, surgery, ophthalmology, dermatology, oncology, neurology, radiology, and preventive medicine.

Private and university affiliated hospitals are the primary source of morbidity and mortality statistics for pets. Death certification is not practiced for pets nor are questions concerning pet ownership included in the census of the human population. Since the pet population at risk of developing a specific disease is usually unknown, it is difficult to accurately calculate incidence rates for diseases. However, to circumvent this problem, several population-based animal cancer registries have been established for well-defined geographic areas (MacVean et al., 1978; Schneider, 1975). In addition, morbidity and mortality data for domestic animals visiting several major veterinary teaching hospitals in the United States have been systematically collected since 1964 (Tjalma et al., 1964). Examples of how this resource could be used for risk assessment are presented in the following section focusing on cancer in pet dogs as the biological endpoint.

PET DOGS WITH CANCER AS SENTINELS: ANALYTICAL EPIDEMIOLOGIC STUDIES

If analytical epidemiologic studies in animals are to be useful for human risk assessment, then they should: (1) be capable of either determining the presence of a known human environmental hazard before it produces adverse effects in people, (2) be able to identify chemicals that cause disease in people but are less likely to be identified by using similar studies in humans because of methodologic or logistical constraints, or (3) clarify the relationship between a set of suspected complex environmental hazards for humans and various disease outcomes. The following three analytic epidemiologic studies in pet dogs illustrate these points.

MESOTHELIOMA AND ASBESTOS EXPOSURE

Epidemiological evidence indicates that asbestos is a causal factor for human mesothelioma and that the latency period for cancer development following occupational exposure usually exceeds 20 years (Selikoff et al., 1980). However, up to 40% of individuals with mesothelioma have no history of exposure to asbestos (McDonald and McDonald, 1977).

Pet dogs with spontaneous mesothelioma were used to identify environmental exposures that might increase their owners' risks of asbestos-related disease (Glickman et al., 1983). These animals were selected because they share people's domicile environment, yet do not indulge in activities (e.g., smoking and working) that confound interpretation of human epidemiologic studies. Eighteen histologically confirmed canine mesotheliomas were diagnosed at the Veterinary Hospital of the University of Pennsylvania, Philadelphia, from April 1977 to December 1981. Sixteen owners of cases and 32 owners of age, breed, and sex-matched controls were interviewed to determine their occupation and their dogs' medical history, life style, diet, and exposure to asbestos. An asbestos-related occupation or hobby of a household member and use of flea repellents on the dog were significantly associated with mesothelioma. In addition, there was a trend indicating an increased risk of mesothelioma with an urban residence. Lung tissue from three dogs with mesothelioma and one dog with squamous cell carcinoma of lung had higher levels of chrysotile asbestos fibers than lung tissue from control dogs (Table 2). These findings indicate that well-designed epidemiological studies of spontaneous tumors in pet animals may provide insight into the role of environmental factors in human cancers and serve as a valuable sentinel model to identify environmental health hazards for humans.

It might be asked how this information could be used to reduce the human risk of mesothelioma? Since the latency period for mesothelioma in the dogs following asbestos exposure is probably less than eight years while human latency exceeds 30 years, and since canine mesothelioma was associated with an asbestos-related occupation or hobby of a household member, one could require that veterinarians report any dog or cat diagnosed with mesothelioma to their local health department. Efforts could then be made to identify the source of asbestos in these households and household members screened for early radiographic signs of asbestos-related diseases. Similar screening programs might be established for owners of dogs or cats with other known or suspected environmentally related diseases (e.g., lead poisoning) if a cost-benefit evaluation proved favorable.

BLADDER CANCER, OBESITY, AND INSECTICIDE EXPOSURE

A case control study of bladder cancer was conducted in pet dogs to (1) assess exposure to insecticides and passive smoking as potential bladder carcinogens, and (2) determine whether obesity affects the risk of bladder cancer in dogs exposed to insecticides (Glickman et al., 1990). This study was prompted by epidemiologic studies in humans indicating a number of risk factors for bladder cancer including industrial chemicals (e.g., alpha- and beta-napthylamine, benzidine, and 4-aminobiphenyl), tobacco smoking, coffee consumption, schistosomiasis, and drug use (e.g., phenacetin and cyclophosphamide) (Matanaski and Elliot, 1981). However, no one factor or

Table 2. Asbestos Fiber Content in Lung Tissue of Dogs

Patient	Category	Age at Diagnosis (Years)	Fiber type (No./g dry wt.) Chrysotile	Amphibole
	Mesothelioma			
12		10	3,100,000	0
14		6	7,200,000	3,300,000
18		6	22,000,000	760,000
	Lung Cancer			
A	Squamous cell carcinoma	12	8,200,000	4,800,000
B	Bronchial-alveolar carcinoma	13	280,000	280,000
C	Bronchial-alveolar carcinoma	8	69,000	0
	Controls*			
D		5	325,000	81,000
E		9	700,000	0
F		6	300,000	200,000
G		6	2,900,000	720,000
H		8	1,100,000	0
I		8	0	340,000

* Nonrespiratory disease and noncancer (From Glickman et al., 1983)

combination of factors satisfactorily explains a majority of human bladder cancer cases.

Eighty-nine pet dogs were diagnosed with histologically confirmed transitional cell carcinoma (TCC) of the bladder between January 1982 and June 1985 at the University of Pennsylvania School of Veterinary Medicine's Surgical Biopsy Service. The 89 dogs with TCC were stratified by year of diagnosis, sex, breed size, and age. Control dogs were selected by randomly searching the records of all other dogs with biopsy reports until a comparable number of dogs with similar characteristics were identified.

The owners of the dogs were contacted by telephone and interviewed to obtain medical, residential, and diet histories for the dogs, exposures to specific household and environmental chemicals, and smoking patterns of household members. Interviews were completed for 58 TCC cases and 71 control dogs. The mean age of the TCC cases and control dogs was 11.1 ± 2.1 years and 10.8 ± 2.4 years, respectively.

Preliminary analyses revealed that the risk of TCC was not related to lifetime exposure to the passive smoking of household members (Table 3).

Table 3. Odds Ratios and Chi Square Test for Trend for Lifetime Pack Year Passive Smoke Exposure at Home

Pack Years at Home	Number of Cases	Controls	Odds Ratio
0	26	32	1.0
1-3,000	18	18	1.3
>3,000	14	21	0.8

$\chi^2 = 0.11$
$P = 0.7$

However, there was a significant dose-response relationship between TCC risk and lifetime exposure to tick and flea dips. An exposure index for each dog was calculated based on the reported pattern of tick and flea dip (none = 0, seasonal or sporadic = 1, continuous = 2) times the number of times per month that dip was applied. The results are shown in Table 4.

Since some insecticides are stored in fat deposits in the body, or fat may act as a tumor promoter, the relationship between the dog's body confirmation one year prior to diagnosis as reported by owner was evaluated in relation to the risk associated with insecticide exposure (Table 5).

Tick and flea dips for dogs contain a variety of active ingredients including organophosphate compounds and synthetic pyrethrins, as well as solvents such as xylene. Therefore, analysis of the data will be extended to include specific commercial products in order to identify a carcinogenic chemical. This has direct human health implications since the ingredients in flea and tick dips are commonly used as insecticides both inside and outside the household. Also, each time a pet animal is treated with a tick and flea dip there is likely to be significant human exposure, primarily by dermal absorption.

Based on an odds ratio of 3.5 for tick and flea dip exposure and a prevalence of use of dips of 42% among the control dogs in this study, the proportion of TCC cases in the pet dog population that can be attributed to dip use is 51%. This does not, however, include the risk of TCC associated with other sources of insecticides including flea shampoos, lawn products, household products, etc. Further studies are in progress to evaluate the importance of these other sources and to better define the role of obesity and the frequency of urination in the pathogenesis of TCC. In the meantime, however, based on these findings, the role of insecticides in the etiology of human bladder cancer should be explored.

Table 4. Odds Ratios and Chi Square Test for Trend for Tick and Flea Dip Exposure

Exposure Index	Number		Odds Ratio
	Cases	Controls	
0	23	41	1.0
1-4	23	24	1.7
>4	12	6	3.6

χ^2 = 5.4
P = 0.02

Table 5. Odds Ratios for TCC Associated with Insecticide Exposure and Body Confirmation

Body Conformation	Flea and Tick Dip Exposure	
	No	Yes
Thin or average (11/24)[a]	1.0 (16/25)	1.4
Overweight or Obese	1.4 (12/18)	9.8 (18/4)[b]

[a](No. cases/No. controls)
[b]P = 0.0003

140

BREAST CANCER, DIET, AND SURVIVAL

What people and pets eat constitutes an integral part of the household environment. In fact, in some instances, people and their dogs consume many of the same foods. For example, Sonnenschein (1988, unpublished) found that on average, 33.7% of the total calories that a dog eats come from human table food. For smaller dogs, the percentage of table food in the diet is considerably higher (up to 100%). For these reasons and because the latency period for most diseases is relatively shorter in dogs than humans, the pet dog may be an ideal sentinel for carcinogens in human foods. Pet dogs can be used to either monitor for carcinogenic risk factors in human food or to evaluate the effect of diet on recurrence rates or survival times following a diagnosis of cancer.

Obesity and increased fat intake have been associated with decreased survival of women with breast cancer. To further define these relations in a spontaneous animal model (Shofer et al., 1989) a cohort of 145 pet dogs with histologically confirmed mammary carcinoma was identified. Information was collected from the owners and veterinarians on medical and reproductive history, management or "life-style" factors, nutritional status (body conformation), treatment, tumor recurrence, and survival. A dietary history using quantitative food frequencies for one year prior to diagnosis was also obtained. A histologic malignancy score was derived for each animal based on seven pathologic criteria. The mean age of the dogs was 10.4 + 2.5 years: 37% had been ovariohysterectomized prior to diagnosis. Product-limit estimates of survival indicated that none of the dietary or nutritional factors alone were statistically significant ($p < 0.05$). However, for dogs in the low fat group (< 39% of calories derived from fat) median survival for those with greater than 27% of total calories derived from protein was 3.0 years, compared with 1.2 years and 6 months for dogs with 23-27% and less than 23% protein, respectively ($p=0.008$). When these data were fitted to a proportional hazards model, recurrence, histologic score and tumor type, percent protein, and history of pseudo-pregnancies were significant. Predicted 1-year survival rates for dogs with 15, 25, and 35% protein in their diets were 17, 69, and 93%, respectively.

SUMMARY

Wildlife and domestic animal sentinels of environmental hazards have several advantages to laboratory rodents for risk assessment. Wildlife and domestic animals can be studied in a nonexperimental setting where exposures to chemical contaminants are more likely to approximate the routes and dosages experienced by people. Epidemiologic methods can be applied to these animal species in a wide variety of natural settings including the home. Alternatively, experiments can be conducted and animals placed in foreign environments such as the workplace, or more precisely, at point sources of toxic discharges from industry--e.g., in streams or rivers. All of the advantages of human epidemiologic studies can be extended to animal populations, plus results may be obtained in a shorter period of time with animals due to their shorter lifespan.

Several disadvantages are associated with the use of wildlife and domestic animals for risk assessment. As with laboratory rodents, there is the problem of extrapolating results to humans. In this regard, more highly outbred animal populations may better predict the heterogeneity of human responses to toxins than do inbred laboratory rodents. In general, human heterogeneity in both exposure and response to toxic agents would be expected to be much greater than that observed in widely used, highly inbred, rodent model strains. This information is necessary to establish the biological basis of the uncertainty (safety) inherent to risk assessment, because of the range of human responses to carcinogens and other toxic substances (Calabrese, 1988).

Other problems presented by using nontraditional animal species include insufficient knowledge of the physiology of the species to interpret the findings, too few veterinarians trained in environmental and analytic epidemiology to conduct the studies, and the mindset of regulatory agencies, which has been slow to incorporate novel approaches to risk assessment. Recognition of the opportunities provided by advances in biochemical markers of exposure, molecular biology of disease, and biotechnology in general need to be coupled with an appreciation of how these opportunities could be effectively used in nontraditional animal species to increase our understanding of the complex interactions between host and environment.

Alternative approaches to the use of laboratory animals for human health risk assessment have traditionally implied in vitro techniques or computer simulation. An expanded concept of an alternative should include epidemiological studies of wildlife and domestic animals with naturally occurring disease. Unlike laboratory animal experiments that primarily benefit humans, research into the cause, prevention, and treatment of environmentally related diseases in wildlife and domestic animals may also benefit the individual animal, the animal species, and ultimately man's living environment.

REFERENCES

American Veterinary Medical Association, 1988, Supply and demand: the chicken or the egg syndrome? Marketing and Practice Strategies for Companion Animal Practitioners, 1:1.

Buck, W.B., 1979, Animals as monitors of environmental quality, Vet. Hum. Toxicol., 21:277-284.

Cypess, R.H., and Hurvitz, A.I., 1974, Animal models, in: "The Handbook of Laboratory Animal Science," E.C. Melby and H.H. Altman, eds., CRC Press, Cleveland, OH, pp.

Glickman, L.T., Domanski, L.M., Maguire, T.G., Dubielzig, R.R., and Churg, A., 1983, Mesothelioma in pet dogs associated with exposure of their owners to asbestos, Environ. Res., 32:305-313.

Glickman, L.T., Shofer, F.S., McKee, L.J., Reig, J.S., Goldschmidt, M.H., 1990, An epidemiologic study of insecticide exposures, obesity, and risk of bladder cancer in household dogs, J. Toxicol. Environ. Hlth. (In press).

Hoel, D., and Hogan, M., 1988, Estimation of population risks, in: "Report of the Subcommittee on Health Effects, Research Strategies Committee, Appendix D: Strategies for Health Effects Research," Y.S. Environmental Protection Agency, Washington, D.C.

Hollstein, M., McCann, J., Angelosanto, F.A., and Nichols, N.N., 1979, Short-term tests for carcinogens and mutagens, Mutat. Res., 65:133-226.

Kendall, R.J., 1988, Wildlife toxicology: A reflection on the past and the challenge of the future, Environ. Toxicol. Chem., 7:337-338.

MacVean, D.W., Monlux, A.W., Anderson, P.S., Solberg, S.L., and Roszel, J.F., 1978, Frequency of canine and feline tumors in a defined population, Vet. Pathol., 15:700-715.

Matanoski, G.M., and Elliott, E.A., 1981, Bladder cancer epidemiology, Epidemiol. Rev., 3:203-229.

McDonald, J.C., and McDonald, A.D., 1977, Epidemiology of mesothelioma from estimated incidence, Prev. Med., 6:426-446.

National Research Council, National Academy of Sciences, Committee on the Institutional Means for Assessment of Risks to Public Health, 1983, Risk Assessment in the Federal Government: Managing the Process, Washington, D.C.

National Research Council, National Academy of Sciences, Committee on the
    Animals as Monitors of Environmental Health Hazards, 1989, Early
    Warnings: Animal Sentinels of Environmental Health Hazards, Washington,
    D.C., (In press).

Schneider, R., 1975, A population based animal tumor registry, in: "Animal
    Disease Monitoring," D.G. Ingram, W.R., Mitchell, and S.W. Martin,
    eds., Charles C. Thomas, Springfield, IL.

Selikoff, I.J., Hammond, E.C., and Seidman, H., 1980, Latency of asbestos
    disease among insulation workers in the United States and Canada,
    Cancer, 46:2736-2740.

Shannon, R.E., 1975, "Systems Simulation," Prentice Hall, Englewood
    Cliffs, NJ, pp.1-35.

Shofer, F.S., Sonnenschein, E.G., Goldschmidt, M.H., Laster, L.L., and
    Glickman, L.T., 1989, Histopathologic and dietary prognostic factors
    for canine mammary carcinoma, Breast Cancer Res. Treat., 13:49-60.

Tjalma, R.A., Puester, W.A., Adelstein, E.H., and Vander Zwagg, R., 1964,
    Clinical records systems and data retrieval function in veterinary
    medicine, A proposal for systematic data programming, J. Amer. Vet.
    Med. Assoc., 145:1189-1197.

SENTINEL ANIMALS (DOGS) AS PREDICTORS OF CHILDHOOD EXPOSURE TO

ENVIRONMENTAL LEAD CONTAMINATION: OBSERVATIONS ON PRELIMINARY RESULTS

Stephanie R. Ostrowski

Surveillance and Programs Branch
Division of Environmental Hazards and Health Effects
Center for Environmental Health and Injury Control
Centers for Disease Control
Atlanta, Georgia 30333

INTRODUCTION

Childhood lead poisoning is perhaps the most common environmentally caused disease of children in the United States and in the developing world. In concept, it is a preventable disease--remove the lead from the child's environment and the disease will disappear. In practice, however, eliminating childhood lead poisoning is a complex public health problem (ATSDR, 1988; Lin-Fu, 1980).

Early screening and detection of exposure and toxicity, along with a nationwide commitment to decreasing the use of lead additives in gasoline, have succeeded in reducing the rates of severe lead poisoning. Effective screening programs, present in many major metropolitan areas, reduce the toxicity and costs of lead poisoning. However, the success of screening programs is limited by the difficulty of locating all children with lead toxicity, largely because many rural smaller urban areas do not have child-hood lead screening programs, nor do all pediatricians include blood lead testing as part of routine care. The 1988 Agency for Toxic Substances and Disease Registry (ATSDR) report to Congress of childhood lead poisoning re-commended the establishment and maintenance of effective screening programs as well as the development of additional environmental measurement techn-iques for field use (ATSDR, 1988).

Several researchers have suggested that, in addition to established childhood lead screening programs, sentinel animal screening programs be used to monitor a variety of environmental toxicants (Schilling and Stehr-Green, 1987; Schull and Cheeke, 1983; Marienfeld, 1979; Mason and Hayes, 1982; Hayes, 1976; Hayes et al., 1981; Buck, 1979; Scharding and Oehme, 1973; Glickman et al., 1983; Reif et al., 1970; Reif and Cohen, 1970). Because of the close association of children and family pets, we propose that dogs are the most appropriate sentinel animal for a variety of child-hood exposures to environmental hazards (Buck, 1979; Scharding and Oehme, 1973; Glickman et al., 1983; Reif et al., 1970). Pets share children's homes and frequently their foods; both growing puppies and toddlers indis-criminately mouth and chew any surfaces with which they come into contact. In this report we compare human and canine blood lead (PbB) data from recent field investigations: two in Jamaica and Alaska and two in

In Situ Evaluations of Biological Hazards of Environmental Pollutants
Edited by S. S. Sandhu *et al.*
Plenum Press, New York, 1990

145

Tennessee and Illinois, all with different levels of environmental lead contamination (Ostrowski et al., 1987a,b; Matte et al., 1988; State of Alaska, 1988; Ostrowski, 1988).

MATERIALS AND METHODS

Control Data

Data from a 1985 study of urban adults living near the Hollywood dump site (Memphis, TN) was used as a source of background PbB values of normal urban adults (Memphis and Shelby County, 1988). Data from a 1983 report on children in rural Idaho were used for recent comparison values for normal background PbB values in the United States (Panhandle District Health Department et al., 1986). Control PbB values for dogs were obtained from a study of 90 unexposed dogs living in a rural, non-industrialized environment in Illinois (Ostrowski et al., 1987a).

Contaminated Communities

In Tennessee and Jamaica, the communities studied contained secondary lead smelters that were the primary source of environmental contamination by elemental lead (Pb) and lead oxides (PbO and PbO_2), both highly bioavailable forms of lead (State of Alaska, 1988; Ostrowski, 1988). In Jamaica, gross contamination of the community was present, but in Tennessee, physical contamination of the nearby community was minimal. By contrast, although gross environmental contamination was present in Alaska, the form of lead in the unrefined ore (galena, PbS) was only minimally bioavailable (State of Alaska, 1988; Ostrowski, 1988).

High Environmental Contamination/Low Bioavailability: Alaska

Alaskan data were obtained from a 1988 public health department investigation of a community with high levels of environmental contamination by lead ore. Potential childhood lead poisoning was the focus for the investigation, although subsequent results indicated that all children and all non-occupationally exposed members of the community had blood lead levels well below the CDC action level (25 ug/dL Pb in whole blood) for childhood lead exposure. Samples were obtained from five dogs as part of environmental data collection.

Low Environmental Contamination/High Bioavailability: Tennessee

A 1987 study in Memphis, Tennessee, compared PbB concentrations in 82 adults and children (<12 years old) and 60 dogs in a community downwind from a secondary lead smelter equipped with modern industrial emission controls. Although contamination of the immediate factory area and nearby industrial park was well-documented by the local health department, enironmental contamination of the surface soil of the nearby residential neighborhoods was minimal.

High Environmental Contamination/High Bioavailability: Jamaica

Data from Jamaica (PbB in 157 adults, 110 children <12 years old, and 19 dogs) was obtained during the course of a 1987 field investigation into a community with severe and ongoing environmental lead contamination. Community-wide lead poisonings were related to backyard lead-acid battery repair and scrap smelting. Although a serious and ongoing childhood lead poisoning outbreak was the focus of the investigation, samples were also obtained from 19 dogs as part of the environmental data collection.

146

RESULTS

Data from these studies are compiled in Table 1.

Control Values

Data from control populations indicate that the geometric mean PbB was 7.7 ug/dL for urban U.S. adults (Memphis, TN, 1985), and 11 ug/dL for rural children (Kellogg, Idaho, 1983). Control dogs in rural Illinois (1987) had a mean PbB of 5.0 ug/dL.

Results by Study Site

Tennessee: For adults, mean PbB was 6.1 ug/dL: for children 0-18 years, 7.4 ug/dL; and for dogs, 13.5 ug/dL.

Alaska: Mean PbB for adults was 4.6 ug/dL; for children 0-12 years, 5.1 ug/dL; and for dogs, 3.3 ug/dL.

Jamaica: Mean PbB for adults was 15.3 ug/dL; for children 0-12 years, 27.0 ug/dL; and for dogs was 43.7 ug/dL.

Comparison of Mean PbB Values: Humans and Sentinel Animals

In both Tennessee and Jamaica, where highly bioavailable forms of lead were present, mean PbBs for dogs were approximately two to four times those for children and adults. In Tennessee, the mean PbB for dogs was 13.5 ug/dL (n=23) compared with children (7.4 ug/dL, n=30) and adults (6.1 ug/dL, n=52) studied, a significant difference ($p < 0.02$). In Jamaica, mean PbB for dogs was 43.7 ug/dL (N=19) compared with 27.0 ug/dL (N=110) for children and 15/3 ug/dL for adults, both significantly different ($p < 0.02$). In Alaska, there was no statistically significant difference between the mean

Table 1. Comparisons of Blood Lead Values in Adults, Children, and Dogs

	CONTROL ug/dl	ALASKA ug/dl	TENNESSEE ug/dl	JAMAICA ug/dl
ADULTS:	*7.7	4.4 (±5.1)	6.1 (±5.1)	13.6 (±1.8)
(N)	(195)	(25)	(52)	(157)
CHILDREN:	**11.0 (±2.0)	4.7 (2414.9)	7.4 (2415.8)	27.0 (±4.4)
(N)	(122)	(115)	(30)	(110)
DOGS:	5.0 (±4.4)	3.3 (±1.0)	13.5 (±8.4)	43.7 (±58.6)
(N)	(90)	(5)	(60)	(19)

* Value from 1985 study of residents near the Hollywood dump site, Memphis, Tennessee (standard deviation not available (Memphis and Shelby County, 1988).

** Value from 1983 survey in Kellogg, Idaho, of rural children < 6 years old (Panhandle District Health Department et al., 1986).

PbB values for dogs (3.3 ug/dL) and for humans (children, 5.1 ug/dL; adults, 4.6 ug/dL). Only in the Jamaican data was a statistically significant difference seen between blood lead values for children and adults (p < 0.02).

DISCUSSION

Human PbB Values: 1976-1980 Versus 1987-1988

The national geometric mean PbB values for urban adults of all races (men, 17.8 ug/dL; women, 12.9 ug/dL) and for urban children between 6 months and 6 years old (18.0 ug/dL) in the 1976-1980 Second National Health and Nutrition Examination Survey (NHANES II) study are more than twice the values reported approximately 10 years later from urban Tennessee and rural Alaska, both areas of supposed environmental contamination (Table 1). This decrease parallels the phasing-out of highly bioavailable tetraethyl lead additives in gasoline, and indicates a reduction in the average PbB among all groups nationwide, a finding previously reported (ATSDR, 1988; Lin-Fu, 1980: Annest et al., 1983). All of the U.S. mean PbB values in these studies were significantly lower (p < 0.02) than those recorded for the corresponding age groups in a heavily contaminated Jamaican lead-smelting community (adult mean PbB, 13.6 ug/dL; children, 27.0 ug/dL).

Human PbB Values: Adults Versus Children

In both the United States and Jamaica, adults had lower mean PbB values than children, and in a setting where severe environmental contamination with a highly bioavailable form of lead existed (Jamaica), the disparity became a statistically significant. These findings are in accord with many previous studies. Screening of adults in a community does not appear to provide a sufficiently sensitive method for detecting existence of a lead hazard nor for initiation of a community childhood lead-screening program.

Canine PbB Versus Human PbB

In data from Tennessee and Jamaica, a significant difference was seen between the mean PbB of dogs and humans. The Alaskan data appear to be discordant, showing no significant difference between the mean PbB for dogs and for people. Additional issues need to be considered, however, when interpreting the data from Alaska. First, the number of dogs tested in Alaska was too small to be statistically significant when compared with the number of children tested. In addition, values below 10 ug/dL are currently believed to be of no medical importance (ATSDR, 1988). When these factors are considered, even this very small sampling of dogs appears qualitatively representative of the lead exposure for children in this Alaskan community, i.e., the mean PbB of dogs predicted that the mean PbB of children would be at or below a level that would be considered cause for concern.

Whether or not these observations can be extrapolated for use in future investigations, the existence of such close agreement between a planned study (Tennessee) and data obtained in field investigations (Jamaica, Alaska) bears additional scrutiny. A number of additional research questions must be answered. The first, and most important, is whether this phenomenon would be seen consistently if additional studies on a larger scale were carried out. Do upper and lower threshold PbB values exist for this relationship? How much does the bioavailability of the form of lead influence this apparent relationship?

The mean PbB of dogs, if consistently found to be higher than or at least as high as the mean PbB of children in a variety of field settings, would be both sufficiently sensitive and specific for lead exposure to be a tool for use in the field. Providing approximately a two-fold margin of safety, it could be used to determine communities where a potential environmental lead hazard exists and where children may reasonably be assumed to be at risk.

CONCLUSIONS

In the past 20 years, knowledge of the effects of lead poisoning has increased. Scientific studies show a progressive decline in the lowest exposure levels of lead at which adverse effects can be detected in children. When national childhood lead poisoning prevention programs began in the early 1970s, lead encephalopathy and other manifestations of severe overt disease were common. Today these outcomes are rare--largely because of the availability and success of childhood lead-screening programs in high-risk areas and because of a national commitment to the reduction of lead in the environment (particularly gasoline, air, and food).

In addition to using traditional childhood lead screening programs, several researchers have suggested screening sentinel animals, particularly dogs. The data examined in this paper indicate that dogs may be useful as biological monitors to indicate the bioavailability of various forms of lead in the environment. Preliminary observations suggest that the mean PbB values of dogs, like those of children, are elevated when similar pathways of environmental exposure exist. If so, incorporation of PbB testing of dogs into ongoing surveillance programs would be merited. Additional research is necessary to determine if these findings can be generalized and used as a public health tool to target populations in need of intervention.

ACKNOWLEDGMENTS

Some of the data used derives from studies supported in part by funds from the Comprehensive Environmental Response, Compensation, and Liability Act trust fund through an inter-agency agreement with the Agency for Toxic Substances and Disease Registry, U.S. Public Health Service.

Field data referred to were compiled from studies and investigations performed with the assistance and active participation of the following persons (who figure elsewhere as coauthors) and agencies: (1) James Fikes, DVM, Val Beasley, DVM, PhD, School of Veterinary Medicine, University of Illinois, Champaign-Urbana, Illinois, (2) J. Peter Figueroa, MD, Principal Medical Officer--Epidemiology: Jamaican Ministry of Health, Kingston, Jamaica, (3) Norman LaChapelle, MSPH, Director of Environmental Health, Memphis and Shelby County Health Department, Memphis, Tennessee, (4) Thomas D. Matte, MD, PhD, National Institute for Occupational Safety and Health, Centers for Disease Control, Atlanta, Georgia, (5) John Middaugh, MD, State Epidemiologist; Michael Jones, MD, Staff Epidemiologist; Michael Beller, MDCM, EIS Officer, Alaska Department of Health and Social Services, Division of Public Health, Section of Epidemiology, Anchorage, Alaska, and (6) Rebecca Schilling, DVM, MPH, Division of Environmental Hazards and Health Effects, Center for Environmental Health and Injury Control, Centers for Disease Control, Atlanta, Georgia.

REFERENCES

Agency for Toxic Substances and Disease Registry (ATSDR), 1988, The nature and extent of lead poisoning in children in the United States: a report to Congress, U.S. Department of Health and Human Services, U.S. Public Health Service, Centers for Disease Control, Atlanta.

Annest, J.L., Pirkle, J.L., Makuc, D., Neese, J.W., Bayse, D.D., and Kovar, M.G., 1983, Chronological trend in blood lead levels between 1976 and 1980, New England Journal of Medicine, 308:1373-1377.

Buck, W.B., 1979, Animals as monitors of environmental quality, Vet. Hum. Toxicol., 21(4):277-84.

Glickman, L.T., Domaski, L.M., Maguire, T.G., Dubielzig, R.R., and Churg, A., 1983, Mesothelioma in pet dogs associated with exposure of their owners to asbestos, Environ. Res., 32:305-13.

Hayes, H.M., 1976, Canine bladder cancer: epidemiologic features, Am. J. Epidemiol., 104(6):673-7.

Hayes, H.M., Hoover, R., and Tarone, R.E., 1981, Bladder cancer in pet dogs: a sentinel for environmental cancer? Am. J. Epidemiol., 114(2)229-33.

Panhandle District Health Department, Idaho Department of Health and Welfare, the Center for Environmental Health, the Centers for Disease Control, and the U.S. Environmental Protection Agency, 1986, Kellogg Revisited--1983, Childhood blood lead and environmental status report.

Lin-Fu, J.S., 1980, Lead poisoning and undue lead exposure in children: history and current status, in: "Low Level Lead Exposure: The Clinical Implications of Current Research," Needleman, H.L., ed., Raven Press, New York, pp.5-16.

Marienfeld, C.J., 1979, Detecting teratogenic substances by watching animal populations, Contr. Epidem. Biostatist., 1:57-70.

Mason, T.J., and Hayes, H.M., Jr., 1982, Diseases among animals as sentinels of environmental exposure, in: "Environmental Epidemiology," Leaverton, P.E., ed., Preager, New York, pp.67-72.

Matte, T.D., Ostrowski, S.R., Burr, G., and Baker, E.L., 1988, Community Survey Near the Jamaica Metal Refining Company Lead Smelter, Interim report to the Ministry of Health, Jamaica, May 25, 1988.

Memphis and Shelby County Health Department, and U.S. Public Health Service Centers for Disease Control, 1988, Health Effects Study of Residents Near the Hollywood Dumpsite, Memphis, TN, Final report, Cooperative agreement #U61-CCU400608-01.

Ostrowski, S.R. 1988, Lead exposure due to ore transport--Skagway, Alaska, 1988, Centers for Disease Control, EPI-89-10-2, U.S. Public Health Service, Department of Health and Human Services, Unpublished data.

Ostrowski, S.R., Schilling, R.J., Fikes, J., and Beasley, V.R., 1987a, Blood Lead in Dogs from a Rural Area: Champaign, Illinois, 1987; Submitted to Veterinary and Human Toxicology.

Ostrowski, S.R., LaChapelle, N., Schilling, R.J., Farrar, J.A., 1987b, Human and canine blood lead levels near a secondary smelter in Memphis, Tennessee, 1987; Submitted to Am. J. Public Health.

Reif, J.S., Cohen, D., 1979, Retrospective radiographic analysis of pulmonary disease in rural and urban dogs, Arch Environ. Health, 20:676-683.

Reif, J.S., Rhodes, W.H., Cohen, D., 1970, Canine pulmonary disease and the urban environment, Arch. Environ. Health 20:676-83.

Scharding, N.N., and Oehme, F.W., 1973, The use of animals for comparative studies of lead poisoning, Clin. Toxicol., 6(3):419-24.

Schilling, R.J., and Stehr-Green, P.A., 1987, Health effects in family pets and 2,3,7,8-TCDD contamination in Missouri: A look at potential animal sentinels, Arch. Environ. Health, 42(2):330-53.

Schull, L.R., and Cheeke, P.R., 1983, Effects of natural and synthetic toxicants on livestock, J. Anim. Sci., 57(2):330-53.

State of Alaska, Department of Health and Social services, Division of Public Health, Section of Epidemiology, 1988, Interim Reports #2 and #3: Skagway Heavy Metal Investigation.

FISH NEOPLASMS FOUND AT HIGH PREVALENCE IN POLLUTED WATERS

John M. Grizzle

Department of Fisheries and Allied Aquacultures
Alabama Agricultural Experiment Station
Auburn University
Auburn, Alabama 36849

INTRODUCTION

High frequencies of neoplasms have been found in several populations of wild fish in polluted environments (Table 1). These epizootics in polluted locations in North America will be reviewed, including studies of Puget Sound, Washington; a pond in Alabama that receives chlorinated sewage effluent; and several areas in the eastern United States and Great Lakes region. The occurrence of neoplasms in fish living in polluted waters has been reviewed by Black (1984a), Couch and Harshbarger (1985), and Mix (1986).

The use of aquatic organisms as sentinels for the presence of chemical carcinogens has been suggested by several authors (Dawe et al., 1964); Dawe and Harshbarger, 1975; Stich and Acton, 1976; Sonstegard, 1978; Hendricks, 1982; Baumann, 1984; Black, 1984a,b; Sonstegard and Leatherland, 1984; Couch and Harshbarger, 1985; Masahito et al., 1988). The research reviewed

Table 1. Fish Neoplasms Found at High Prevalence in Polluted Waters in North America.

Location	Species	Neoplasm	Maximum Prevalence (percent)
Hudson River	Tomcod	Hepatocellular carcinoma	25
Boston Harbor	Winter flounder	Hepatic carcinoma	8
Fox River, IL	Brown bullhead	Hepatocellular carcinoma	13
Lake Ontario	Yellow perch	Testicular leiomyoma	8
Oakville, Ont.	White sucker	Oral papilloma	50
Buffalo, NY	Freshwater drum	Dermal neoplasms	16
	White suckers	Oral papilloma	8
	Brown bullhead	Liver or skin neoplasms	17
Black River, OH	Brown bullhead	Cholangiocarcinoma	44 (in age-4)
Torch Lake, MI	Sauger	Hepatocellular carcinoma	100
	Walleye	Hepatocellular carcinoma	27
Tuskegee, AL	Black bullhead	Oral papilloma	73
Puget Sound	English sole	Hepatocellular carcinoma	32

*See text for references.

In Situ Evaluations of Biological Hazards of Environmental Pollutants
Edited by S. S. Sandhu *et al.*
Plenum Press, New York, 1990

151

provides evidence that wild fish populations can be used as sentinels. In some cases, chemical carcinogens were not known to be present, or at least were not recognized by the public, until the fish neoplasms were discovered.

In this review, prevalence indicates the percent of the population being studied that is affected by a particular disease at a given time. In wild fish populations, the prevalence of a disease is determined by examining a sample of the population and determining the percent affected. By contrast, incidence indicates the rate of occurrence of new cases of a particular disease in a population being studied. Accurate determination of the incidence of neoplasms in wild fish populations is not feasible.

In this review, only reports of neoplasms will be considered. A neoplasm is a heritably altered, relatively autonomous growth of tissue (Pitot, 1986). In this review, tumor will be used as a synonym for neoplasm, and cancer will be considered synonymous to malignant neoplasm. The terms benign and malignant have been problematic in fish oncology because neoplasms usually considered to be malignant seldom have been demonstrated to metastasize, and most of the histologic differences between benign and malignant neoplasms are relative (i.e., growth rate, mitotic index, anaplasia). Invasiveness is frequently a characteristic of malignant neoplasms in fish, but other features, such as degree of cellular differentiation, are also important.

FISH POPULATIONS WITH HIGH FREQUENCY OF NEOPLASMS

Atlantic Coastal Region

Philadelphia. Lucké and Schlumberger (1941) described carcinomas (termed epitheliomas by Lucké and Schlumberger) in brown bullheads, Ictalurus nebulosus, from the Delaware and Schuylkill rivers near Philadelphia. The lesions were usually in the skin of the head or elsewhere on the body. Over a 2-year period, 166 brown bullheads with neoplasms, most located on the lips or in the oral cavity, were obtained. Although no data were presented, reports from fishermen indicated that 1 of 300 to 500 fish had tumors, with higher frequencies in some locations. Neoplasms were not found in small fish (<25 cm long), and no seasonal variation was noted.

Hudson River. Grossly visible liver tumors were observed in 25% ($\underline{N}$=264) of the Atlantic tomcod, Microgadus tomcod, collected from the estuarine portion of the Hudson River (Smith et al., 1979). Histologically these lesions were identified as hepatocellular carcinomas. The ultrastructure of these carcinomas was described by Cormier (1986). The Hudson River is contaminated with many types of industrial, municipal, and agricultural wastes, and Atlantic tomcod from this estuary contain elevated concentrations of polychlorinated biphenyls (PCB) (Klauda et al., 1981).

Chesapeake Bay Area. White perch, Morone americana, with hepatic neoplasms have been reported in two recent studies, but types of pollution in the collection areas were not determined. In a survey of 15 estuarine tributaries of the Chesapeake Bay, 10% ($\underline{N}$=254) of the adult white perch had cholangiomas and 5% had hepatocellular neoplasms (May et al., 1987). Some of the hepatocellular neoplasms were classified as clear-cell adenomas and others were termed basophilic neoplasms. The relationship of these types of neoplasms to those found in other studies of fish hepatocellular neoplasms is not clear. In another study, 2 of 21 white perch collected from Chesapeake Bay had cholangiomas (Bunton and Baksi, 1988).

Thiyagarajah and Bender (1988) examined oyster toadfish, Opsanus tau, collected near a petroleum refinery outfall in the lower York River, a tributary of Chesapeake Bay. One of the fish had grossly visible neoplasms that were diagnosed histologically as pancreatic ductal adenomas and hepatic adenoma. Only 10 oyster toadfish were examined, so an accurate assessment of prevalence was not possible. This report is significant because neoplasms had not been reported previously in this species, which has potential as a sentinel for environmental pollutants.

Eight of 398 Fundulus heteroclitus examined from the Elizabeth River, Virginia, had neoplasms (Hargis et al., 1989). Six of the neoplasms were oral papillomas, one was a schwannoma, and one was a hemangioendothelioma. The Elizabeth River has sediments with high concentrations of organic chemicals and metals (O'Connor and Huggett, 1988). The prevalence of neoplasms in Fundulus heteroclitus from unpolluted areas is not known. This commonly studied species has been collected extensively from unpolluted areas, and reports of neoplasms are rare, so a 2% prevalence of neoplasms in this species is noteworthy.

Boston Harbor. Murchelano and Wolke (1985) found that the winter flounder, Pseudopleuronectes americanus, from Boston Harbor had an 8% ($\underline{N}$=200) prevalence of hepatic neoplasms. This fish has both hepatocellular carcinomas and cholangiocarcinomas, as well as a variety of preneoplastic and nonneoplastic lesions. In Boston Harbor, grossly visible lesions were present only in winter flounder from the southern shore of Deer Island. In other polluted areas, both winter flounder and windowpane flounder, Scophthalmus aquosus, from New Haven Harbor, Connecticut, and upper Narragansett Bay, Rhode Island, had hepatic neoplasms (3.4% prevalence, $\underline{N}$=119). Winter flounder examined from unpolluted sites on the south shore of central and eastern Long Island, New York; Casco Bay, Maine; and Georges Bank did not have neoplasms ($\underline{N}$=93).

Great Lakes Region

Several areas in or near the Great Lakes have high concentrations of organic pollutants (Baumann and Whittle, 1988). The Fox River, near Chicago, has been included in the Great Lakes region because of its proximity to the Great Lakes and similarity in fish species, even though it is actually in the Mississippi River drainage. Neoplasms found in fish from the Great Lakes area have been reviewed previously by Baumann (1984) and Black (1984b).

Fox River, Illinois. Lesions of fish collected from the Fox River near Chicago were compared to those of fish from Lake of the Woods, Ontario (Brown et al., 1973; Brown et al., 1977). Water from the Fox River contained several organic compounds, such as organophosphates, crude oil, toluene, benzene, benzanthracene, naphthalene, triazine, benzoic acid, and ethyl ether, which were not detected in the reference lake. Frequencies of neoplasms were reported for 1967-1972 and for 1972-1976 (Brown et al., 1977); results were similar for these two periods. During 1972-1976, 5.3% ($\underline{N}$=113) of the northern pike, Escox lucius, and muskellunge, Esox masquinongy, in the Fox River had reticular cell carcinoma, compared to 0.48% ($\underline{N}$=410) in Lake of the Woods. During the same time period, 13.37% ($\underline{N}$=284) of the brown bullheads in the Fox River had hepatocellular neoplasms, compared to 1.15% ($\underline{N}$=87) in the reference lake. Other species had neoplasms with prevalences from 2.2 to 4.97%.

Lake Ontario. Male yellow perch, Perca flavenscens, collected from South Bay, Lake Huron, Ontario, over a 6-year period, had an 8.4% prevalence ($\underline{N}$=3579) of testicular neoplasms diagnosed as leiomyomas (Budd et al., 1975). The female yellow perch in these collections had a 0.3% prevalence ($\underline{N}$=1320) of ovarian neoplasms. Prevalence of testicular neoplasms in yellow perch collected from six lakes in Ontario (including Lakes Erie,

Superior, and Ontario) was 0 to 12.7%. Four other types of neoplasms found in yellow perch during these surveys were represented by single specimens. Pollutants in the locations where fish were collected were not determined.

A survey of white suckers, Castotomus commersoni, from western Lake Ontario indicated that fish from Burlington Harbor and Oakville Creek had prevalences of oral papillomas of 35.1% and 50.8%, respectively (Sonstegard, 1977). The frequency of these lesions decreased in samples of fish collected away from the Oakville-Burlington area, with frequencies as low as 2.2% and 0.8% at Toronto, Ontario, and Rochester, New York, respectively. In comparison, white suckers in Lake Huron and Lake Superior had an overall prevalence of papillomas between 0 and 0.7%. Also, fish in the Oakville-Burlington area tended to have multiple papillomas. The frequency of these neoplasms was age related: they first appeared in 4-year-old fish and increased in frequency to 50-70% in 6- to 9-year-olds from areas with highest prevalence. The papillomas were also larger in the older fish, indicating that the tumors grow with increasing age. Electron microscopy revealed C-type particles in the papillomas, and reverse transcriptase activity was associated with the particulate fractions separated on sucrose gradients.

Buffalo, New York, area. Black (1983) reported dermal neoplasms in freshwater drum, Aplodinotus grunniens, and oral papillomas in white suckers from the Niagara River and adjacent areas of Lake Erie, including the locations where the Buffalo River and Smokes Creek flow into Lake Erie. Freshwater drum had dermal neoplasms with frequencies as high as 16.67% ($\underline{N}$=30) in Lake Erie near Wanakah, New York, and 13.3% ($\underline{N}$=15) at the confluence of Frenchmans Creek and the Niagara River. The neoplasms of freshwater drum were more common in larger fish. White suckers over 30 cm long had oral papillomas with an overall frequency of 8.46% ($\underline{N}$=520). Although a high prevalence of neoplasms was observed at some locations with relatively low concentrations of polycyclic aromatic hydrocarbons (PAH) in sediment, freshwater drum and white suckers can move freely from areas of high sediment concentration of PAH to nearby areas with low concentration. Various types of neoplasms were found in five additional species of fish (Black, 1983), including a 17% prevalence ($\underline{N}$=28)) of grossly visible skin or liver neoplasms in large adult brown bullheads in the Buffalo River (Black et al., 1985).

An extract of sediment from the Buffalo River was fed to brown bullheads and also applied to the skin of brown bullheads and mice (Black et al., 1985). Control fish and mice did not develop neoplasms. One of six fish fed the extract for 7 months had a cholangioma and two other fish had clear-cell and eosinophilic nodules in the liver. After 18 months, 8 of 22 fish with skin exposure (1 exposure/week) to the extract developed papillomas. After 30 weeks, 30% ($\underline{N}$=50) of the mice exposed (5 exposures/week) to Buffalo River sediment extract had developed papillomas. An undetermined number of the mouse skin lesions became squamous cell carcinomas. In the same experiment, only 12% of mice exposed to 66 ug/mL benzo[$\underline{a}$]pyrene developed papillomas.

Black River. The Black River, in Lorain County, Ohio, flows into Lake Erie near Cleveland. Brown bullheads, collected 4-8 km upstream from Lake Erie, had a high prevalence of liver, skin, and lip neoplasms (Baumann et al., 1987). The results of this survey expanded and confirmed the results or earlier studies of Black River fish (Baumann et al., 1982; Harshbarger et al., 1984; Baumann and Harshbarger, 1985). Most of the liver neoplasms in brown bullheads were cholangiocarcinomas, approximately 60% of the skin and lip neoplasms were papillomas, and the remaining skin and lip tumors were squamous cell carcinomas. No evidence of viruses in the lesions was found with electron microscopy.

In brown bullheads from the Black River, the prevalence of neoplasms was age dependent (Baumann et al., 1987). Skin and lip neoplasms occurred in less than 1% (N=263) of 2-year-olds, but frequencies in age-4 fish were as high as 32% for lip neoplasms and 18% for skin neoplasms (N=50). Prevalence of liver tumors was less than 2% (N=263) in 2-year-olds, exceeded 11% (N=92) in 3-year-olds, and was 28% (N=56) to 44% (N=50) in age-4 fish. Prevalence was even higher in 4-year-olds sampled in September (54%, N=11) and in 5-year-old fish (60%, N=5), but few fish survived to this age. Brown bullheads collected from a reference lake (Buckeye Lake) had no liver or skin neoplasms and a 1.5% frequency of lip tumors in 3-year-olds.

Ten PAH were identified in brown bullheads from the Black River, and concentrations of these PAH were much higher than in reference fish (Baumann et al., 1987). There were also 1,300 ng/g wet weight of PCB in Black River fish, compared to 50 ng/g in reference fish. The sediment of the Black River contains high concentrations of PAH (Baumann et al., 1982). An organic fraction extracted from Black River sediment was tested in the rodent experiment previously described for Buffalo River sediment. After 30 weeks of exposure (5 exposures/week), 80% (N=25) of the mice had papillomas (Black et al., 1985).

Torch Lake, Michigan. This lake is located in the Keweenaw Peninsula near the Portage Lake Ship Canal of southern Lake Superior (there is another Torch Lake in Michigan, east of Lake Michigan, but it was not involved in this study). Copper mine wastes were dumped into the lake for several years, filling at least 20% of the original volume of the lake. Hepatic neoplasms were observed in all of the sauger, Stizostedion canadense, (N=23) and a minimum of 27% of the walleye, Stizostedion vitreum, (N=22) collected from Torch Lake (Black et al., 1982). The prevalence of the sauger neoplasms was determined by the number of grossly visible tumors, and some were examined histologically. The prevalence of walleye neoplasms was based on lesions confirmed histologically. All of the liver lesions examined histologically were hepatocellular carcinomas. Dermal fibromas, with variable ossification, and mesotheliomas (differential diagnosis, lymphangioma) were also present in both species. The high prevalence of neoplasms in sauger may have been related to the old age of the specimens; juvenile sauger had not been noted in the lake for several years. The organic chemicals used in copper extraction and herbicide manufacturing have been suggested as possible causes of the neoplasms (Couch and Harshbarger, 1985).

Southeastern United States

Tuskegee, Alabama, wastewater treatment plant. In late 1979 and early 1980, 73% (N=133) of the black bullheads, Ictalurus melas, in the final oxidation pond of the Tuskegee wastewater treatment plant (WWTP) had oral papillomas (Grizzle et al., 1981). This pond received chlorinated effluent from the WWTP. No neoplasms were found in black bullheads from a nearby pond filled by runoff from pasture and woodland (unpublished data). Viruses have been associated with some fish papillomas, but there was no evidence that viruses were present in the oral papillomas on black bullheads in the Tuskegee WWTP pond (Grizzle et al., 1984).

This WWTP received only domestic wastewater, and, except for low concentrations of chloroform (9.0 to 13.5 ug/l) and bromodichloromethane (0.7 ug/l) present in the water, chemicals suspected to be carcinogens were not detected in the water or sediment of the WWTP pond. Some organic extracts of the wastewater tested positive for mutagenicity in Ames tests; extracts were most mutagenic during the summer (Grizzle et al., 1984). Tan et al. (1981) presented evidence for induction of mixed function oxidase systems and for hepatic dysfunction in black bullheads from this pond. After the neoplasms were discovered, treatment plant personnel reduced the amount of chlorine used for effluent disinfection; the total residual chlorine concentration entering the pond decreased from 1.0-3.1 mg/l (monthly averages) to 0.25-1.2 mg/l (monthly averages). Three years after the chlorination rate was reduced, the prevalence of neoplasms had decreased to 23% ($\underline{N}$=23) (Grizzle et al., 1984).

Papillomas developed on 4 or 5 black bullheads that were exposed in situ for 537 days by confining the fish to a cage in the WWTP pond (Grizzle et al., 1984). In addition, 1 of 12 black bullheads stocked at a later time developed a papilloma after 57 days in a cage. These papillomas were grossly and histologically similar to the papillomas on the wild black bullheads in this pond. Papillomas did not develop in control fish, nor in exposed brown bullheads, yellow bullheads (Ictalurus natalis), and channel catfish (Ictalurus punctatus). Compared to exposed black bullheads and control channel catfish, in situ exposed channel catfish had increased levels of hepatic glucuronosyltransferase that could conjugate and thereby reduce the effects of carcinogens. The advantages and limitations of in situ exposure of caged fish in polluted environments were discussed by Grizzle et al. (1988).

Pacific Coastal Region

Puget Sound, Washington. A few areas in Puget Sound have sediments with high concentrations of anthropogenic chemicals, but most of Puget Sound is less polluted, allowing comparisons of fish collected from locations with different levels of sediment contamination. Puget Sound fish with hepatic neoplasms have been studied extensively, and this research was reviewed by Malins et al. (1987a, 1988). In addition to neoplasms, several types of non-neoplastic lesions are present in fish from Puget Sound. Myers et al. (1987) found that certain types of these non-neoplastic lesions of the liver had high frequencies of co-occurrence with hepatic neoplasms and were useful indicators of exposure to carcinogens.

English sole, Parophrys vetulus, from various polluted areas of Puget Sound had high prevalences of hepatic neoplasms: 32% ($\underline{N}$=62) (Pierce et al., 1978) and 16.2% ($\underline{N}$=136) (Malins et al., 1984) in the Duwamish Waterway, 26.7% ($\underline{N}$=75) in Eagle Harbor (Malins et al., 1985b), 7.5% ($\underline{N}$=66) at Mukilteo (Malins et al., 1985a), and 12.1% ($\underline{N}$=66) at Port Gardner in Everett Harber (Malins et al., 1984). No fish with neoplasms were found in several minimally polluted locations in Puget Sound (Pierce et al., 1978; Malins et al., 1984; Malins et al., 1985a). Site of capture and fish age were the most important risk factors for neoplasms as well as other hepatic lesions (Rhodes et al., 1987). Examination of 896 English sole from Commencement Bay revealed 2.6% with hepatic neoplasms; no neoplasms were found in fish from the reference area, Carr Inlet (Becker et al., 1987). Prevalence of neoplasms was positively correlated with fish age but not with sex. Becker et al. (1987) compared their data to an earlier study of English sole from the same bay (Malins et al., 1984) and found that the spatial patterns of lesion prevalences were similar.

Over 900 different organic compounds were identified in sediments of Commencement Bay (Malins et al., 1984). Aromatic hydrocarbons were also found in invertebrate animals recovered from the stomach of English sole from Puget Sound (Malins et al., 1985a,b), indicating that the organic chemicals present in sediment are bioavailable through the diet. There was a correlation ($\underline{r}_s$=0.48, $\underline{P}$=0.003) between sediment concentration of aromatic hydrocarbons and prevalence of hepatic neoplasms in English sole from several locations (Malins et al., 1984). There was also a correlation ($\underline{r}_s$ 0.37, $\underline{P}$=0.019) between a concentration of metals in sediment and prevalence of neoplasms. There was a higher correlation between the concentration

of bile metabolites of aromatic compounds and prevalence of hepatic neoplasms in English sole (r_s=0.853, P<0.002) (Krahn et al., 1986).

Los Angeles. White croaker, Genyonemus lineatus, from the vicinity of
sewage outfalls near Los Angeles had cholangioma (1 of 25 fish from Hyper-
ion outfall), cholangiocellular carcinoma (1 of 25 fish from Queensway
Bay), and hepatocellular carcinoma (1 of 25 fish from near Reservation
Point) (Malins et al., 1987b). Preneoplastic lesions also were present in
fish from these locations. No neoplasms were present in white croaker from
the reference area. Sediments from the sewage outfall areas contained high
concentrations of aromatic hydrocarbons, PCB, and DDT. Concentrations of
these toxicants were generally even higher in the food organisms from the
white croaker's stomach than in sediment.

DISCUSSION

Causes of neoplasms in fish are presumed to be similar to those of mammals: chemical carcinogens, oncogenic viruses, radiation, and inherited
conditions. Several of the reports reviewed indicate an association between neoplasms of wild fish and chemicals in the environment. Positive
correlations of hepatic neoplasm prevalence with sediment concentrations
and bile metabolites of aromatic hydrocarbons, as discussed by Malins et
al. (1987a, 1988), is strong evidence for a role of these chemicals in the
occurrence of neoplasms. Development of skin neoplasms in brown bullheads
and mice exposed to organic extracts from the Black River and the Buffalo
River (Black et al., 1985), and corroborating field data (Baumann et al.,
1987), indicate that the relationship between fish neoplasms in Puget Sound
and carcinogenic pollutants is not an isolated event in a marine environment, but rather, occurs in polluted fresh water as well.

Development of papillomas in black bullheads confined to cages in a
wastewater treatment pond provides convincing evidence that the environmental conditions in this pond triggered the neoplasms observed in wild
fish in the pond (Grizzle et al., 1984). Mutagenicity of organic extracts
of the water and enzyme induction in exposed fish indicate that chemical
contaminants were involved. Tumor prevalence decreased after the chlorination rate was reduced, suggesting that chlorine was involved in the formation of carcinogens or as a promoter of tumor formation. No evidence was
found for viruses in these tumors, and fish from a different genetic stock
were used in the in situ exposures to eliminate the possibility that the
neoplasms were genetically transmitted.

Certain pollutants seem to be involved in increasing the prevalence of
neoplasms in resident fish, but whether these chemicals act as carcinogens,
promoters of carcinogenesis, or activators of oncogenic viruses is not
known. Chemical carcinogens interact with DNA to cause genetic changes in
a cell; however, some chemicals also serve as promoters that stimulate
transformed cells to increase in number so that a tumor develops. Some
chemicals are probably both carcinogens and promoters; an initial exposure
causes genetic change, and continuing exposure stimulates development and
growth of the neoplasm. Chemicals present in polluted waters could contribute to more than one phase of oncogenesis.

Typically, pollution-associated neoplasms occur in fish species that
are demersal, suggesting that contact with contaminated sediments or consumption of benthic organisms is related to the development of neoplasms.
Most of the external neoplasms of demersal species are on or near the
mouth, which would be exposed to chemicals in sediments or food organisms.
The liver is a likely target of ingested carcinogens, and elevated levels
of carcinogens were found in the food organisms in the stomach of English

157

sole from Puget Sound (Malins et al., 1985a,b). However, consumption of contaminated food is not required for hepatic carcinogenesis in fish; many laboratory experiments have demonstrated that exposure of fish to water-borne carcinogens can result in hepatic tumors (Hendricks, 1982). Pelagic species, such as the sauger and walleye in Torch Lake (Black et al., 1982), can also be exposed to carcinogens, from either the water, sediment, or food chain.

Not all pollutants would be expected to cause an increase in tumor pre-valence. Noncarcinogenic pollutants will not cause neoplasms in wild fish; rather, these pollutants may kill fish or deplete fish populations by des-troying habitat or by interfering with reproduction. These effects reduce species diversity and remove sensitive species that could have been senti-nels for other pollutants that are carcinogenic.

In human populations, an increased incidence of neoplasms can indicate exposure to carcinogens. The expected incidence of commonly occurring neo-plasms is known for many human populations, so that an increased incidence can be recognized. Although the expected prevalence of neoplasms in fish populations has not been adequately determined, neoplasms in most fish populations are rare and even one neoplasm is often noteworthy.

After a high frequency of neoplasms in wild fish has indicated the pre-sence of a carcinogen, the risk to human health should be evaluated. Such an assessment should consider the following possibilities: (1) edible por-tions of aquatic animals from the polluted location contain carcinogens, (2) consumption of water from the polluted lake or river increases exposure to carcinogens, and (3) disturbance of contaminated sediments could in-crease the exposure of aquatic life and humans to carcinogens.

Of course, not all tumors of fish are caused by carcinogenic pollut-ants. Neoplasms occurring at a high frequency but not caused by chemical carcinogens often develop from the nervous system, chromatophores, or mesenchymal cells. Examples of such neoplasms include neurofibromas and schwannomas in bicolor damselfish, Pomacentrus partitus, in south Florida (Schmale et al., 1986), and genetically transmitted melanomas in Xiphophorus (Anders et al., 1984). Sonstegard (1977) found a high pre-valence of gonadal neoplasms (up to 100% in older males) in hybrid goldfish (Carassius auratus) X common carp (Cyprinus carpio) in the Great Lakes. These gonadal neoplasms may have been related to genetic factors rather than exposure to carcinogens. Ornamental carp, Cyprinus carpio, with com-plex genetic histories, also develop ovarian neoplasms that may be inherit-ed (Ishikawa and Takayama, 1977).

Most neoplasms in fish from polluted waters originate from liver or epi-thelial tissue in the oral cavity or skin. The relationship between hepa-tic neoplasms in fish and chemical carcinogens is strong. The liver is the most common site of neoplasms in fish exposed to chemical carcinogens in laboratory experiments (Couch and Harshbarger, 1985). Experimentally in-duced hepatic neoplasms are histologically similar to those found in wild fish from polluted waters (Myers et al., 1987).

Fish sentinels will not replace chemical analysis to determine the types of pollutants present in aquatic environments. But there are thou-sands of anthropogenic compounds present in some environments, and accurate quantification of all chemicals is not practical. Although it would be desirable to have qualitative and quantitative information on all compounds present in environments of concern, such extensive information is expensive and the quantity of pollutants varies between localities and over time. Additional methods, such as the use of sentinel species or in situ expo-sures, are needed to monitor potential effects of pollutants.

ACKNOWLEDGMENTS

    Support was provided by the Southeastern Cooperative Fish Disease
Project.

REFERENCES

Anders, F., Schartl, M., and Barnekow, A., 1984, *Xiphophorus* as an in vivo
    model for studies on oncogenes, Natl. Cancer Inst. Monogr., 65:97.
Baumann, P.C., 1984, Cancer in wild freshwater fish populations with
    emphasis on the Great Lakes, J. Great Lakes Res., 10:251.
Baumann, P.C., and Harshbarger, J.C., 1985, Frequencies of liver neoplasia
    in a feral fish population and associated carcinogens, Mar. Environ.
    Res., 17:324.
Baumann, P.C., Smith, W.D., and Ribick, M., 1982, Hepatic tumor rates and
    polynuclear aromatic hydrocarbon levels in two populations of brown
    bullhead (*Ictalurus nebulosus*), in: "Polynuclear Aromatic Hydro-
    carbons: Physical and Biological Chemistry," M. Cooke, A.J. Dennis,
    and G. Fisher, eds., Battelle Press, Columbus, Ohio, pp. 93-102.
Baumann, P.C., Smith, W.D., and Parland, W.K., 1987, Tumor frequencies and
    contaminant concentrations in brown bullheads from an industrialized
    river and a recreational lake, Trans. Am. Fish. Soc., 116:78.
Baumann, P.C., and Whittle, D.M., 1988, The status of selected organics in
    the Laurentian Great Lakes: An overview of DDT, PCBs, dioxins, furans,
    and aromatic hydrocarbons, Aquatic Toxicology (Amsterdam) 11:241.
Becker, D.S., Ginn, T.S., Landolt, M.L., and Powell, D.B., 1987, Hepatic
    lesions in English sole (*Parophrys vetulus*) from Commencement Bay, Mar.
    Environ. Res., 23:153.
Black, J.J., 1983, Field and laboratory studies of environmental carcino-
    genesis in Niagara River fish, J. Great Lakes Res., 9:326.
Black, J.J., 1984a, Aquatic animal bioassays for carcinogenesis, Trans-
    plantation Proceedings, 16:406.
Black, J.J., 1984b, Environmental implications of neoplasia in Great Lakes
    fish, Mar. Environ. Res., 14:529.
Black, J.J., Evans, E.D., Harshbarger, J.C., and Zeigel, R.F., 1982,
    Epizootic neoplasms in fishes from a lake polluted by copper mining
    wastes. J. Natl. Cancer Inst., 69:915.
Black, J., Fox, H., Black, P., and Bock, F., 1985, Carcinogenic effects of
    river sediment extract in fish and mice, in: "Water Chlorination:
    Chemistry, Environmental Impact and Health Effects, Vol. 5," R.L.
    Jolly, R.J. Bull, W.P. Davis, S. Katz, M.H. Roberts, Jr., and V.A.
    Jacobs, eds., Lewis Publishers, Chelsea, MI, pp. 415-427.
Brown, E.R., Hazdra, J.J., Keith, L., Greenspan, I., Kwapinski, J.B.G.,
    and Beamer, P., 1973, Frequency of fish tumors found in a polluted
    watershed as compared to nonpolluted Canadian waters, Cancer Res.,
    33:189.
Brown, E.R., Sinclair, T., Keith, L., Beamer, P., Hazdra, J.J., Nair, V.,
    and Callaghan, O., 1977, Chemical pollutants in relation to diseases in
    fish, Ann. N.Y. Acad. Sci., 298:535.
Budd, J., Schroder, J.D., and Dukes, K.D., 1975, Tumors of the yellow
    perch, in: "The Pathology of Fishes," W.E. Rebelin and G. Migaki, eds.,
    Univ. of Wisconsin Press, Madison, WI, pp. 895-906.
Bunton, T.E., and Baksi, S.M., 1988, Cholangioma in white perch (*Morone
    americana*) from the Chesapeake Bay, J. Wildl. Dis., 24:137.
Couch, J.A., and Harshbarger, J.C., 1985, Effects of carcinogenic agents
    on aquatic animals: An environmental and experimental overview,
    Environ. Carcinogenesis Review, 3:63.
Cormier, S.M., 1986, Fine structure of hepatocytes and hepatocellular car-
    cinoma of the Atlantic tomcod, *Microgadus tomcod* (Walbaum), J. Fish
    Dis., 9:179.

159

Dawe, C.J., Stanton, M.F., and Schwartz, F.J., 1964, Hepatic neoplasms in native bottom-feeding fish of Deep Creek Lake, Maryland, Cancer Res., 24:1194.

Dawe, C.J., and Harshbarger, J.C., 1975, Neoplasms in feral fishes: Their significance to cancer research, in: "The Pathology of Fishes," W.E. Rebelin and G. Migaki, eds., University of Wisconsin Press, Madison, WI, pp. 871-894.

Grizzle, J.M., Schwedler, T.E., and Scott, A.L., 1981, Papillomas of black bullheads, Ictalurus melas (Rafinesque), living in a chlorinated sewage pond, J. Fish Dis., 4:345.

Grizzle, J.M., Melius, P., and Strength, D.R., 1984, Papillomas on fish exposed to chlorinated wastewater, J. Natl. Cancer Inst., 73:1133.

Grizzle, J.M., Horowitz, S.A., and Strength, D.R., 1988, Caged fish as monitors of pollution: Effects of chlorinated effluent from a waste-water treatment plant, Water Resour. Bull., 24:951.

Hargis, W.J., Jr., Zwerner, D.E., Thoney, D.A., Kelly, K.L., and Warinner, J.E., III, 1989, Neoplasms in mummichog from the Elizabeth River, Virginia, Journal of Aquatic Animal Health, in press.

Harshbarger, J.C., Cullen, L.J., Calabrese, M.J., Spero, P.M., Baumann, P.C., and Parland, W.K., 1984, Epidermal, hepatocellular and cho-langiocellular carcinomas in brown bullheads, Ictalurus nebulosus, from industrially polluted Black River, OH, Mar. Environ. Res., 14:535.

Hendricks, J.D., 1982, Chemical carcinogenesis in fish, in: "Aquatic Toxi-cology, Vol. 1," L.J. Weber, ed., Raven Press, New York, pp. 149-211.

Ishikawa, T., and Takayama, S., 1977, Ovarian neoplasia in ornamental hybrid carp (nishikigoi) in Japan, Ann. N.Y. Acad. Sci., 298:330.

Klauda, R.J., Peck, T.H., and Rice, G.K., 1981, Accumulation of poly-chlorinated biphenyls in Atlantic tomcod (Microgadus tomcod) collected from the Hudson River estuary, New York, Bull. Environ. Contam. Toxicol., 27:829.

Krahn, M.W., Rhodes, L.D., Myers, M.S., Moore, L.K., MacLeod, W.D., Jr., and Malins, D.C., 1986, Associations between metabolites of aromatic compounds in bile and the occurrence of hepatic lesions in English sole (Parophrys vetulus) from Puget Sound, Washington, Arch. Environ. Contam. Toxicol., 15:61.

Lucké, B., and Schlumberger, H., 1941, Transplantable epitheliomas of the lip and mouth of catfish: I. Pathology. Transplantation to anterior chamber of eye and into cornea, J. Exp. Med., 74:397.

Malins, D.C., McCain, B.B., Brown, D.W., Chan, S., Myers, M.S., Landahl, J.T., Prohaska, P.G., Friedman, A.J., Rhodes, L.D., Burrows, D.G., Gronlund, W.D., and Hodgins, H.O., 1984, Chemical pollutants in sedi-ments and diseases of bottom-dwelling fish in Puget Sound, Washington, Environ. Sci. Technol., 18:705.

Malins, D.C., Krahn, M.M., Brown, D.W., Rhodes, L.D., Myers, M.S., McCain, B.B., and Chan, S.-L., 1985a, Toxic chemicals in marine sediment and biota from Mukilteo, Washington: Relationships with hepatic neoplasms and other hepatic lesions in English sole (Parophrys vetulus), J. Natl. Cancer Inst., 74:487.

Malins, D.C., Krahn, M.M., Myers, M.S., Rhodes, L.D., Brown, D.W., Krone, C.A., McCain, B.B., and Chan, S.-L., 1985b, Toxic chemicals in sedi-ments and biota from a creosote-polluted harbor: Relationships with hepatic neoplasms and other hepatic lesions in English sole (Parophrys vetulus), Carcinogenesis, 6:1463.

Malins, D.C., McCain, B.B., Myers, M.S., Brown, D.W., Krahn, M.M., Rougal, W.T., Schiewe, M.H., Landahl, J.T., Chan, S.-L., 1987a, Field and lab-oratory studies of the etiology of liver neoplasms in marine fish from Puget Sound, Environ. Health Perspect., 71:5.

Malins, D.C., McCain, B.B., Brown, D.W., Myers, M.S., Krahn, M.M., Chan, S.L., 1987b, Toxic chemicals, including aromatic and chlorinated hydro-carbons and their derivatives, and liver lesions in white croaker (Genyonemus lineatus) from the vicinity of Los Angeles, Environ. Sci. Technol., 21:765.

Malins, D.C., McCain, B.B., Landahl, J.T., Myers, M.S., Krahn, M.M., Brown, D.W., Chan, S.-L., and Roubal, W.T., 1988, Neoplastic and other diseases in fish in relation to toxic chemicals: an overview, Aquatic Toxicology, 11:43.

Masahito, P., Ishikawa, T., and Sugano, H., 1988, Fish tumors and their importance in cancer research, Jpn. J. Cancer Res. (Gann), 79:545.

May, E.B., Lukacovic, R., King, H., Lipsky, M.M., 1987, Hyperplastic and neoplastic alterations in the livers of white perch (Morone americana) from the Chesapeake Bay, J. Natl. Cancer Inst., 79:137.

Mix, M.C., 1986, Cancerous diseases in aquatic animals and their association with environmental pollutants: A critical literature review, Mar. Environ. Res., 20:1.

Murchelano, R.A., and Wolke, R.E., 1985, Epizootic carcinoma in the winter flounder, Pseudopleuronectes americanus, Science, 228:587.

Myers, M.S., Rhodes, L.D., and McCain, B.B., 1987, Pathologic anatomy and patterns of occurrence of hepatic neoplasms, putative preneoplastic lesions, and other idiopathic hepatic conditions in English sole (Parophrys vetulus) from Puget Sound, Washington, J. Natl. Cancer Inst., 78:333.

O'Connor, J.M., and Huggett, R.J., 1988, Aquatic pollution problems, North American coast, including Chesapeake Bay, Aquatic Toxicology (Amsterdam), 11:163.

Pierce, K.V., McCain, B.B., and Wellings, S.R., 1978, Pathology of hepatomas and other liver abnormalities in English sole (Parophrys vetulus) from the Duwamish River estuary, Seattle, Washington, J. Natl. Cancer Inst., 60:1445.

Pitot, H.C., 1986, "Fundamentals of Oncology," 3rd ed., Marcel Dekker, New York.

Rhodes, L.D., Myers, M.S., Gronlund, W.D., and McCain, B.B., 1987, Epizootic characteristics of hepatic and renal lesions in English sole, Parophrys vetulus, from Puget Sound, J. Fish Biol., 31:395.

Schmale, M.C., Hensley, G.T., and Udey, L.R., 1986, Neurofibromatosis in the bicolor damselfish (Pomacentrus partitus) as a model of von Recklinghausen neurofibromatosis, Ann. N.Y. Acad. Sci., 486:386.

Smith, C.E., Peck, T.H., Klauda, R.J., and McLaren, J.B., 1979, Hepatomas in Atlantic tomcod Microgadus tomcod (Walbaum) collected in the Hudson River estuary in New York, J. Fish Dis., 2:313.

Sonstegard, R.A., 1977, Environmental carcinogenesis studies in fishes of the Great Lakes of North America, Ann. N.Y. Acad. Sci., 298:261.

Sonstegard, R.A., 1978, Feral aquatic organisms as indicators of waterborne environmental carcinogens, in: "Aquatic Pollutants: Transformation and Biological Effects," O. Hutzinger, I.H. Van Lelyveld, and B.C.J. Zoeteman, eds., Pergamon Press, Oxford, England, pp. 349-358.

Sonstegard, R.A., and Leatherland, J.F., 1984, Comparative epidemiology: The use of fishes in assessing carcinogenic contaminants, in: "Contaminant Effects on Fisheries," V.C. Cairns, P.V. Hodson, and J.O. Nriagu, eds., John Wiley and Sons, New York, pp. 223-231.

Stich, H.F., and Acton, A.B., 1976, The possible use of fish tumors in monitoring for carcinogens in the marine environment, Prog. Exp. Tumor. Res., 20:44.

Tan, B., Melius, P., and Grizzle, J.M., 1981, Hepatic enzymes and tumor histopathology of black bullheads with papillomas, in: "Polynuclear Aromatic Hydrocarbons Chemical Analysis and Biological Fate. 5th International Symposium," M. Cooke and A. Dennis, eds., Battelle Press, Columbus, OH, pp. 377-386.

Thiyagarajah, A., and Bender, M.E., 1988, Lesions in the pancreas and liver of an oyster toadfish, Opsanus tau (L.), collected from the lower York River, Virginia, USA, J. Fish Dis., 11:359.

BIOLOGICAL MARKERS IN ANIMAL SENTINELS: LABORATORY STUDIES IMPROVE

INTERPRETATION OF FIELD DATA

John F. McCarthy, Braulio D. Jimenez, Lee R. Shugart,
Frederick V. Sloop

Environmental Sciences Division
Oak Ridge National Laboratory
Oak Ridge, Tennessee

Aimo Oikari

University of Joensuu
Joensuu, Finland

INTRODUCTION

A number of approaches have been used to evaluate the biological hazards of environmental pollution. Chemical analysis of the concentrations of toxic compounds in environmental media is clearly an important component of such an evaluation. Advanced analytical procedures are specific, quantitative, and exquisitely sensitive and precise. However, the biological significance of the chemical concentrations is not at all clear. We understand the toxic action of but a few of the thousands of chemicals in the environment and have almost no information on the toxicity of complex mixtures of chemicals. Furthermore, a chemical survey is a snapshot in time and space. Variations in concentrations over time resulting from intermittent releases of effluents by industries or from storm events, changes in winds, etc., cannot be accounted for without repeated analyses. Spatial patchiness of contaminant patterns also requires extensive and expensive sampling and chemical analyses.

One approach to addressing these problems is to monitor animals in the environment as sentinels of environmental contamination. Animals integrate chemical exposures over time and across their spatial range and indicate the integrated dose of bioavailable contaminants in soil, water, sediment, or the food chain. We have been measuring biological markers in fish from contaminated streams at government facilities around Oak Ridge as part of a larger biological monitoring program. Biological markers (biomarkers) provide information that cannot be obtained from direct measurement of body burdens and that can be relevant to evaluating the biological significance of exposure (e.g., measures of genotoxic effects). Biomarkers provide evidence of exposure to compounds that do not bioaccumulate or that are rapidly metabolized and eliminated. Furthermore, biomarkers integrate the toxicological and pharmacokinetic interactions resulting from exposure to complex mixtures of contaminants and present a biologically relevant measure of toxicant action at target tissues and the cumulative adverse effect of the exposure.

In Situ Evaluations of Biological Hazards of Environmental Pollutants
Edited by S. S. Sandhu *et al.*
Plenum Press, New York, 1990

This paper describes some of the results of our studies of biomarker responses of fish. We report results of field surveys of animals from contaminated and unpolluted reference streams, as well as laboratory studies of biomarker responses. The goal of this presentation is to emphasize two points that are critical to the successful application and interpretation of biomarker responses in environmental species.

1. It is necessary to measure a suite of biomarkers. No single biomarker can provide all the information necessary to evaluate exposure or its significance. Furthermore, the responses of one biomarker provide information that improves interpretation of other biomarkers.

2. The responses of biomarkers reflect a complex biological response to the entire exposure scenario. Cumulative stress from the environment, including temperature, crowding, food availability, and parasites, as well as low-level insults from nontoxic concentrations of complex mixtures of chemicals, can significantly affect the response of an organism to toxic contaminants. Damage from environmental pollutants may not be a simple dose-dependent function of how much of what chemical enters an animal; the same exposure may cause more toxic damage to a stressed animal. However, the power of biomarkers in evaluating in situ hazards is that they provide biologically significant insights into cumulative effects of integrated situations.

METHODS AND MATERIALS

Field Survey of Fish from East Fork Poplar Creek

Biological markers are being examined in fish as part of a Biological Monitoring and Abatement Program stipulated under the National Pollutant Discharge Elimination System permit issued to the Oak Ridge Y-12 Plant. This comprehensive monitoring program includes tasks that evaluate the toxicity of the ambient stream water, the bioaccumulation of contaminants in fish and molluscs, and the structure and function of the aquatic community, in addition to the measurement of biomarkers (Loar et al., 1988). Effluent discharges from the Y-12 Plant enter the headwaters of East Fork Poplar Creek (EFPC) 23.7 km above its confluence with Poplar Creek Effluent discharges of 388 L/s from the Y-12 Plant and 227 L/s from the City of Oak Ridge Wastewater Treatment Facility (ORWTF) at East Fork Kilometer (EFK) 13.4 together constitute 39% of the mean annual flow in EFPC at a gauging station of EFK 5.3. The stream also receives urban runoff and some agricultural runoff between EFK 22.7 and 7.7.

Water and sediments in EFPC downstream of the Y-12 Plant contain metals, organic chemicals, and radionuclides discharged over many year of operation. Sediment samples contained 10 priority pollutant organics (including seven polycyclic aromatic hydrocarbons, polychlorinated biphenyls, and total phenols) and seven metals (As, Cd, Pb, Hg, Ni, Ag, and Zr) at concentrations above background (TVA, 1986). Monitoring of the discharge into EFPC identified potential toxicity problems due to copper, ammonia, residual chlorine, perchloroethylene, total nitrogen, and oil and grease. Of these, the first four could have been toxic at the maximum concentration reported, although wide variability in concentrations was observed (Loar et al., 1988). However, monthly testing of 24-h composite samples collected at the outfall of the Y-12 Plant (EFK 23.4) provides no evidence of chronic toxicity, based on bioassays that measured Ceriodaphnia survival and reproduction and fathead minnow larval survival and growth (Loar et al., 1988). Accumulation of mercury and PCBs in fish and molluscs in EFPC decreases with increasing distance from the Y-12 Plant (Loar et al., 1988).

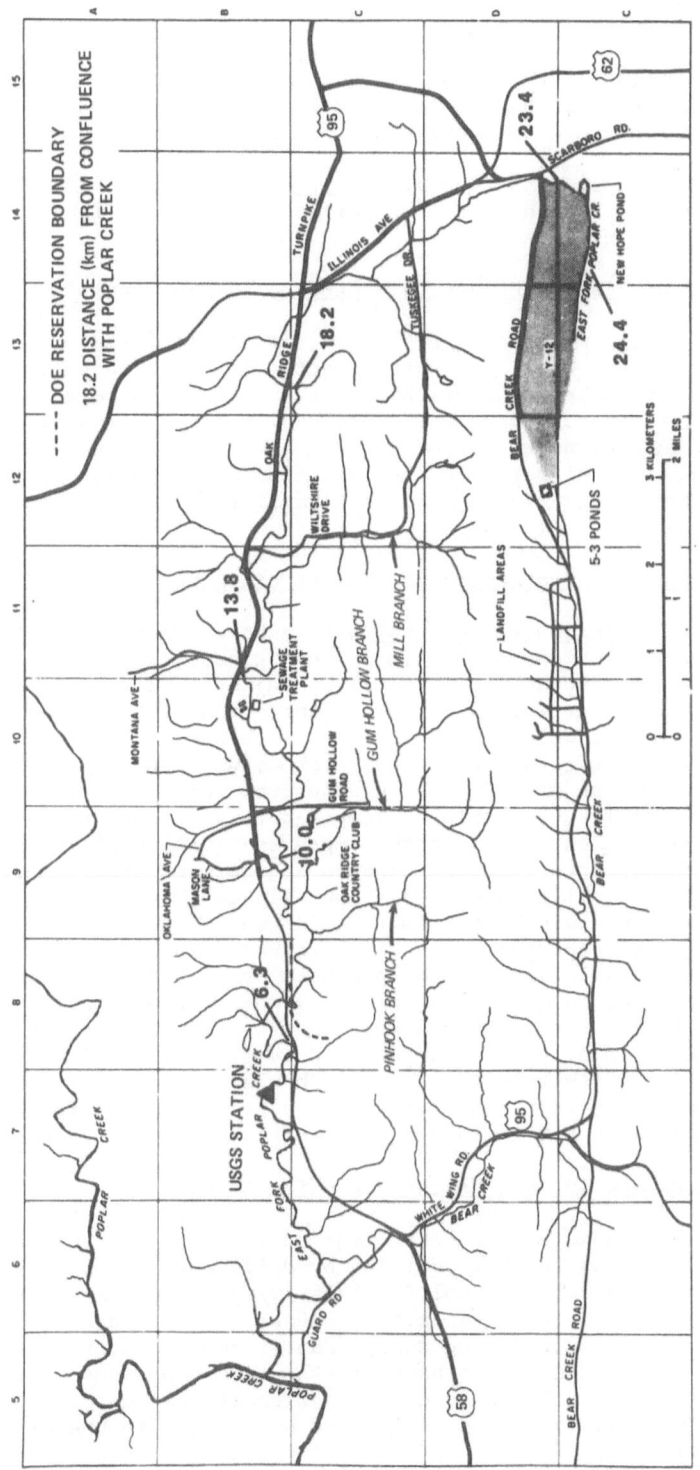

Figure 1. Map of the East Fork Poplar Creek watershed showing the locations of the six primary study sites.

This paper discusses results of measurements of biomarkers at four sites on EFPC and a site on Brushy Fork, an ecologically similar stream that is used as an uncontaminated reference site. Brushy Fork has no source of industrial or municipal inputs, but may receive some agricultural runoff. The sites on EFPC are reported as Station 1 (at the industrial discharge from the Y-12 Plant, EFK 23.4), Station 2 (at EFK 18.6), Station 3 (just above and below the discharge of the ORWTF. EFK 13.4), and Station 5 (at the gauging station above the confluence with Poplar Creek, EFK 5.3).

Adult male redbreast sunfish (_Lepomis auritus_) were collected by electroshocking at each site. Fish were transported alive to the laboratory for dissection and analysis.

Laboratory Exposure of Fish to Hepatotoxins and Benzo(a)pyrene

Fish (hybrid _Lepomis_) were dosed by gavage with allyl formate, a known hepatotoxic agent, and one day later were injected intraperitoneally with benzo(a)pyrene (BaP; gold label, Aldrich) in corn oil at a dose of 8 mg/kg body weight. Vehicle controls were not treated with allyl formate, but were injected with an equivalent volume of corn oil without BaP. BaP controls received injections of BaP in corn oil but no allyl formate. Fish were killed five days after injection and assayed for enzyme activities and genotoxic damage.

Laboratory Exposure of Fish to Aqueous Solutions of Benzo(a)pyrene (BaP)

Bluegill sunfish (_Lepomis macrochirus_) were exposed under flow-through conditions to aqueous solutions of BaP in a 100-L glass aquarium. Solutions were generated continuously either (1) by injection of a solution of BaP dissolved in dioxane carrier (dioxane concentration of 0.1 mL/L) into a mixing bottle with flowing water (McCarthy and Jimenez, 1985) or (2) without carrier by passing water through a generator column containing glass beads coated with an excess of BaP. These exposures were conducted with a flow rate of approximately 200 mL/min. "Temperature-stressed" animals were acclimated to 10°C and, then placed in the exposure aquarium at 20°C. All other fish were acclimated to 20°C and exposed at the same temperature. Concentrations of BaP in all exposures was approximately 1 ug/L (approximately 25% of the aqueous solubility limit) and remained constant over the exposure period. In one experiment (results presented in Table 1), fish were exposed to aqueous solutions of BaP produced by a generator column, using a static-renewal protocol in which solutions were renewed daily; concentrations decreased from 0.9 to 0.3 ug/L over the 24 h between renewals. The aqueous concentration of BaP was determined by passing a volume of water through a C-18 SepPak (Waters Assoc.), which was eluted with methanol. The BaP in the methanol extract was quantified by using high-performance liquid chromatography with a C-18 column and fluorescence detection.

Measurement of DNA Damage by an Alkaline Unwinding Assay

DNA was isolated by homogenizing the liver of fish in 1N NH_4OH/0.2% Trition X-100. The DNA was further purified by differential extraction with chloroform/isoamyl alcohol/phenol (24/1/25, v/v) and passage through a molecular sieve (Sephadex G-50). DNA strand breaks were measured in the isolated DNA by an alkaline unwinding assay (Kanter and Schwartz, 1982; Shugart, 1988a,b). The technique is based on the time-dependent partial alkaline unwinding of DNA followed by determination of the duplex:total DNA ratio (F value). The relative number of strand breaks ($\underline{N}$) in DNA of animals in experimental groups (or from EFPC stations) can be compared to those in control groups (or from the Brushy Fork reference site) as follows (Kanter and Schwartz, 1982; Shugart, 1988a,b):

$$\underline{N} = (\ln F_E/\ln F_C) - 1, \text{ (Equation 1)}$$

Table 1. Changes in EROD activity in sunfish (<u>Lepomis</u> sp) exposed to aqueous solutions of BaP

Exposure Time	BaP + Dioxane	BaP (no Carrier)
Control	12 + 2.5 (8)	7.5 + 2.5 (10)
7 d	26 + 2.1 (4)	5.5 + 1.8 (5)
14 d	2.0 + 0.9 (4)	31 + 11 (4)
18 d	3.1 + 0.7 (3)	25 + 3.5 (3)

Fish were exposed either to a flowing solution of BaP (approximately 1 ug/L) with dioxane carrier (0.1 mL dioxane/L water) or to aqueous BaP solution produced without carrier with a generator column. In the latter exposure, BaP solutions were not flow-through, but were renewed daily; concentrations decreased from 0.9 to 0.3 ug BaP/L between renewals. EROD activity is in units of pmol/min/mg microsomal protein. Values are reported as means ± s.e.m. (number of fish).

where F_E and F_C are the mean F values of DNA from the experimental and control groups, respectively. $\underline{N}$ values greater than 0 indicate that DNA from the experimental groups had more strand breaks than DNA from control groups; an $\underline{N}$ of 5, for example, indicates that the DNA from the experimental group had five times more strand breaks than the DNA from the control group. Statistical tests of the significance of differences in the number of strand breaks were calculated by using the values of ln F; if the ln F of two groups differed by $P < 0.05$, then the ratios, $\underline{N}$, are also significantly different.

Ethoxyresorufin-O-deethylase (EROD) Activity in Hepatic Microsomes

    Fish hepatic microsomes were prepared by differential centrifugation. Fish were killed by severing their spinal cords, and the livers were immediately removed and placed in ice-cold 0.1 M Tris buffer at pH 7.4. The minced tissues were homogenized in five volumes of buffer and centrifuged at 3000 x g for 20 minutes with a Beckman J-21B centrifuge. The resulting supernatants were centrifuged at 105,000 x g for 60 minutes in 0.1 M Tris buffer, 20% glycerol, pH 7.4 (Jimenez et al., 1988). Microsomes were frozen in liquid nitrogen and stored at -120° until assayed.

    The activity of 7-ethoxyresorufin-O-deethylase in the hepatic microsomal fraction was determined fluorometrically at 25°C by measuring the production of resorufin from 7-ethoxyresorufin (Molecular Probes, Oregon), using an Aminco SPF500 spectrofluorometer with excitation at 530 nm and emission measured at 582 nm (Burke and Mayer, 1974). The final reaction buffer contained 150 mM Tris at pH 7.7, 200 nmol NADPH, and 1 nmol 7-ethoxyresorufin in the cuvette. The activity is expressed as picomoles of resorufin per minute per milligram of microsomal protein (Jimenez et al., 1988).

Metallothionein Assay

    The concentration of metallothionein (MT) was measured in the high-speed supernatant resulting from isolation of the hepatic microsomal fraction with a Chelex-100/[109]Cadmium assay described in Sloop et al. (1989). Briefly, the supernatant was acidified to pH <2 and passed through a Sephadex G-10 column to remove endogenous metals bound to the MT. Proteins of large molecular weight were removed by centrifugation after isoelectric precipitation at pH 5.5. An excess of [109]Cd was added, the pH was raised to approximately 60, and the solution was passed through a Chelex-100 ion exchange resin column to remove ^{109}Cd not bound to protein. MT is quantified in units of nanomoles of ^{109}Cd bound per gram of total liver protein.

Blood Chemistry

Levels of lactic dehydrogenase (LDH) were measured in serum of field-collected fish or in plasma of laboratory fish by using a Cobas Fara centrifugal analyzer (Gay et al., 1968).

Parasite Loads in Fish Livers

Parasites were removed from livers of field-collected fish by homogenization in a solution containing 0.25 M sucrose and 0.1 M KOH. Cysts remaining after solubilization of the liver were counted.

RESULTS AND DISCUSSION

Biomarker Responses of Fish in East Fork Poplar Creek

The biomarker responses of bluegills collected at different stations are shown in Figure 2., for EROD, DNA strand breaks, and MT. Fish analyzed for EROD and DNA strand breaks were collected in May 1988, and fish analyzed for MT were collected in August 1988. Although EROD activity in livers of fish from the upper stations of EFPC is significantly higher ($P < 0.05$) than in fish from the reference stream, levels are much higher in fish collected near the ORWRF outfall (Station 3) than at the industrial outfall (Station 1). EROD activity in fish further downstream (Station 5) is similar to that in fish from the reference site. A somewhat similar pattern is observed for MT levels; significant elevation of MT levels ($P < 0.05$) is observed only at Station 3. However, genotoxic stress, expressed as strand breaks in the DNA, is very high in each of the three upstream stations and decreases significantly ($P < 0.05$) downstream at Station 5 (compared to values at Station 3). It should be noted that EROD activity (but not the other two biomarkers) was assayed repeatedly over a 2-year period at these sites, and the general pattern was very similar to that observed in this sampling (Loar et al., 1988).

The significance of the biological response of the fish from the stream is difficult to interpret. Two different patterns are observed. Genetic damage suggests that toxic exposure and effects are severe near the industrial outfall and continue to be a problem down to the municipal outfall, while EROD and MT suggest that the fish from the upstream reaches are not very different from those at the reference sites.

The genetic damage data are reinforced by the general pattern of biological responses examined by the Biological Monitoring Program in EFPC over the last two years. These data suggest that fish in the upper reaches of the stream are much more severely affected than those downstream. Condition factors, liver and visceral somatic indices, and other indicators of physiological and bioenergetic status of the fish clearly suggest that fish near the industrial outfall are severely affected by the discharges and that the situation improves downstream (Loar et al., 1988). Parasite infestation is much more severe in animals near the industrial outfall (Figure 3), and histological examination of liver tissue demonstrated that fish from the upstream stations have much more of the total liver volume occupied by parasitic and necrotic lesions (data analyzed by D. Hinton, reported in Loar et al., 1988). Species richness of fish (number of different species observed over 12 months) and diversity of insects (number of

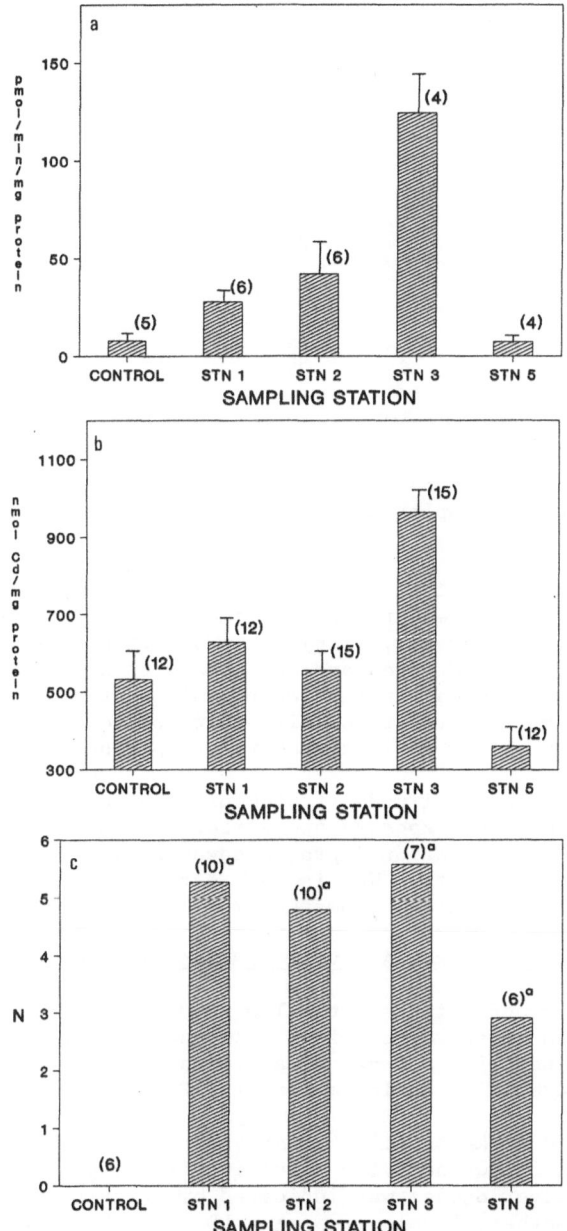

Figure 2. Biomarker responses of redbreast sunfish collected from East Fork
Poplar Creek and a reference (control) site (Brush Fork). (A) EROD
activity of hepatic microsomes (picomoles per minute per milligram
protein). (B) Metallothioein levels in fish livers (nanomoles
^{109}Cd bound per milligram total liver protein). (C) DNA strand
breaks ($\underline{N}$, the relative number of breaks per alkaline unwinding
unit). Values for EROD and MT are reported as means ± s.e.m.
(number of fish). Because of the calculation of $\underline{N}$ (Equation 1),
variance in DNA strand breaks cannot be expressed as an s.e.m.; the
statistical significance of differences between the stations is
discussed in the text; stations with strand breaks significantly
different from the reference (control) site are indicated by "a".

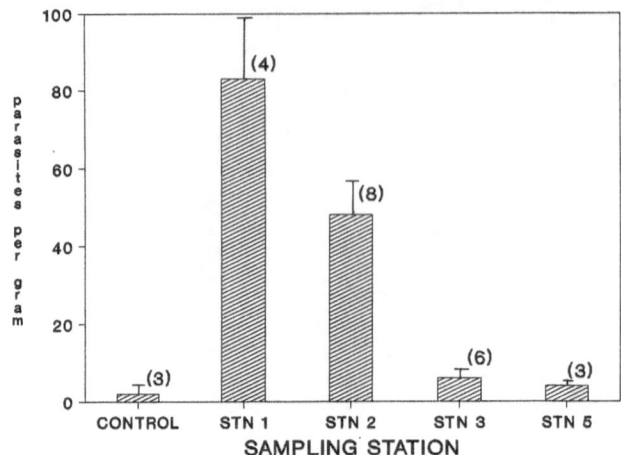

Figure 3. Numbers of parasites in livers of redbreast sunfish collected from East Fork Poplar Creek. These fish were sampled in June 1987.

species per taxon) are low at the industrial outfall, improve downstream, and are very similar to values at the reference site at Station 5 near the mouth of the stream (Loar et al., 1988). Reproductive competence, measured as the number of oocytes recruited and the number of atretic oocytes, was significantly lower ($P < 0.05$) in redbreast sunfish at Station 1, than in fish downstream or from the reference site (data collected in spring, 1988; M.S. Greeley, Jr., Oak Ridge National Laboratory personal communication).

In general then, the total body of biomonitoring data suggests a clear pattern indicating that the exposure and effects of the industrial outfall are severe and that the situation improves downstream, presumably due to dilution of industrial effluents. Why is this pattern not uniformly reflected in the pattern of biochemical biomarkers?

Effects of Hepatotoxins on EROD Activity and Genetic Damage

We explored the hypothesis that the low EROD activity at the industrial outfall was due to an hepatotoxic effect from the effluents. Fish in the laboratory were treated with a known hepatotoxic agent (allyl formate) and subsequently challenged with an injection of BaP.

Treatment with 50 uL/kg of allyl formate significantly ($P < 0.05$) reduced the induction of EROD activity resulting from exposure to BaP (Table 2). The extent of liver damage was indicated by an assay that can be considered another biomarker, levels of plasma LDH, which increased significantly ($P < 0.05$) in the animals treated with allyl formate. The presence of this cytoplasmic enzyme in blood is assumed to result from liver cell damage from the hepatotoxic agent.

Damage to the integrity of the DNA, expressed as strand breaks, is also exacerbated by treatment with the hepatotoxin (Table 2). The number of strand breaks in DNA from animals treated with BaP alone is significantly less ($P < 0.05$) than that from animals treated with allyl formate prior to injection of BaP.

Effects of Stress of DNA Strand Breaks

Potent cytotoxic agents are not the only stressors that can alter the extent of genetic damage. Exposure of fish in the laboratory to aqueous

170

Table 2. Effect of hapatotoxins on EROD activity and genetic damage

Treatment[a]		EROD	LDH	DNA Damage ($\underline{N}$)
[ALF]	[BaP]			
(Vehicle Control)				
0	0	6.6 ± 0.64 (11)	95 ± 18.6 (11)	--[b] (9)
(BaP Control)				
0	8	113 ± 31 (5)	174 ± 31.6 (4)	0.948 (7)
(Experimental)				
50	0	44* ± 19 (4)	408* ± 109 (5)	2.30* (7)

[a] Fish were injected with benzo(a)pyrene (BaP) in corn oil (8 ug/g body weight) one day after administration of 50 uL/kg body weight allyl formate (ALF) by gavage. EROD activity is in units of picomoles per minute per milligram microsomal protein. LDH activity is in units of picomoles per minute per milligram microsomal protein. LDH activity is reported as units/per minute per milligram microsomal protein. LDH activity is reported as units/per minute per milligram microsomal protein. DNA strand breaks are expressed as $\underline{N}$, the number of breaks per alkaline unwinding unit. Values for EROD and LDH are reported as means ± s.e.m. (number of fish). Because of the calculation of $\underline{N}$ (Equation 1), variance in DNA strand breaks cannot be expressed as an s.e.m.; statistically significant differences in $\underline{N}$ are indicated by an asterisk.

[b] The vehical control group was used as the control for calculating the value of $\underline{N}$ (see Equation 1).

solutions of BaP (1 ug/L) results in markedly different levels of strand breaks, depending on the conditions of the exposure (Table 3). Exposure to BaP for 16 days with no dioxane present as a carrier (BaP solution generated by flowing water over glass beads coated with BaP) resulted in very little damage to the DNA. Much higher levels of damage were observed when the BaP was introduced into the water with dioxane as a carrier (dioxane concentration of 0.1 mL/L in water). Animals stressed by transfer from a 10°C acclimation temperature to a 20°C exposure aquarium (temperature stressed) with no carrier, and with BaP solution produced by a generator column, also demonstrated significantly higher levels of DNA damage than did animals that had been previously acclimated to the exposure temperature; in the absence of BaP, temperature stress alone produced no damage to the DNA (data not shown). Thus, the significance of exposure to BaP, a known genotoxic agent, can be very different, depending on the conditions of the exposure. Temperature stress and co-exposure to dioxane produced high levels of damage, comparable to levels observed in fish from the most affected sites at East Fork Poplar Creek, while exposure to BaP with no additional stresses produced little damage to DNA.

The high levels of DNA strand breaks in the (BaP plus dioxane) exposures were not associated with the induction of high levels of EROD activity. Fish exposed to BaP with or without dioxane carrier demonstrated a small but significant increase in EROD activity (Table 1). The increase occurred approximately seven days earlier in fish exposed to BaP with dioxane. The concentration of dioxane in the water (0.01%) is too low to alter the activity of the BaP in water and thus should not affect the bioconcentration this hydrophobic contaminant by the fish (Rand and Petrocelli, 1985). Nevertheless, the presence of the dioxane does appear to alter either the rate of uptake of the BaP or its interaction with molecular targets. It should be noted that the levels of EROD activity resulting from continuous exposure to BaP at a concentration close to its aqueous solubility limit (Table 1) are much lower than those observed in affected sites on EFPC and are comparable to levels in fish from the reference stream (Figure 2a).

Table 3. DNA damage, measured as strand breaks, by an alkaline unwinding assay (Shugart, 1988a,b) in livers of Sunfish exposed to aqueous solutions of BaP

Exposure[a]	$\underline{N}$		
	Day 4	Day 7	Day 14
BaP Only	0.25	0.45	0.85
BaP + Dioxane	2.00*	3.60*	4.95*
Dioxane Control	0.15	0.10	--b
Temperature stressed	0.65	0.50	3.90*

[a] All exposures were at 20°C. Fish, other than those in the temperature-stressed exposure, were acclimated to 20°C prior to exposure. Treatments were as follows: BaP only exposures contained BaP solutions generated by passing the water over glass beads coated with an excess of BaP; BaP +Dioxane exposures contained BaP solutions produced by using dioxane as a carrier to dissolve the BaP in the water (dioxane concentration of 0.1 mL/L water); dioxane control exposures did not contain BaP, but did contain flowing solutions with the dioxane carrier at the same concentration as in the dioxane-carrier exposure; and temperature-stressed exposures contained BaP solution produced with the generator column (no dioxane or other carrier), but animals were acclimated to 10°C and introduced into exposure water at 20°C. DNA strand breaks are expressed as $\underline{N}$, the number of breaks per alkaline unwinding unit. Asterisks indicate significant differences ($P < 0.05$) from the control (from Shugart et al., 1988).

[b] The day 14 dioxane control group was used as the control for calculating the value of $\underline{N}$ (see Equation 1).

Integration of Field and Laboratory Results: Interpretation of Biomarkers

Biological markers are sensitive responses that indicate the magnitude of the host response to exposure to environmental contaminants; however, interpreting the significance of the biomarker responses is not always straightforward. In this paper, we have attempted to illustrate several points that must be considered in interpreting biomarkers in field-collected animals.

1. Biomarker measurements provide information on how the organism responds within the context of the entire exposure scenario and may not be strictly dose dependent. Cellular damage, presence of a co-contaminant (such as dioxane), and temperature stress altered the response of enzyme and genetic biomarkers to a well-characterized contaminant, BaP. This may make it difficult or impossible to use biomarker responses to infer the "dose" of a contaminant to which an organism was exposed. However, since most real-world exposures usually involve complex mixtures of chemicals, concepts of dose-dependent responses to individual chemicals are somewhat irrelevant. Biomarker responses provide significant insights into the cumulative effects of the integrated exposure situation. While laboratory studies can demonstrate the effects of chemical mixtures or temperature stress on toxic damage, it is unrealistic to even attempt to understand and predict how real-world exposure scenarios will modulate effects of a large number of toxic chemicals. Biomarkers measured in animal sentinels from affected sites give a direct measure of the integrated response to the entirety of the exposure. Our challenge, then, is not to develop simple dose-response relationships, but rather to understand the relationship between biomarker responses, such as measures of genotoxicity, and long-term deleterious effects to the individual and its population.

2. Measurement of a single biomarker can lead to misinterpretation of the extent or severity of exposure. EROD activity and MT levels at Stations 1 and 2 are only slightly greater than those observed in fish from the reference stream (Figures 2a and 2b). The low level of response may result from gross impairment of liver function due to parasitic and necrotic lesions observed in fish from these stations (Loar et al., 1988). The laboratory study on the effects of an hepatotoxin demonstrates the significance of cellular damage on the capacity of the fish liver to respond to a known inducer of EROD activity. While it is possible that Stations 1 and 2 do not contain chemicals that induce EROD or MT, the overall pattern of effects at those sites argues against this interpretation: (1) EROD and MT respond to very different types of contaminants, yet both of these markers respond only marginally above responses at reference sites, (2) Stations 1 and 2 exhibit significant toxic exposure based on two years of biomonitoring data, demonstrating increased parasitic and necrotic lesions in livers, impaired physiological and bioenergetic measures of fitness and condition, reduced species richness and diversity, and impaired reproductive competence, and (3) genetic damage, measured as DNA strand breaks, is higher these stations.

3. Results from one biomarker can improve interpretation of other biomarkers. The histological observations of increased parasitic and necrotic lesions and data on increased levels of parasites in livers of animals from Station 1 and 2 are both biomarkers that demonstrate damage from contaminants, but that also improve interpretation of EROD and perhaps MT biomarkers. Low EROD activity in animals with extensive hepatic lesions cannot be inferred to demonstrate low concentrations of chemical inducers such as BaP. MT has a number of biological functions, including metal homeostasis and metal detoxification, as well as secondary roles such as maintenance of intracellular redox potential, and defense against physical and inflammatory stress, starvation, and infection (Brady, 1982; Webb and Cain, 1982; Hamer, 1986; Dunn et al., 1987). While MT is used in biomarker studies as a measure of the biological response to excessive amounts of heavy metals, it may be equally or more useful as a less specific indicator of environmental and chemical stresses that modify the organism's response to toxic chemicals.

Measurement of DNA strand breaks appears to be a sensitive and informative biomarker that responds relatively rapidly to chemical exposures. It is significant that continued exposure of fish to BaP with dioxane (Table 3) results in an eventual decrease in the number of breaks back to control values after 30 days of continuous exposure, perhaps because of induction of DNA repair capacity (Shugart, 1988b). Nevertheless, field-collected animals showed high levels of DNA strand breaks (Figure 2c); if repair mechanisms were induced in the animals from the stream, they were insufficient to cope with the continued insult and cumulative stress in the environment. Even more information and improved interpretation could be provided if the longer-term consequences of genotoxic damage, such as hypomethylation of the DNA (Shugart, 1989) or micronuclei formation (Hose et al., 1987) were measured simultaneously. Combined data reflecting exposure to EROD inducers, histopathological damage, induced MT, and DNA strand breaks, as well as longer-term consequences of the cumulative insult to DNA integrity, could form a solid foundation to begin to evaluate the biological significance of the environmental contaminants and their overall effect on exposed organisms.

4. Regardless of the difficulties in interpreting the biomarker data, it is important to note that the biomarkers were much more informative than the chronic toxicity tests with Ceriodaphnia and fathead minnow larvae. The monthly chronic bioassays found no indication of toxic effect even at Station 1 (Loar et al., 1988), while the community and population surveys

clearly demonstrated adverse ecological effects in the upper portions of the stream. The biomarker responses, especially as interpreted within the context of the full suite of measurements discussed in this paper, clearly warned of toxic effects in the upstream sites. The chronic toxicity bio-assays do not appear to be sensitive enough to be relied on to protect the ecological integrity of aquatic communities. Biomarkers measured in animals collected from potentially affected sites offer great promise in evaluating the environmental risks posed by pollutant discharges.

SUMMARY AND CONCLUSIONS

Biological markers measured in fish from EFPC are consistent with the laboratory observation that histopathological lesions in livers impair the capacity of the organ to induce EROD activity and perhaps affects MT concentrations. DNA integrity is also compromised in the upper three stations of the streams. The cumulative effects of industrial discharges, combined with stresses and perhaps the presence of nontoxic levels of mixtures of chemicals from the ORWTF discharge, may account for the high level of bio-marker responses at Station 3. Biomarker responses may be complex and can be best interpreted if a suite of responses of several biological endpoints are evaluated together.

Since environmental stress and chemical mixtures can modify toxicolog-ical response to a chemical exposure, biomarkers are informative and biolog-ically significant measures of the adverse effects of environmentally real-istic exposure scenarios. Ultimately, improved understanding of biomarker responses may make it possible to unequivocally interpret the significance of these measurements in animals from sites of suspected contamination. In this study the comprehensive multi-task biomonitoring program provided a more readily interpretable body of data than did the biomarker data alone.

However, the biomonitoring program is a very costly endeavor, especial-ly the labor-intensive studies of community structure. The biomarker com-ponent comprises only 10-15% of the total cost of the monitoring program. Although the state of the science is still in the early phases of a learning curve for interpreting biomarker responses, field studies of bio-markers offer great potential as sensitive, cost-effective tools to evalu-ate the biological hazards of environmental contamination. Success in this endeavor can significantly reduce the expense and time required for site evaluations, without compromising the goal of environmental protection.

ACKNOWLEDGMENTS

We thank M.K. Gustin and T. Petanen for technical assistance and data analysis. This work was supported by the Exploratory Studies Program, Oak Ridge National Laboratory, by the Oak Ridge Y-12 Plant, Department of Environmental Management, and by the Academy of Finland/Research Council for Environmental Sciences (project 06/133). The Oak Ridge Y-12 Plant and the Oak Ridge National Laboratory are operated by Marietta Energy Systems, Inc., under contract DE-AC05-84OR21400 with the U.S. Department of Energy. Publication No. 3365, Environmental Sciences Division, Oak Ridge National Laboratory.

REFERENCES

Brady, F.O., 1982, The physiological function of metallothionein, TIBS, 7:143-145.
Burke, M.D., and Mayer, R.T., 1974, Ethoxyresorufin: Direct fluorometric

assay of a O-dealkylation which is preferentially inducible by
    3-methylcholanthrene, Drug Metab. Dispos., 2:583-588.
Dunn, M.A., Blalock, T.L., and Cousins, R.J., 1987, Metallothionein, Proc.
    Soc. Exp. Biol. Med., 185:107-119.
Gay, R.J., McComb, R.B., and Bowers, G.H., 1968, An assay for activity of
    lactic dehydrogenase, Clin. Chem., (Winston-Salem, NC) 14:740.
Hamer, D.H., 1986, Metallothioneins, Annu. Rev. Biochem., 55:913-951.
Hose, J.E., Cross, J.N., Smith, S.G., and Diehl, D., 1987, Elevated
    circulating micronuclei in fishes from contaminated sites off southern
    California, Mar. Environ. Research, 22:167-176.
Jimenez, B.D., Burtis, L.S., Ezell, G.H., Egan, B.Z., Lee, N.E.,
    Beauchamp, J.J., and McCarthy, J.F., 1988, The mixed function oxidase
    system of bluegill sunfish, Lepomis macrochirus: Correlation of
    activities in experimental and wild fish, Environ. Toxicol. Chem.,
    7:623-634.
Kanter, P.M., and Schwartz, H.S., 1982, A fluorescence enhancement assay
    for cellular DNA damage, Mol. Pharmacol., 22:145-151.
Loar, J.M., Adams, S.M., Boston, H.L., Jimenez, B.D., McCarthy, J.F.,
    Smith, J.G., Southworth, G.R., and Stewart, A.J., 1988, First annual
    report on the Y-12 Plant Biological Monitoring and Abatement Program,
    ORNL/TM, Oak Ridge National Laboratory, Oak Ridge, TN.
McCarthy, J.F., and Jimenez, B.D., 1985, Reduced bioavailability to blue-
    gills of polycyclic aromatic hydrocarbons bound to dissolved humic
    material, Environ. Toxicol. Chem., 4:511-521.
Oikari, A., and Jimenez, B.D., 1989, Effects of hepatotoxicants on the
    induction of microsomal monooygenase activity in sunfish liver by
    Beta-naphthoflavone and benzo(a)pyrene, Toxicol. Appl. Pharmacol.,
    (submitted).
Rand, G.M., and Petrocelli, S.R., 1985, "Fundamentals of Aquatic Toxi-
    cology," Hemisphere Publishing Corp., New York, p.666.
Shugart, L.R., 1988a, An alkaline unwinding assay for the detection of
    DNA damage in aquatic organisms, Mar. Environ. Res., 24:321-325.
Shugart, L.R., 1989b, Quantitation of chemically induced damage to DNA of
    aquatic organisms by alkaline unwinding assay, Aquatic Toxicology,
    13:43-52.
Shugart, L.R., 1989, 5-Methyl deoxycytidine content of DNA from bluegill
    sunfish (Lepomis machrochirus) exposed to benzo(a)pyrene, Environ.
    Toxicol. Chem., (submitted).
Shugart, L.R., Jimenez, B.D., and McCarthy, J.F., 1988, DNA damage as a
    biological marker in aquatic organisms exposed to benzo[a]pyrene, in:
    "Eleventh International Symposium on PAH's," W.E. May and M. Cooke,
    eds., Battelle Press, Columbus, OH, (in press).
Sloop, F.V., Shugart, L.R., McCarthy, J.F., and Jacobson, K.B., 1989,
    Assay of metal binding proteins using Chelex-100 and cadmium, Anal.
    Biochem., (submitted).
Tennessee Valley Authority (TVA), 1986, Instream Contaminant Study, Task
    5: Summary Report, Report to U.S. Department of Energy, Oak Ridge
    Operations Office, Office of Natural Resources and Economic Develop-
    ment, Tennessee Valley Authority, Knoxville, TN.
Webb, M., and Cain, K., 1982, Functions of metallothionein, Biochem.
    Pharmacol., 31:137-142.

DNA-CARCINOGEN ADDUCTS IN FISH AS A TOOL FOR MEASURING THE EFFECTIVE

BIOLOGICAL DOSE OF AQUATIC CARCINOGENS

Bruce P. Dunn

British Columbia Cancer Research Centre
Vancouver, British Columbia
Canada V5Z 1L3

INTRODUCTION

The majority of chemical carcinogens act by inducing genetic damage in
cells. The most common type of such damage results when the carcinogen (or
its metabolite) covalently binds to DNA, forming a DNA-carcinogen adduct.
Such adducts are generally thought to be relatively benign in nondividing
cells, and may be removed from the DNA by a variety of enzymatic DNA repair
mechanisms. However, if repair is not complete when cells with DNA-
carcinogen adducts attempt to divide, the damaged DNA may be used as a tem-
plate for DNA synthesis. Attempts to replicate adducted DNA can lead to a
variety of types of heritable genetic damage, including point mutations,
deletions, insertions and translocations. Such mutations are capable of ac-
tivating oncogenes and producing cells that are genetically programmed to
form a neoplasm.

DNA adducts of environmental carcinogens occupy a central position on
the route leading from the presence of carcinogens in the environment to
the eventual production of cancer. Before it can form a DNA adduct capable
of inducing heritable mutations, an environmental carcinogen must be taken
up by an organism, must be distributed to the tissue in question, and in
most cases must be metabolically activated. In order to survive until it
is detected by analytical procedures, the adduct must escape DNA repair
processes. Thus, when DNA adduct levels are measured in a wild organism
that has been exposed to environmental carcinogens, the adduct level that
is seen is a function not only of the environmental level of a carcinogen,
but also of its bioavailability, uptake, tissue distribution, metabolism,
and the repairability of the adduct species that are formed. The events
that occur during the period up to adduct formation can be characterized as
"things that occur to the carcinogens." Once an adduct causes genetic damage in a cell, the original chemical nature of the carcinogen becomes irrelevant, and the events leading to a neoplasm can be characterized as "things
that occur to genetically damaged cells."

In a variety of experimental animal systems, it has been shown that
while adduct formation and genetic damage by a carcinogen are necessary for
the induction of cancer, such damage may not always be sufficient to cause
cancer. On a simple level, the genetic damage induced in a single cell by
the carcinogen may be in regions of the DNA not involved in the processes
by which a cell becomes cancerous. Alternately, a mutation may be of a

In Situ Evaluations of Biological Hazards of Environmental Pollutants
Edited by S. S. Sandhu *et al.*
Plenum Press, New York, 1990

type that does not lead to neoplastic transformation. Furthermore, a single genetic change may not by itself be capable of generating the full neoplastic phenotype, and further genetic changes in the same cell may be necessary. These changes may be either helped or directly caused by tumor promoters. Finally, even if all necessary genetic changes in a cell have occurred, the nascent tumor must escape immune surveillance and progress to form a neoplasm. Thus, the detection of DNA-carcinogen adducts in the tissues of a wild organism indicates that genetic damage of cells has occurred, but adduct levels cannot by themselves predict whether such damage will result in the production of tumors.

PROCEDURES FOR MEASURING DNA ADDUCTS

Until recently, DNA-carcinogen adducts were measurable only when radioactively labeled carcinogens were used in laboratory experiments. New techniques have now become available which overcome this limitation, and allow the measurement of selected classes of DNA-carcinogen adducts with great sensitivity (Santella, 1988). For fluorescent carcinogens, DNA adducts can be measured by fluorometric techniques, using either intact or hydrolyzed DNA. For both fluorescent and nonfluorescent carcinogens, DNA adducts can be detected by immunochemical techniques based on monoclonal or polyclonal antibodies. A third technique, ^{32}P-postlabeling, is even more sensitive than antibody-based techniques and needs less DNA, but is at the moment restricted largely to the analysis of certain classes of large bulky adducts (Watson, 1987). In this procedure, DNA is hydrolyzed to mononucleotides and their adducts. Adducted nucleotides are purified by HPLC (Dunn and San, 1988), solvent/solvent extraction, or nuclease P1 destruction of normal nucleotides (Watson, 1987). Adducted nucleotides are labeled in the laboratory with ^{32}P. Adducts are then purified, and separated in two dimensions by thin-layer chromatography (Dunn and San, 1988). Adducts are detected by autoradiography, with individual adducts separated by thin-layer chromatography appearing as dark spots on the X-ray film.

DNA-CARCINOGEN ADDUCTS FOR IN SITU MEASUREMENT OF ENVIRONMENTAL CARCINOGENS

Because adducts are directly involved in the carcinogenic process, their measurement is highly relevant. Only chemicals which can cause genetic damage in a given species are observable. The assay ignores both environmental chemicals that are not carcinogens, and carcinogens that are not metabolically activated. Finally, adduct measurement can be performed on different tissues from animals to investigate tissue-specific genetic damage.

DNA-carcinogen adducts in wild organisms can be used for both in situ measurement of exposure to environmental carcinogens and as a measure of the effects of environmental carcinogens. When used to estimate exposure of a given species to a carcinogen, adducts have a number of very useful properties. Only carcinogens that are biologically available are measured. This requirement eliminates carcinogens that are tightly bound to inorganic matrices, or those present in food organisms not eaten by the species. Adducts have an appreciable lifetime in organisms (days to weeks) that is generally much longer than that of the parent carcinogen. This is particularly important for the monitoring of exposure to readily metabolized carcinogens such as polycyclic aromatic hydrocarbons. For this class of compounds, body burdens in vertebrate species are essentially zero, and thus body burden cannot be used to estimate exposure (Dunn, 1989).

Measurement of adduct levels also has advantages for in situ evaluation of the biological effects of environmental carcinogens. In measuring effects, the adduct values are interpreted not for what they can say about the level of carcinogen exposure, but for what they can say about the genetic consequences of exposure. The primary advantage of using adduct levels as an endpoint is that they are measured on organisms as they exist in the wild. Effects do not have to be extrapolated from laboratory experiments. A second major advantage when monitoring for carcinogen effects is that each individual organism analyzed can provide a numerical value of adduct levels. Only a small number of organisms (typically five to ten) need to be used in any analysis of effects, and the analysis of even a single individual may be helpful. This feature is in direct contrast to the examination of wild organisms for tumors, where hundreds of organisms may have to be examined to obtain a statistically reliable measure of tumor frequencies.

If the desired endpoint of a field study is to monitor exposure to and effects from carcinogens, adduct measurements are highly relevant. If, however, the desired endpoint is cancer, then the usefulness of adduct measurements is lessened by the fact that there are many intervening biological effects between adduct formation and the development of neoplasms.

RESULTS OF MEASURING DNA ADDUCTS IN FISH FROM POLLUTED AREAS

A variety of species of fish from carcinogen-contaminated areas have been shown to have elevated frequencies of tumors (Mix, 1986). Using ^{32}P-postlabeling we have examined DNA from livers of brown bullheads from the Buffalo and Detroit rivers, where sediments are contaminated with polycyclic aromatic hydrocarbons and bullheads suffer from pollution-related liver tumors (Dunn et al., 1987). We detected elevated levels of aromatic DNA-carcinogen adducts in fish from the polluted areas relative to fish raised in clean aquaria. The adducts seen appeared as a diffuse diagonal radioactive zone on two-dimensional thin-layer chromatograms. We have interpreted this as resulting from multiple overlapping adduct spots. While it is not possible to directly determine the chemical nature of the adducts seen in these fish populations, the data are consistent with their identification as adducts of polycyclic aromatic hydrocarbons.

FUTURE POTENTIAL OF ADDUCT MEASUREMENTS

A major advantage of using adduct levels as an indicator for carcinogen exposure is that in all organisms the target molecule (DNA) is the same, and the range of adducts formed is frequently the same for exposure to a given carcinogen. Adduct levels can thus provide a basis for comparing the effects of carcinogens on different species, exposed through different routes. This facility is particularly important when considering the possible human health effects of contaminated environments. The technology is now available for comparing adduct levels in humans (either biopsies or tissues from autopsy) with levels in animals seen after environmental or laboratory exposure (Santella, 1988; Watson, 1987). Adduct measurements can thus be used to aid in extrapolating animal data for use in human risk estimates.

Using modern methods for detecting DNA-carcinogen adducts, it is easily possible to quantitate adduct levels in laboratory organisms exposed to environmental samples or extracts. These data may then be compared with data obtained on wild organisms from polluted areas, and with data from laboratory organisms exposed to known quantities of pure carcinogens. In doing this, adduct levels can provide a common denominator between different types of laboratory and field exposures. An even more generic approach is to apply environmental extracts to a standardized laboratory cell system,

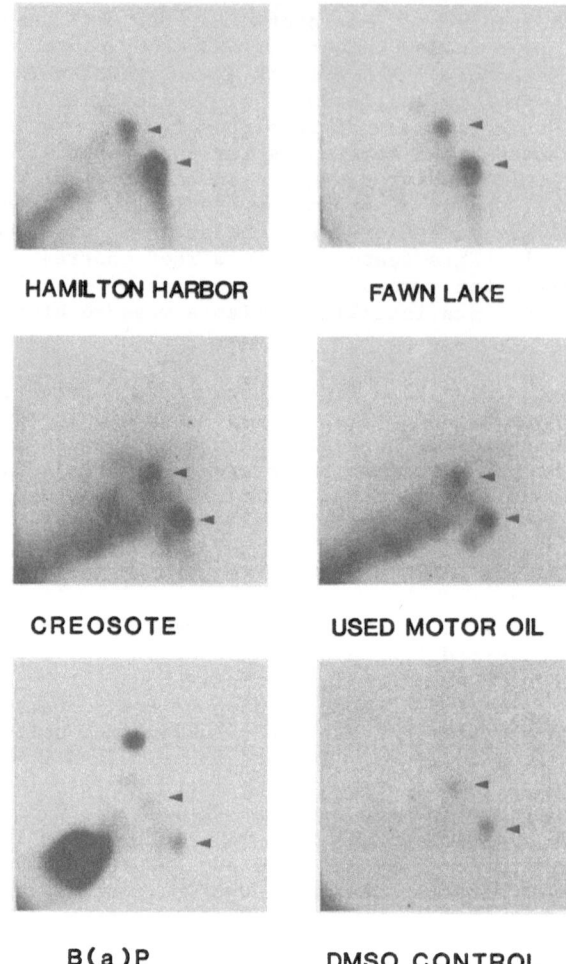

HAMILTON HARBOR   FAWN LAKE

CREOSOTE   USED MOTOR OIL

B(a)P   DMSO CONTROL

Figure 1.   DNA adducts in cultured cells treated with environmental
            extracts.

and to then measure genetic damage as manifested by adduct formation. This
approach is analogous to the widespread use of bacterial mutagenesis tests
for determining the presence of mutagens in crude environmental mixtures.

In preliminary experiments, we have used ^{32}P-postlabeling to detect
aromatic DNA-carcinogen adducts in cultured C3H/10T1/2 cells grown in culture. This mammalian cell line can metabolize polycyclic aromatic hydrocar-
bon carcinogens to DNA-reactive metabolites that form DNA-carcinogen adducts. Cells were exposed to extracts of sediments from a PAH-contaminated
fresh water harbor area (Lake Ontario, Hamilton, Ontario) and a noncontami-
nated reference lake (Fawn Lake, Ontario) (Figure 1). Samples from the control lake exhibited two assay-related background spots (marked with arrows
on the figure), while cells exposed to an extract of the sediment from the
harbor of Hamilton, Ontario, showed a diagonal zone, representing the presence of multiple aromatic DNA-carcinogen adducts. This chromatographic
pattern was similar to that seen in fish from the harbor (Dunn, unpublished
results) and from other PAH-contaminated areas (Dunn et al., 1987). For
comparative purposes, cells were exposed to extracts of commercial creosote
and of used crankcase oil. Both materials contain DMSO-extractable polycyclic aromatic hydrocarbons, and are contributors to the aquatic burden of

these carcinogens. In both cases, a similar diagonal radioactive zone was produced, suggesting the similarity of the carcinogens in the sediments and in the pollutant sources. The ready induction of adducts in cultured cells by environmental samples suggests that testing protocols based on this type of experiment may be useful in evaluating the genotoxicity of complex environmental mixtures.

Acetone extracts of sediments from two Ontario lakes were provided by Dr. C. Metcalfe, Trent University, Peterborough, Ontario. Hamilton harbor samples are from an area known to be contaminated by polycyclic aromatic hydrocarbons, while Fawn Lake samples were from a remote, pristine lake. Extracts were dissolved in DMSO at a concentration equivalent to 1 g sediment/ml DMSO. Used automobile engine oil and commercial creosote from a hardware store were diluted with 9 volumes of hexane and then extracted with 10 volumes of DMSO. C3H/10T1/2 cells were treated for 3 hours with 1% DMSO extracts (equivalent to 10 mg sediment or 1 ul creosote or used engine oil per ml culture medium) or 10^{-6}M benzo(a)pyrene or DMSO alone, then were harvested, and DNA adducts determined by the nuclease P1 version of the ^{32}P-postlabeling assay (Dunn and San, 1988). Arrows indicate background spots that commonly occur in DNA from a variety of sources.

ACKNOWLEDGMENTS

I would like to thank Dr. C. Metcalfe and Mr. Gordon Balch of Trent University, Peterborough, Ontario, for generously providing extracts of lake sediments. This work was supported by Grant A3403 from the Natural Sciences and Engineering Research Council of Canada. The expert technical assistance of Daniel Twa is gratefully acknowledged.

REFERENCES

Dunn, B.P., 1989, Carcinogen adducts as an indicator for the public health risks of consuming carcinogen-exposed fish and shellfish, Environ. Health Perspect. (in press).
Dunn, B.P., Black, J.J., and Maccubbin, A., 1987, ^{32}P-Postlabeling analysis of aromatic DNA adducts in fish from polluted areas, Cancer Res. 47:6543.
Dunn, B.P., and San, R.H.C., 1988, HPLC enrichment of hydrophobic DNA-carcinogen adducts for enhanced sensitivity of ^{32}P-postlabeling analysis, Carcinogenesis, 9:1055.
Mix, M.C., 1986, Cancerous diseases in aquatic animals and their association with environmental pollutants: a critical literature review, Mar. Environ. Res., 20:1.
Santella, R.M., 1988, Application of new techniques for the determination of carcinogen adducts to human population monitoring, Mutat. Res. 205:271.
Watson, W.P., 1987, Post-labeling for detecting DNA damage, Mutagenesis, 2:319.

IN SITU MONITORING OF ENVIRONMENTAL CLASTOGENS USING TRADESCANTIA-

MICRONUCLEUS BIOASSAY

Te-Hsiu Ma

Department of Biological Sciences and
Institute for Environmental Management
Western Illinois University
Macomb, Illinois 61455

INTRODUCTION

In order to safeguard environmental quality and to detect health hazards directly in the real life situation, effective in situ monitoring bioassays are needed. Based upon a recent review, 17 major groups of organisms across the plant and animal kingdoms can be used in the in situ
monitoring of environmental mutagens and clastogens (Ma and Harris, 1985).
About 200 articles involving some 40 different bioassays have been published since the late 1960s; most of these were published in the last decade.

Among the eukaryotic bioassays, plant systems using chromosome damage
as the end point have the highest sensitivity to clastogens and the con-
gruity with data obtained by other bioassays. Of these, the Tradescantia-
microncleus (Trad-MCN) bioassay is a highly sensitive test owing to the
high degree of synchrony of both the target cells (early meiotic prophase
I) and of the cells that are evaluated for micronuclei (early tetrads of
meiosis) (Ma, 1981, 1983).

The high sensitivity of the meiotic chromosomes is also attributed to
the crossingover during the pachytene/diplotene stages of the early prophase I of meiosis. Chromosome breakages at this stage can be created
either by excision of the DNA helices or the interference of fusion at the
time of crossingover.

The results obtained from this biomonitor together with those of the
chemical monitoring efforts would provide much better information of the
real environmental status under the climatic and edaphic-conditions. If
this biomonitoring is used before chemical monitoring to determine the
environmental impacts on living systems, both chemical monitoring efforts
and expenditures could be drastically reduced.

Since the development of the Trad-MCN bioassay in 1978 (Ma et al.,
1978), this system has been used for in situ monitoring of air and water
pollutants at various locations in the United States, Canada, the People's
Republic of China and Mexico. This paper reviews the Trad-MCN monitoring
data obtained by our laboratory and other laboratories in the United States
and other countries.

In Situ Evaluations of Biological Hazards of Environmental Pollutants
Edited by S. S. Sandhu *et al.*
Plenum Press, New York, 1990

183

MATERIALS AND METHODS

Two major clones (#4430 and #03) of _Tradescantia_ plants were used for Trad-MCN bioassay. Clone #4430 is more sensitive to gaseous agents in the air, and clone #03 is more sensitive to agents in the water. In either case, the plant cuttings that bear young inflorescences are the samples for the test. Due to the relatively short duration of early prophase I in the target cells (about 6 h), exposure to liquid agents by absorption through the vascular system of the plant cuttings could be any length from 6 h to 30 h. Exposure to gaseous agents could be as short as 1 h or as long as 30 h, depending upon the concentration and potency of the gaseous agents. Since _in situ_ exposures are usually chronic, longer exposure durations are always desirable to evaluate the possible cumulative effects of the agent in the plant system. However, extended exposures, up to 48 h or more, are unnecessary since the portion of the meiotic cycle vulnerable to clastogens lasts only 6 h. Extended exposure periods may affect the chromosomes of a separate meiotic group, but may not be cumulative in the meiotic cells of the same population.

A regular sample of 15 plant cuttings was maintained in a cup of tap-water (if not heavily polluted) or nutrient solution (one-third diluted Hoagland solution). For the monitoring of gaseous agents, each sample unit was housed in a small cage and a number of cages were loaded into a clean-air box (Ma, et al., 1982) for transport to the sites of pollution. Individual caged samples were left at each of the chosen sites for a desirable duration. The exposed samples were brought back in the clean-air box for a 24-hour recovery time, and the inflorescences from each of the plant cuttings were fixed in aceto-alcohol (1:3 ratio). The inflorescences were stored in 70% ethanol after 24 hours of fixation. The procedure for slide preparation, scoring and data analysis are described in earlier publications (Ma et al., 1978, 1983).

For monitoring water pollutants, the plant cuttings were carried by a floating device called "Aquatoon" on the lake, river or reservoir for a period of 6 to 24 h. Exposed samples could be fixed with or without recovery time, depending upon the duration of exposure. A modified _in situ_ monitoring procedure was used in which the water samples were collected from various sites and tested in the laboratory (Ma et al., 1984, 1987a). This was referred to as "_in situ_ sampling" test. Most of the data on water monitoring were derived from this kind of test.

For both gaseous and liquid pollutant monitoring processes, the samples were transported to and from the sites in a clean-air box. Each monitoring trip maintained a laboratory control and a field control group.

RESULTS AND DISCUSSION

A typical set of data obtained from a series of validation experiments (Ma, 1979) is presented in Table 1 as an example of general treatment and interpretation of the test results. These results were derived from scoring 5 to 9 slides of each of the experimental groups that received either X-rays, or EMS or sodium azide. Statistical analyses were applied to determine the responses of each treatment at a 5% significance level. To simplify this presentation, the consolidated data of each group are presented only by a positive (+), or negative (-), or, in some cases, a borderline (+/-) response.

In situ air monitoring should be done repeatedly under the tolerable range of climatic factors. However, extremely hot or cold conditions could result in extraordinarily high frequencies of micronuclei since the meiotic

Table 1. Representative Data on Micronucleus Frequencies Induced by
          X-rays, or EMA or Sodium Azide

Clastogens	Dosages	Treated groups MCN/100 tetrads	S.D.	Control groups MCN/100 tetrads	S.D.	Responses
X-rays	20 R	22.8	4.8	4.1	1.3	+
EMS(liquid)	50 mM	13.2	0.8	2.9	0.3	+
(gaseous)	1000 ppm*	17.4	3.3	5.2	1.3	+
Sodium(liquid) azide	0.2 mM	10.1	1.8	3.9	1.4	+
(gaseous)**	136 ppm	31.2	8.2	5.2	1.3	+

*ppm denotes the concentration of the gas generated by a known volume of
volatile liquid utilized to convert it into gaseous volume according to
Avogadro's gas law.

**Hydrozoic acid fumes.

pollen mother cells are highly sensitive to temperature changes (Ho and
Fang, 1983; Ho, 1985). Consolidated results of air monitoring at various
parking garages, truck stops and bus stops in the United States (Ma et al.,
1982, Ma et al., 1984a) and the People's Republic of China (PRC), (Fang,
1981) are summarized in Table 2.

Inaccessibility to the industrial sites in the United States is a major
limiting factor in the development of the in situ monitoring program. In
some cases, the pollutants are major health hazards to the industrial workers and nearby residents, especially when the workers and their families
reside at the industrial sites. The consolidated data of on-site monitoring at various industrial sites in the United States (Ma et al., 1982) and
the People's Republic of China (Fang, 1981) are summarized in Table 3.

Table 2. Results of Air Monitoring at Parking Garages, Truck Stops and
          Bus Stops

Sites	Freq.	Duration(h)	Responses	Remarks	References
Parking garages					
Chicago, IL	2	4 - 6	-	clear day	Ma et al., 1982
Decatur, IL	4	1 - 4.5	-		
Peoria, IL-I	3	2 - 6	+		Ma et al., 1984a
Peoria, IL-II	2	2.5 - 4	+/-	few cars	
Truck/bus stops					
Dixie - I	2	2 - 4	-		Ma et al., 1982
Dixie - II	2	2 - 3	-		
Woodhull, IL	2	2.5 - 5	+		Ma et al., 1984a
Peoria, IL	2	0.5 - 1	+/-	rained	
Qingdao, PRC	1	5	-	gasoline	Fang, 1981

Table 3. Results of Air Monitoring at the Industrial Sites

Sites	Freq.	Duration(h)	Responses	Remarks	References
Granite City, MO	2	3.5	+/-	foundry	Ma et al., 1982
Wood River, IL	1	3.5	+	petro. co.	
Qingdao, PRC					
Ag chem co-I	1	4	+		Fang, 1981
Ag chem co-I	1	6	+		
Rubber co.	1	6	+/-		Ma et al., 1982
Dry cleaner	1	15	-	at night	
Farmer's market	4	2.5 - 5	+/-		
Animal farm	1	6	-		

Although some epidemiological data on occupational exposure to air pollutants are available, on-site monitoring data from factories and laboratories and other common workplaces are lacking. Farmhouse air monitoring for fumes from pesticides is an urgent need, and recently, radon has become of increasing concern. Attempts have been made to correlate chromosome damage and radon concentration. The Trad-MCN assay would be the most adequate monitor which goes hand-in-hand with the physical and radiochemical monitors. Consolidated results of air monitoring in laboratories, residential houses and common workplaces in the United States (Ma et al., 1982) and the People's Republic of China (Ma et al., 1982) are summarized in Table 4.

The on-site monitoring of open water (lake, river, reservoir) is not a simple matter because of the variation of climatic conditions and the quality and quantity of the water pollutants. A limited amount of data were obtained from the "aquatoon" monitoring technique directly at the site and shown in Table 5.

Table 4. Results of Air Monitoring of Laboratories, Residential Houses and Common Workplaces

Sites	Freq.	Duration(h)	Responses	Remarks	References
Residential houses, PRC	2	4.5	-	coal fire	Ma et al., 1982
Farmhouse #1-I	1	16	+		
Farmhouse #1-II	1	17	+		
Farmhouse #2	1	17	-		
Smoker's office	2	10 - 24	+		Ma and Harris, 1987
Smokerium	1	10	+	lounge	
Trailer #1	1	16	-		
Trailer #2	1	16	-		
Office #1	1	24	+	new paint	
Laboratory	1	24	+	new paint	
Herbarium	2	24	-	dichloro-benzene	

Table 5. Results of "Aquatoon" Monitoring on Site

Sites	Freq.	Duration(h)	Responses	Remarks	References
Reservoir, Macomb, IL	2	3	-		Ma, unpublished
Lake Superior, Canada	6	24	+/-	+5 out of 6 sites	Grant et al., 1989

The "in situ sampling" approach to monitoring drinking water is superior to the Aquatoon approach because the influence of climatic factors can be avoided. Bioassays accompanied with a chemical analysis of drinking water would be a more meaningful approach. Results of "in situ sampling" tests of water samples from various sites of the United States (Ma et al., 1984, Ma et al., 1987a), the People's Republic of China (Lo, 1985) and Mexico (Ruiz et al., 1987) are summarized in Table 6.

In most cases, industrial wastewater can be monitored only through the in situ sampling approach because the Trad-MCN assay cannot tolerate high concentrations of pollutants. Proper dilution of the wastewater samples had to be done before treating the plant cuttings. Clastogenicity of the industrial wastewater samples collected from various sites of PRC (Chen and Fang, 1981; Fang, 1981a; Chen and Zhang, 1982; Chen and Xiang, 1983; Chen et al., 1983; Zheng, 1985) and Mexico (Ruiz et al., 1987) are given in Table 7.

Trad-MCN tests were applied to soil solutions derived from the contaminated soil samples collected near a steel mill in Qingdao, PRC (Ho et al., 1983) and a battery factory near Moline, IL (Ma et al., 1984a). Positive responses were obtained from all of the samples collected at different distances from the steel mill. Positive response was obtained only from the alkaline fraction of the extracts of the contaminated soil near the battery factory. Results of the soil tests are summarized in Table 8.

CONCLUSIONS

The Tradescantia-Micronucleus (Trad-MCN) bioassay is a genotoxicity test using chromosome damage in the meiotic pollen mother cells of Tradescantia as the endpoint. Its efficacy for health hazard assessment should not be overlooked because the test uses plant chromosomes. With the universality of DNA structure and damage and repair mechanisms,

Table 6. Results of "In Situ Sampling Tests of Drinking Water

Sites	Freq.	Duration(h)	Responses	Remarks	References
Reservoir #1-I Macomb, IL	13	30	+/-	Seasonal	Ma et al., 1984a
Reservoir #1-II Macomb, IL	26	30	+/-	Seasonal	
Shallow well Lewistown, IL	22	30	+/-	Seasonal	Ma et al., 1987a
Deep well	22	30	-		
Reservoir #2 Sichuan, PRC	1	30	+/-	Seasonal	Lo, 1985

Table 7.   Clastogenicity of Industrial Waste Water Samples

Sites	Freq.	Duration(h)	Responses	Remarks	References
Sea water #1 Qingdao, PRC	3	12	+	Effluent	Chen and Fang, 1981
Sea water #2 Qingdao, PRC	1	12	+	Effluent	Fang, 1981a
Sea water Dagang, PRC	1	6	+/-	Effluent	Chen and Zhang, 1982
Sea water Xiaogang, PRC	1	6	+	High Cr	
Sea water Er Zhung, PRC	1	6	+	Effluent	
Brackish water leather factory	1	6	+	Effluent	Chen and Xiang, 1983
Brackish water paint factory	1	6	+	Effluent	
Brackish water paper mill	1	6	+	Effluent	
Brackish water plating co.	1	6	+	Effluent	
Brackish water dying co.	1	6	+	Effluent	
Brackish water ag chem. factory	1	6	+	Effluent	
Brackish water steel mill	1	12	+	Effluent	
Brackish water printing/dying co., Fuzhou, PRC	1	12	+	Effluent	Zheng, 1985
Waste water Queretaro, Mexico	1	12	+/-	Effluent	Ruiz et al., 1987
Sea water Hungdao, PRC	1	8	+	Oil Polluted	Chen, et al., 1983

clastogenicity revealed in plant chromosomes should have the same validity as those exhibited in mammalian chromosomes, including those in humans.

Trad-MCN is a efficient short-term in situ and in vivo bioassay that can provide a first alert for the presence of hazardous agents in the air or in drinking water. Application of this bioassay in monitoring environmental pollutants will benefit humans greatly as well as other living beings.

All pollutants in the air and in drinking water are in the form of mixtures. The Trad-MCN test can be applied directly to the mixture without condensation, purification or sterilization. The synergistic or antagonistic effects of the mixtures could be detected under a variety of climatic conditions. The interaction of the agents with the climatic factors can also be detected.

Table 8.   Clastogenicity of Soil Solutions

Sites	Freq.	Duration(h)	Responses	Remarks	References
Steel mill Qingdao, PRC				Fly ash	Ho et al., 1983
Soil sample #1 500 m dist.	1	6	+	Fluoride	
Soil sample #2 1000 m dist.	1	6	+	Fluoride	
Soil sample #3 1500 m dist.	1	6	+	Fluoride	
Battery factory Moline, IL				Cu,Pb,Cd.	Ma et al., 1984a
Soil fraction #1	1	6	-	Acid	
Soil fraction #2	1	6	+	Alkaline	
Soil fraction #3	1	5	-	Neutral	

REFERENCES

Chen, D., and Fang, T., 1981, A preliminary study on the use of
   Tradescantia-Micronucleus technique in monitoring marine pollution,
   J. Shandong Coll. Oceanology, 11:81 (in Chinese).
Chen, D., and Zhang, Y., 1982, Tradescantia-Micronucleus tests on water
   samples from several seashore areas of Qingdao and Jiaozhouwan, J.
   Environ. Sci., 3:35 (in Chinese).
Chen, D., and Xiang, D., 1983, Preliminary results of Tradescantia-
   Micronucleus tests on the wastewater samples from industrial factor-
   ies in Qingdao, J. Environ. Sci., 4:45 (in Chinese).
Chen, D., Dai, H., Yang, X., 1983, Tradescantia-Micronucleus tests on sea
   water contaminated by crude oil, Environ. Pollut. and Protection,
   2:39 (in Chinese).
Fang, T., 1981, A report on the studies of effects of environmental pol-
   pulion on chromosomes - A Sino-American collaborated research project, II. Tradescantia-Micronucleus bioassay on environmental mutagens in the air and water samples from some industrial areas of
   Qingdoa, PRC and on the pesticide-DDV, J. Shandong Coll. Oceanology,
   11:9.
Grant, W.F., Lee, H.G., Logan, D.M., Salamone, M.F., 1989, The use of
   Tradescantia and vicia faba bioassays for the in situ detection of
   mutagens in a aquatic environment, Environ. Molecular Mutagenesis,
   14(Suppl. 15):75.
Ho, J., Zhou, R., Fang, T., 1983, Tradescantia-Micronucleus tests on
   fluoride contaminated soil, Chinese Soil Utilization and Conserva-
   tion, in: "Proc. 5th Ann. Meeting Chinese Soil Sci. Soc.," 2:325.
Ho, J., and Fang, T., 1983, Induction of micronuclei in Tradescantia
   under sudden temperature changes, Chinese J. Heredity, 6:15 (in
   Chinese).
Ho, J., 1985, Application of Tradescantia-Micronucleus tests under low
   temperature seasons, J. Environ. Sci., 4:35.
Lo, M., 1985, Tradescantia-Micronucleus tests on drinking water, Sichuan
   Environment, 4:45.
Ma, T.H., Sparrow, A.H., Schairer, L.A., and Nauman, A.F., 1978, Effect
   of 1,2-dibromoethane (DBE) on meiotic chromosomes of Tradescantia,
   Mutat. Res., 58:251.
Ma, T.H., 1979, Micronuclei induced by X-rays and chemical mutagens in

meiotic pollen mother cells of Tradescantia - A promising mutagen test system, <u>Mutat</u>. <u>Res</u>., 64:307.

Ma, T.H., 1981, Tradescantia MCN-in-Tetrad mutagen test for on-site monitoring and further validation, U.S. EPA Report, EPA-600/SI-81-019.

Ma, T.H., Anderson, V.A., and Ahmed, I., 1982, Environmental clastogens detected by meiotic pollen mother cells of Tradescantia, <u>in</u>: "Genotoxic Effects of Airborne Agents," R. Tice, R.D.L. Costa, and K.M. Schaich, eds., Plenum Press, New York, p.141.

Ma, T.H., 1983, Tradescantia-Micronucleus (Trad-MCN) test for environmental clastogens, <u>in</u>: "<u>In</u> <u>vitro</u> Toxicity Testing of Environmental Agents, Pt. A.," A.R. Kolber, T.K. Wong, L.D. Grant, R.S. DeWoskin, and T.J. Hughes, eds., Plenum Press, New York, p.191.

Ma, T.H., and Harris, M.M., 1985, <u>In</u> <u>situ</u> monitoring of environmental mutagens, <u>in</u>: "Hazard Assessment of Chemicals, Current Development," 4:77.

Ma, T.H., Anderson, V.A., Harris, M.M., Neas, R.E., and Lee, T.S., 1984, Mutagenicity of drinking water detected by the Tradescantia-Micronucleus test, <u>Can</u>. <u>J</u>. <u>Genet</u>. <u>Cytol</u>., 27:143.

Ma, T.H., Harris, M.M., Anderson, V.A., Ahmed, I., Mohammad, K., Bare, J.L., and Lin, G., 1984a, Tradescantia-Micronucleus (Trad-MCN) tests on 140 health-related agents, <u>Mutat</u>. <u>Res</u>., 138:157.

Ma, T.H., and Harris, M.M., 1987, Tradescantia-Micronucleus (Trad-MCN) bioassay - A promising indoor pollution monitoring system, <u>in</u>: "Proc. 4th Internl. Conf. Indoor Air Quality and Climate," W. Berlin, p.243.

Ma, T.H., Neas, R.E., Harris, M.M., Xu, Z., Cook, C., and Swofford, D., 1987a, <u>In</u> <u>vivo</u> tests (Tradescantia- and Mouse-Micronucleus) and Chemical analysis on drinking water of rural communities, <u>in</u>: "Short-term Bioassays in the Analysis of Complex Mixtures V.," S.S. Sandhu, D.M. DeMarini, M.J. Mass, M.M. Moore and J.S. Mumford, eds., Plenum Press, New York, p.189.

Ruiz, L.E., Valtierra, E.R., and Ma, T.H., 1987, Presencia de agentes genetoxicos en aquas residuales empleadas para riego utilizando el sistema de micronucleos en cellulas gameticas de Tradescantia clone #4430, <u>in</u>: "III Semana dela Ecologia y Proteccion del Ambiente," Universidad Autonoma de Queretaro, QRO, Mexico, p.50.

Zheng, D., 1985, Tradescantia-Micronucleus tests on the industrial wastewater from a printing and dying factory in Fuzhou City, J. Fujian Normal University, 21:5.

THE USE OF DROSOPHILA MELANOGASTER FOR IN SITU BIOMONITORING

Lawrence G. Harshman

Department of Entomology and
Department of Genetics
University of California
Davis, California 95616

Bruce D. Hammock

Burroughs Wellcome Toxicology Scholar
Department of Entomology and
Department of Environmental Toxicology
University of California
Davis, California 95616

INTRODUCTION

Drosophila melanogaster was the principal organism used for the development of diploid transmission genetics at the beginning of the century and more recently has been the subject of pivotal studies on eucaryote gene regulation, development and behavior. D. melanogaster is potentially valuable for in situ biomonitoring because it is convenient to test different life stages, it can be used for a multi-faceted analysis of environmental genotoxins and a rapidly expanding base of information on its molecular genetics facilitates the development of new methods for bioassays.

LIFE CYCLE ASSAYS

Drosophila melanogaster is relatively easy and inexpensive to rear which is important for genotoxicity tests that often require large sample sizes. Mass handling procedures and automatic counting devices will become increasingly useful in this regard. The short generation time of the flies means that investigators could rapidly identify health hazards.

With D. melanogaster it is possible to conduct toxicological surveys with different life stages. There are various ways to expose D. melanogaster to toxic compounds. These include adding compounds to fly food, leaving a compound residue in an otherwise empty vial, topical application, injection and absorption of airborne agents. Techniques are also available for mass collection of eggs and larvae (Roberts 1986). Numerous pupal stages can be identified (Roberts 1986) and it is possible to monitor metamorphosis after exposing either larvae or pupae to potentially teratogenic environments.

The adult stage can be used for tests of toxins on male and female reproduction, viability and senescence. It may be useful to determine

In Situ Evaluations of Biological Hazards of Environmental Pollutants
Edited by S. S. Sandhu *et al.*
Plenum Press, New York, 1990

reproductive effects since insect chemosterilants often are mutagens (Borkovec, 1973) and/or carcinostatic agents (Hayes 1968). Sterility can be assessed by the inability of females to lay fertile eggs or males to successfully inseminate females. Effects on female reproductive output (fecundity) can be measured by transferring inseminated females to fresh fly medium daily and counting the number of progeny that emerge from the set of transfers. Male reproductive output (virility) is more difficult to assess. One method is to place a male in a vial containing medium and virgin females. After one day the females can be transferred to individual vials with medium to determine their fecundity. The male can be repeatedly confined with females to measure reproductive output.

Adult mortality is a convenient parameter for in situ monitoring because it is possible to transport large numbers of flies to test sites and the proportion that die after exposure can be quickly determined. A more speculative approach would be to hold adults in suspect environments for several weeks and then examine them for evidence of accelerated aging. In this manner it may be possible to rapidly identify subtle physiological risks in the environment.

Various life stages of D. melanogaster can be used to evaluate terrestrial or aquatic samples from the field. Often contaminated materials are a complex mixture of compounds which are rarely evaluated (Pereira, 1983). Successive solvent fractions of the samples can be tested for cytotoxicity or genotoxicity with different life stages of D. melanogaster to monitor a range of possible biological effects.

GENETICS AND TOXICOLOGY

The widespread availability of commonly used laboratory stocks of D. melanogaster makes it possible to standardize the results of different investigations. In many in situ bioassays the genetic differences between the test organisms used by the same or different investigators are ignored. Yet there may be substantial genetically-based differential sensitivity to the effects of toxins. This makes it difficult to compare results from different studies or to be sure the differences between treatment and control organisms are due only to the environment.

The extensive genetics of D. melanogaster may be useful for toxicology studies. For instance, chromosomes with multiple inversions and a dominant mutation can be used to control the transmission of homologous chromosomes. It would be possible to place chromosomes with high or low detoxication enzyme activity into a common genetic background and thus construct lines with a range of detoxication potential. Such lines could facilitate study of detoxication or activation of toxins. Another possibility is to use low or no activity mutations in detoxication genes as a way of creating genotypes which are more sensitive or less sensitive to the presence of genotoxins or cytotoxins (Harshman et al., submitted manuscript).

There are numerous mutation tests available for D. melanogaster. This includes tests for sex-linked lethals, autosomal recessive lethals, sex-linked lethal mosaicism, recessive visible mutants, sex-chromosome aneuploidy, translocations, position effects produced by chromosome rearrangements and somatic mutations (Abrahamson and Lewis, 1971; Valencia et al., 1984; Zimmering et al., 1986). One D. melanogaster chromosome mutation test alone can detect non-disjunction, chromosome exchange and deletion (Valencia et al., 1984; Zimmering et al., 1986). In contrast, for most organisms chromosome aberrations must be detected by tedious cytological methods. The availability of efficient chromosome mutation tests in D. melanogaster may be particularly significant because half of the spontaneous human abortions and many birth defects are a result of chromosome aberrations (Sankaranarayan, 1979). Chromosome mutation tests have been used to test volatile genotoxins (Sharkarnis, 1969; Verburgt and Vogel, 1977; Abraham et al., 1979; Zimmering and Kammermeyer, 1983). For assessment of human health hazards D. melanogaster may be particularly valuable when chromosome mutation tests are used to evaluate volatile genotoxins in situ.

A recent advance is to develop chromosome mutation tests in D. melanogaster based on defined molecular changes. The white-ivory eye mutation is a result of a 2.9 Kb duplication of a portion of this gene and reversions are the corresponding deletion. Green et al. (1986) constructed a quadruplication of the white-ivory locus to increase the frequency of reversions. Using this test they found that compounds which specifically promote deletions produce a high mutation rate (Green et al., 1986).

BIOASSAY DEVELOPMENT

There are opportunities to use the extensive genetics of D. melanogaster and molecular biology to increase the utility of this species as an indicator of human health hazards. For instance, the Pelement transformation system of D. melanogaster should prove useful for constructing new indicator strains. It might be possible to transform D. melanogaster with the human TCDD receptor and monitor P450 response for a dioxin bioassay (Jones et al., 1985). In general, it may be possible to construct D. melanogaster lines with mammalian receptors, regulatory elements and detoxication enzymes to parallel the human response to environmental health hazards.

It would be useful to develop mutation tests based on insecticide resistance. One way to employ insecticide resistance might be to expose adults in the field then bring them back to the laboratory for mass collection of eggs. The number of eggs collected could be determined by an automatic counting device or by weight. The eggs would then be transferred onto medium with insecticide and the few individuals surviving could easily be counted to determine a "mutation rate" to resistance per given number of eggs. The progeny of these individuals would also be tested to ensure the resistance is heritable. This type of test would allow the investigator to assess environmental genotoxicity without having to examine a large number of flies.

Induction of detoxication enzymes is a potential biological indicator of exposure to toxic substances. In D. melanogaster it has been possible to indirectly select for induction of a glutathione S-transferase activity (Harshman et al., submitted manuscript). The implication of this work is that it should be possible to select for modified inducibility of detoxication enzyme activities in D. melanogaster and other species to increase the specificity or sensitivity of the induction response.

In summary, D. melanogaster is potentially useful for in situ biomonitoring. In this capacity it may be especially valuable for monitoring volatile genotoxins in the environment. Drosophila melanogaster may also be valuable as a model for the development of new bioassay methods.

REFERENCES

Abraham, S.K., Goswami, V. and Kesavan, P.C., 1979, Mutagenicity of inhaled diethyl sulfate vapour in Drosophila melanogaster and its implications for the utility of the system for screening air pollutants, Mut. Res. 66:195-198.

Abrahamson, S. and Lewis, E.B., 1971, The detection of mutations in
    Drosophila melanogaster, In: "Chemical Mutagens, Vol. 2," A. Hollaender
    (ed.), Plenum Press, N.Y.
Borkovec, A.B., 1973, Insect chemosterilants as mutagens, In: "Chemical
    Mutagens, Vol. 3," A. Hollaender (ed.), Plenum Press, N.Y.
Green, M.M., Todo, T., Ryo, N. and Fujikawa, K., 1986, Genetic-molecular
    basis for a simple Drosophila melanogaster somatic system that detects
    environmental mutagens, P.N.A.S. 83:6667-6671.
Harshman, L.G., Ottea, J.A. and Hammock, B.D., Environment-dependent
    expression of detoxication enzyme activity in a Drosophila melanogaster
    selection experiment, Evolution, submitted manuscript.
Harshman, L.G., Green, M.M., MacKay, W., Bewley, G. and Edlin, G., Relative
    survival of catalase deficient genotypes on irradiated Drosophila food,
    Drosophila Information Service, submitted manuscript.
Hayes, W.J., 1968, Toxicological aspects of chemosterilants, In:
    "Principles of Insect Chemosterilization," G.C. Labrecque and C.N.
    Smith (eds.), Appleton-Century-Crafts, New York.
Jones, P.B.C., Galeazzi, D.R., Fisher, J.M. and Whitlock, J.P., 1985,
    Control of cytochrome P1-450 gene expression by dioxin, Science
    227:1499-1502.
Pereira, M.A., 1983, International symposium on tumor production, Envir.
    Health Perspect. 50:3-370.
Roberts, D.B., 1986, Basic Drosophila care and techniques, In: "Drosophila,
    A Practical Approach," D.B. Roberts (ed.), IRL Press, Washington D.C.
Sankaranarayanan, K., 1979, The role of non-disjunction in aneuploidy in
    man: An overview, Mut. Res. 61:1-28.
Shakarnis, V.F., 1969, Induction of X chromosome non-disjunction and
    recessive sex-linked mutations in females of Drosophila melanogaster by
    1,2-dichloroethane, Sov. Genet. 5:1666-1671.
Valencia, R., Abrahamson, S., Lee, W.R., Von Halle, E.S., Woodruff, R.C.,
    Wurgler, F.E. and Simmering, S., 1984, Chromosome mutation tests for
    mutagenesis in Drosophila melanogaster, Mut. Res. 134:61-88.
Verburgt, F.G. and Vogel, E., 1977, Vinyl chloride mutagenesis in
    Drosophila melanogaster, Mut. Res. 48:327-336.
Vogel, E. and Sobels, F.N., 1976, The function of Drosophila in genetic
    toxicology testing, In:"Chemical Mutagens, Vol. 4," A. Hollaender
    (ed.), Plenum Press, N.Y.
Zimmering, S. and Kammermeyer, K.L., 1983, Comparison of excision repair-
    deficient mei-9 and mus 201 females in the test for paternal sex chromo-
    some loss in Drosophila with procarbazine and diethylnitrosamine (DEN),
    Environ. Mutagen. 5:235-237.
Zimmering, S., Mason, J.M. and Osgood, C., 1986, Current status of aneuplo-
    idy testing in Drosophila, Mut. Res. 167:71-87.

CHEMICAL ETIOLOGY OF ENDEMIC DISEASES: A GLOBAL PERSPECTIVE

P.J. Peterson

Monitoring and Assessment Research Centre
The Old Coach House
Campden Hill
London W8 7AD, United Kingdom

INTRODUCTION

Human health, especially of rural populations in developing countries, is still directly influenced by the environment, whether from vector-borne diseases or from the geochemical abundance or deficiency of elements. With the exception of specific minority groups, usually those with unusual dietary habits, populations in the developed world are no longer markedly affected by such factors. Nevertheless, in global terms, as the majority of peoples live in rural surroundings, the environmental influence is all-important. In recent years vector-borne disease and diarrheal-type diseases, especially in children, have assumed prominence while nutritionally related diseases are often neglected.

In this paper, several of the major diseases with a chemical etiology are discussed with an emphasis on element concentration and occurrence rather than on the medical aspects. Iodine-deficient goitre, fluorosis, arsenism, selenosis, Keshan Disease and Kaschin-Beck Disease form the basis of the paper. In many countries where these diseases occur, reliable epidemiological and statistical data are often sparse and dose-response relationships poorly quantified, thus compounding estimates of the magnitude of the disease.

As a group these diseases are often referred to as "endemic diseases," although the term is misleading, for they are not restricted to any one country. In a general sense the phrase distinguishes them from vector-borne or other communicable diseases.

Deficiencies in zinc and copper or even iron and calcium, which are usually of lesser importance, as well as health effects of heavy metals, which are commonly reviewed, have been omitted from this paper.

ENDEMIC GOITRE

Goitre can be induced by either a deficiency or an excess of iodine. This has been clearly shown in data for the iodine concentration in drinking water from 201 locations in the People's Republic of China where the "U"-shaped relationship is well established (Wang Mingyuan and Zhang Shen, 1986). The presence of goitrogenic substances, cyanogenetic glyco sides and others in food is a complicating factor, but discussion of this has been omitted from the paper.

In Situ Evaluations of Biological Hazards of Environmental Pollutants
Edited by S. S. Sandhu *et al.*
Plenum Press, New York, 1990

Iodine is enriched in the thyroid gland, which accounts for approximately 80 percent of the total body store, and is an essential constituent of the thyroid hormones thyroxine and triiodothyronine. High iodine levels cause thyrotoxicosis and hence iodine-induced goitre due to reduced thyroid uptake of iodine, although this situation is much less common than iodine-deficient goitre.

Iodine-deficient goitre is global in scope with some 800 million to 1 billion people living in deficient areas [World Health Organization (WHO) 1988]. Some 190 million people have goitre and over three million people suffer from iodine-deficient cretinism. The incidence of cretinism can be 10 percent in severely affected areas in China, India and Indonesia. More than 43 million people in Southeast Asia suffer mental and physical impairment due to iodine deficiency (Clugston and Bagchi, 1986). The disease is an ancient one, having been described in the fourth century B.C. in China. Seaweed was used as a documented cure around 280 A.D. (Anon, 1988). In China alone, goitre occurs in 28 provinces, in some 1,550 counties with around 18 million people exhibiting clinical symptoms. In India the number affected is 40 million (Paul, 1986).

The incidence rate of goitre varies with source and dietary concentrations as well as with age and gender, commonly ranging up to 40 percent in severely affected areas. Ten to 20 percent of dietary iodine comes from drinking water, the remainder from food. The critical intake is around 50-100 ug day^{-1}. With an intake of less than 50 ug day^{-1} health risks are apparent. Concentrations of iodine in the urine or in hair are ready markers to monitor iodine exposure. Values of around 0.5 ± 0.25 ug g^{-1} iodine in hair have been reported from endemic areas, whereas "normal" populations in China show values of around 1 ± 0.2 ug^{-1} (Zhu Wenyu, 1989).

The incidence of goitre is greatest in areas of high mountains and plateaus and in lands far from the ocean's large reservoir of iodine. Although the incidence of goitre varies between and within countries, Iceland is the only country reported to have no endemic goitre. High-iodine induced goitre is typical of coastal areas where high intakes of seaweeds or vegetables pickled in kelp salt are consumed, although high-iodine water is significant in many areas.

"Normal" soils range around 5-10 ug g^{-1} iodine, with plant values being less than 1 ug g^{-1} (Fuge, 1988). The incidence of endemic goitre rises significantly with a decrease in soil iodine. Data from a range of countries show that the incidence rate is light around 1 ug g^{-1} iodine in soil but is serious, say, 40 percent, when the value is around 0.4 ug g^{-1}.

Similarly there is a strong negative correlation with the iodine content of drinking water and the prevalence of goitre around the world. The threshold varies with food iodine intakes, but values of less than 5 ug L^{-1} can be found in affected areas while values greater than 10 ug L^{-1} are often adequate.

Basic preventive measures using iodized salt are commonly implemented in many countries but the coverage globally still gives rise for concern. Oral iodine oil or intramuscular injections have also been used as a longer acting preventive medicine.

FLUOROSIS

Fluoride is one of the essential elements, a lack of which is manifested as dental caries throughout many parts of the world. Yet excess

fluoride is the more serious disease. High intakes of dietary fluorides
have been shown to cause fluorosis in man and animals resulting in increased dental caries, osteosclerosis, ligamentous and tendinous calcifica-
tion and extreme bone deformity (WHO, 1984). The problem is global in
scope, occurring on all continents and affecting many millions of people,
either as dental fluorosis or the more serious skeletal fluorosis that occurs mainly in tropical countries. The latter disease has been reported in
13 of the 25 states of India, affecting over one million people (Susheela
and Das, 1988). In China, 18 million exhibit dental fluorosis and around
one million people exhibit skeletal fluorosis (Anon, 1988). In the province of Jiangxi, an examination of over one million people revealed a prevalence rate of 6.5 percent, but this ranged from 30 to 72 percent in some
villages. Similarly, in some villages in Tanzania the incidence rate is
very high. In one village, marked skeletal fluorosis was seen in children
as well as in parts of India and South Africa (WHO, 1984).

Fluorosis is an ancient disease. Teeth in the skull of the Chinese
Xujiayao Man, dated at around 100,000 B.C., show evidence of dental fluorosis. Many ancient Chinese texts also refer to "yellow teeth" now known
to be caused by fluorosis (Anon, 1988).

High-fluoride water is the principal cause of fluorosis around the
world. Three other exposure pathways are known including dietary intake
from high-fluoride soils, high-fluoride teas and the inhalation of fluoride
concentrations from indoor high-fluoride coal combustion. In the latter
case indoor air reached 0.026 ppm compared with fuel-wood concentrations of
0.0025 ppm (Anon, 1988). A prevalence rate of 75 percent was reported in
some villages in China (Li Fucheng et al., 1988).

The most common form of dental fluorosis arises from the drinking of
high-fluoride well waters, or lake waters whose concentrations range up to
10 mg L^{-1}. Various studies have shown a relationship between increasing
fluoride concentrations and increasing incidence of dental fluorosis (Li
Ribang, 1989).

Concentrations of fluorides in "normal" soil range around 200 to 400 ug
g^{-1} but can exceed several thousand ug g^{-1}. Typical values for plants,
on the other hand, are 5 to 10 ug g^{-1}, except for tea, which is around
100 ug g^{-1}. A poor correlation between soil and plant fluoride concentrations has been shown in many studies.

Several "special" teas from China may contain up to 600 ug g^{-1},
giving rise to a fluoride concentration in tea water of 3 mg L^{-1}, daily
consumption of up to 15 mg day^{-1} thereby resulting. Dental fluorosis
occurs above 1 mg day^{-1} fluoride intake but the incidence varies with
dietary factors. Another unusual dietary source is the consumption of
high-fluoride salt where 4 to 5 mg day^{-1} fluoride is consumed. Typical
areas are parts of China, Thailand, and Vietnam.

Endemic fluorosis can be prevented. The usual measure adopted is to
provide an alternative source of drinking water where the fluoride concentration is low. Elevated fluoride exposure still can occur, however, by
the consumption of local foods grown on high-fluoride soils or on soils
irrigated with the high-fluoride water.

ARSENISM

Long-term exposure to inorganic arsenic can give rise to health
effects. The most characteristic effects are hyperkeratosis of the palms
and soles of the feet together with hyperpigmentation, particularly in

areas not exposed to the sun [International Agency for Research on Cancer (IARC), 1980; WHO, 1981]. Skin tumors have also been commonly reported on the hands and feet, and peripheral vascular disturbances resulting in gangrene, termed "blackfoot disease," have been reported from Taiwan (Tseng, 1977).

The most important cause of environmental arsenism relates to the consumption of high-arsenic drinking water, usually well water. Argentina, Taiwan, the United States, Canada, Chile, Japan and China have all reported chronic exposures. In the case of Chile it has been estimated that some quarter of a million people have been affected.

In villages in Xinjiang Providence of China, the incidence of arsenism from deep wells was 46 percent of the population, the incidence increasing with age (Wang Lian-fang et al., 1988). Mean arsenic concentrations in water reached 0.75 mg L^{-1} in the most seriously affected villages. In other villages in the same province where some 50,000 people were affected by continuous exposure over 10 years to a concentration of 0.12 mg L^{-1}, mild arsenic poisoning resulted (Wang Guoquan et al., 1988). A dose-response relationship was also revealed.

Elevated levels of urinary arsenic and hair arsenic are found in people exposed to high concentrations of arsenic. Urine levels are usually used as an index of occupational exposure to arsenic. This and other data have been reviewed by WHO (1981) where the relationship between hair arsenic concentrations and well water arsenic concentrations has been shown.

SELENOSIS

Endemic human selenosis has been reported from several countries, including China, the United States and Venezuela but is relatively uncommon, affecting only small numbers of people. In such situations hair is lost and nails are shed. The major exposure route is via food grown on soils containing elevated levels of selenium, or via the consumption of food items in which selenium has been bioaccumulated. Seleniferous soil occurs in parts of Hubei and Shaanxi provinces in China. In one site the dietary intake was 4,900 ug day^{-1} compared with an intake from deficient areas of around 10 ug day^{-1}. Selenium in blood or in hair provided a useful biological monitoring technique. Selenium concentrations in hair from such areas can exceed 50 ug g^{-1} compared with the average values of around 0.4 ug g^{-1}. Hair from people in Keshan Disease areas contains only around 0.07 ug g^{-1} selenium (Tan Jianan, Zieve and Peterson, 1989).

KESHAN DISEASE

An endemic cardiomyopathy, Keshan Disease, has been reported since the early 1930s from many areas in China ranging from the Southwest to the Northeast and extending into Inner Mongolia and North Korea (Democratic People's Republic of Korea). Early records and stone tablets also refer to the characteristic signs from different areas of China.

The disease affects children under 10 years of age especially, and women of child-bearing age living in rural areas, although the incidence rate among age classes varies between Northeast and Southwest China. There are pronounced annual fluctuations of the disease that have been related to climatic factors, with both summer- and winter-type diseases (Anon, 1988). An incidence rate of up to 11 percent in susceptible age groups and a fatality rate of up to 80 percent in earlier years has been reported. The disease occurs in 15 provinces and in 309 counties of China but has declined significantly in recent years.

Epidemiological studies show that the disease occurs in low selenium environments, resulting in a low dietary intake. Critical daily intakes are around 15 ug day^{-1} selenium. Preventive trials, especially with sodium selenite tablets or selenized salt, have been carried out for a number of years with population groups of various sizes. In one large trial, 1.05 million people received treatment, compared with 0.6 million controls in Sichuan Province between 1984 to 1986 (Cheng Yunyu and Qian Pengchu, 1988). In the controls the incidence rate rose 33 percent while the rates in the treated areas fell by 62 percent.

Selenium concentrations in topsoils in affected areas range around 0.06-0.09 ug g^{-1} for red-brown earths, purplish soils and drab earths, giving rise to a concentration in rice of around 0.015 ug g^{-1} compared with 0.07 ug g^{-1} in nondisease areas. Selenium in blood or hair is a good biological indicator of the disease states. Values of 0.12 ug g^{-1} in hair from people living on purplish soils can be compared with 0.38 ug g^{-1} from nondisease yellow-brown soil area.

Further nutritional factors including other trace elements, vitamin C and unknown virus infections have also been implicated in the disease in earlier years, but their current relevance is unclear.

KASCHIN-BECK DISEASE

Kaschin-Beck Disease (deformans endemica) is regarded as an endemic chronic high-incidence and degenerative osteoarthrosis whose etiology is still incompletely understood (Anon, 1988). It has been reported from 15 provinces and 303 countries of China extending into North Korea, Inner Mongolia and the U.S.S.R., where it is often referred to as Urov Disease. Over two million people exhibit clinical symptoms and 30 million live in affected areas in China alone.

Historical records indicating the joint deformations and dwarfism show that the disease has been known for over 300 years. The disease area overlaps with many of the Keshan disease areas often occurring side by side in some villages. Individuals can also exhibit both diseases, but this is uncommon. Similarly the disease incidence fluctuates annually, although new outbreaks of Kaschin-Beck Disease have been reported in further areas in recent years. The disease occurs on low selenium soils, there is a low dietary intake of the element, and hair and blood selenium are very low and are comparable to Keshan Disease values.

Many selenium supplementation trials have been carried out including X-ray examination of hand/bone development. As the hair and blood selenium concentrations increase, the incidence rate of the disease decreases. In one study involving 437,000 three- to six-year-old children in seven endemic counties in Heilongjiang Province, the incidence rate measured after X-ray analysis dropped from 41 percent to 9 percent over a 3-year supplementation period (Zhang Boachu and Niu Guaquan, 1988). Furthermore, the treatment led to an improvement in the epiphyseal plate development.

Various reports have implicated two other factors in the etiology of the disease, namely humic acids in the drinking water and the presence of oxysporum toxins from _Fusarium_-contaminated grains from the affected areas. A change of well water away from the humic acid type was reported to result in a decrease of Kaschin-Beck Disease from 39.6 percent to 6.6 percent over a 5-year period (Wang Yongzhen et al., 1988). Nevertheless, many more detailed epidemiological surveys and supplementation trials are required before the disease can be successfully brought under control.

CONCLUSIONS

A study of the distribution of the so-called endemic diseases indicates that they are global in scale with pronounced health effects measurable in terms of millions of people. Data relevant to the individual disease are poorly documented in many countries thereby masking their true extent. In some countries, with Kaschin-Beck Disease for example, the chemical etiology is incompletely understood, while some reports indicate its increasing prevalence in new areas. With iodine- deficient goitre, on the other hand, the number of cases is decreasing in some countries where supplementation therapy is in use.

There is an urgent need for epidemiological training in many countries and a better infrastructure in the environmental health field so appropriate investigations and preventive action can be taken. The decline in economic growth, coupled with the increasing population growth throughout most of the developing countries, indicates that many of these major human health problems will continue unnecessarily into the next century.

REFERENCES

Anon, 1988, "Atlas of Endemic Diseases and their Environments in the People's Republic of China," Beijing, China Map Press (in press).
Cheng Yunyu and Qian Pengchu, 1988, Effect of Selenium table salt supplementation in prevention of Keshan disease among 1.05 million population, in: "Abstracts, International Symposium on Environmental Life Elements and Health," Beijing, November, 1-5, 1988, p.241.
Clugston, G.A., and Bagchi, K., 1986, Tackling iodine deficiency in Southeast Asia, World Health Forum, 7:39.
Fuge, R., 1988, Sources of halogens in the environment, influences on human and animal health, Environ. Geochem. Health, 10:51.
IARC, 1980, Arsenic and Its Compounds, IARC Monographs 23, International Agency for Research on Cancer, Lyon, p.39.
Li Fucheng et al., 1988, A report of aluminum-fluoride combined toxicosis by food chain, in: "Abstracts, International Symposium on Environmental Life Elements and Health," Beijing, November 1-5, 1988, p.24.
Li Ribang, 1989, Environmental Fluoride and its Effects on Health in China and Other Countries, MARC Report, Monitoring and Assessment Research Centre, King's College London, University of London, in press.
Paul, S., 1986, Lessons of India's goitre control programme, World Health Forum, 7:39.
Susheela, A.K., and Das, T.K., 1988, Fluoride toxicity and fluorosis: Diagnostic test for early detection and preventive measures adopted in India, in: "Abstracts, International Symposium on Environmental Life Elements and Health," Beijing, November 1-5, 1988, p.89.
Tan Jianan, Zieve, R., and Peterson, P.J., 1989, Environmental selenium in China and other countries, MARC Report, Monitoring and Assessment Research Centre, King's College London, University of London, in press.
Tseng, W.P., 1977, Effects and dose-response relationships of skin cancer and Blackfoot disease with arsenic, Environ. Health Perspect., 19:109.
Wang Lianfang, Sun Xingzhi, Ai Haidi, Xu Xunfeng, Lin Fafu, Liu Hongdi, and Feng Zhaoyue, 1988, An investigation of endemic arsenism on the Southwest part of the Dzungaria Basin, Xinjiang, in: "Preprint, International Symposium on Environmental Life Elements and Health," Beijing, November 1-5, 1988.
Wang Guoquan et al., 1988, Endemic arsenism in Kuytun, Xinjiang, in: "Abstracts, International Symposium on Environmental Life Elements and Health, Beijing, November 1-5, 1988, p.30.
Wang Mingyuan, and Zhang Shen, 1986, Biogeochemical provinces and endemia, Scientia Sincia, B29:389.

Wang Yongzhen, Shang Ke, Zhu Yuyao, Zhao Yuhua, and Yao Jingtong, 1988, Efficient studies on control of Kaschin-Beck Disease with decreasing humic acid in deep wells in different types of Kaschin-Beck Disease regions, in: "Abstracts, International Symposium on Environmental Life Elements and Health," Beijing, November 1-5, 1988, p.240.

World Health Organization, 1981, Arsenic, Environmental Health Criteria 18, World Health Organization, Geneva, p.174.

World Health Organization, 1984, Fluorine and Fluorides, Environmental Health Criteria 36, World Health Organization, Geneva, p.136.

World Health Organization, 1988, Specific deficiencies-iodine, World Health, May 1988, p.30.

Zhang Baochu, and Niu Guaquan, 1988, The preventive effects of selenium on the radiological changes of Kaschin-Beck Disease, in: "Preprint, International Symposium on Environmental Life Elements and Health," Beijing, November 1-5, 1988.

Zhu Wenyu, 1989, Environmental Iodide and Endemic Goitre in China and Other Countries, MARC Report, Monitoring and Assessment Research Centre, King's College London, University of London, in press.

INTEGRATION OF DATA
FOR EFFECTIVE PROBLEM
SOLVING AND ASSESSMENT

CHEMICAL ANALYSIS FOR ASSESSMENT AND EVALUATION OF ENVIRONMENTAL

POLLUTANTS: FACT OR ARTIFACT

M. Wilson Tabor

Institute of Environmental Health
Environmental Analytical Chemistry Research Laboratory
University of Cincinnati Medical Center
Cincinnati, Ohio 45267-0056

INTRODUCTION

Due to the presence of both known and unknown sites and sources of environmental pollutants in many industrialized regions, there exists the possibility for releases of hazardous anthropogenic organics into the environment that present a potential hazard to human health. Many of these releases remain undetected until after the fact; i.e., endpoints of exposure surface in the indigenous population, chemical monitoring/biological monitoring or survey programs show a contamination or threat of release, or the pollutants affect the flora of fauna at the site of the contamination or release. An approach to early detection of environmental pollutants or release of these pollutants is the use of in situ monitoring methodologies via sentinel organisms. These situations require an approach by the environmental analytical chemist that provides data appropriate to assess and evaluate the situation so that effective measures can be instituted. The acquisition of such data requires analyses at trace levels, i.e., ppb or lower, thereby increasing the probability of interferences from artifacts of isolation or inadequately resolved components in the complex mixture under analysis. Considering these problems, an overall analytical strategy and an understanding of the analytical process is required for such cases of assessment and evaluation of environmental pollutants.

A major goal of environmental analytical chemistry is to provide analytical data on environmental pollutants as to their identity and concentration in order to assess and evaluate the biological hazards associated with these compounds. To achieve this goal, the environmental analytical chemists needs to develop a strategy for the chemical investigation and analysis, and to consider the problems associated with the acquisition of analytic data on environmental and biological samples in order to produce data that are scientifically sound. In the planning of the strategy for the chemical investigation and analysis, the analyst needs to consider the following issues: (1) sampling strategy and required samples; (2) sample stability and storage; (3) sample preparation for analysis; (4) methods of detection of analytes and required sensitivity; and (5) quality assurance/quality control. These issues, external to the actual analytical process, but an integral part of the overall analytical strategy, were addressed recently in a review of chemical analysis for assessment and evaluation of human exposure to hazardous anthropogenic compounds (Tabor, 1988).

In Situ Evaluations of Biological Hazards of Environmental Pollutants
Edited by S. S. Sandhu *et al.*
Plenum Press, New York, 1990

Although the intrinsic issues to the analytical process are implicit in a sound quality assurance plan (e.g., Hall, 1984, Taylor, 1987), they may become a major source of problems in chemical analyses due to a lack of understanding or experience on the part of chemists as to the lability of biologically reactive pollutants, the biologic mechanism of action of these pollutants and other biological factors affecting the analysis. Likewise, similar problems may arise with biologist/toxicologists due to their lack of understanding of or experience with the chemical reactivity of labile (i.e., bioreactive) compounds, under extremes of temperature and pH or with physical/chemical processes (e.g., irreversible adsorption or catalysis of chemical reactions on glass and metal surfaces) that can occur during he analytical process. The purpose of this presentation is to consider some of the important intrinsic issues thereby preventing artifacts in the analytical process.

SAMPLE STABILITY AND STORAGE

One primary intrinsic issue in the analytical process is associated with sample stability and sample storage. Veillon (1986) has stated that the primary causes of analytical error in trace element analyses of biological samples are improper sampling, especially contamination during sample collection, and contamination from, or adsorption loss to, the sample container. Additionally, this was emphasized by Aitio (1984), who pointed out that a portion of the scatter seen in the results of analyses of samples for a particular analyte is due to changes that occurred during sample collection and storage.

A number of problems, related to how sample handling can affect the introduction of artifacts into the analytical process, have been discussed by Albro (1979). The four categories of problems detailed in this overview of problems in analytical methodology are: (1) contamination; (2) loss; (3) desiccation; and (4) alteration. Some of these issues are addressed in the following sections.

Containers for Collection and Storage of Samples

Veillon (1986) has stated that the collection of a sample without contaminating it is the first step in a successful analysis. This first step involves not only the choice of the proper container but also the use of proper procedures in cleaning the containers for use in sampling. Containers used for the collection and storage of sample are often the initial source of interfering contaminants, analyte loss or sample alteration. Samples collected for the analysis of both metals and organic constituents can be affected by the type of container.

Generally, both glass and plastic containers are used for the collection of samples for metal analysis (Meranger et al., 1981). The proper cleaning of both of these types of containers to be used for trace metal analysis can minimize extraneous and interfering contaminants. To prepare these containers, typical cleaning procedures, which involve nitric acid soaking, rinsing and storage, have been described (Tabor, 1985a). Also, the type of materials used in the manufacture of glass containers can be a source of metal contamination or can contribute to losses that will affect many types of metal analyses. For example, in lead analyses, the problem of contamination has been overcome now that lead-free sampling tubes and containers are available (Tabor, 1987).

For the analysis of organics, samples are typically collected in glass or stainless steel containers (Tabor, 1988). These containers should be scrupulously cleaned, usually via procedures that insure the removal of greases, oils and other contaminants that would interfere with trace organics analyses. In general, sample contamination from containers can be minimized by thorough cleaning and rinsing of the containers followed by storage of the containers in a dust-free area prior to use. For example, one convenient method of decontaminating glassware is to heat these containers at temperatures > 500°C overnight in a muffle furnace or a high temperature vacuum oven. It should be noted that materials, other than glass, cleaned easily include stainless steel and Teflon[TM]. Examples of cleaning procedures for sample collection and storage containers have been published (Fed. Regist., 1984; QueHee et al., 1983; Tabor, 1985a; Tabor, 1988).

Not only have sample containers been found to be sources of contaminants but also the screw caps and cap liners have been shown to be major sources of contaminants and sample losses that interfere with trace organic analyses. De Zeeuw et al. (1975) reported that plastic cap liners were a major source of plasticizer contamination with concentrations of contaminants in the range of 0.1 to 5.0 ppm. Methods to ascribe sources of plasticizer contamination have been published by Peterson and Freeman (1982). Foil-backed cap liners used on containers to transport and store samples were shown by Albro (1979) to contribute significant levels of contaminants to trace organic analyses. A procedure to correct for contaminants arising from foil liners has been published by Clement et al. (1980). Also, Levine et al. (1983) reported that Teflon[TM]-lined silicone cap liners for center-hole screw caps used in containers for water samples taken for the analysis of purgeable organic compounds allowed for the contamination, cross-contamination and/or analyte loss via the diffusion of volatile organics through these septum seals. Therefore, care must be taken in the choice of container closures and proper quality control procedures must be followed to ensure that sample integrity is not compromised during transport and storage of samples.

Storage of Samples

Samples should be stored in such a way as to minimize analyte losses, analyte deterioration or other changes in the sample that would compromise the sample. One example of sample loss and/or change was described above in the study of Levine et al. (1983) regarding purgeable organics analysis. Subsequently, sample holding times for these and similar analytes were reduced by 50% (Fed. Regist., 1984). Other related studies have led to additional changes in these procedures. These changes have been detailed in the final promulgation of guidelines establishing test procedures for the analysis of pollutants under the Clean Water Act, in terms of holding times and the use of preservatives. A second type of problem related to holding times for samples for trace analyses has been described by Veillon (1986). Although some sample types, like aqueous samples for trace metal analyses, are better stored in plastic containers, the use of these containers presents another possible source of error for samples or stored frozen for long periods. Plastic materials like polyethylene or polypropylene are not completely impermeable to water vapor, and it should be noted that even frozen water has a finite vapor pressure. Therefore, this mode of storage can lead to a gradual loss of moisture over time via lyophilization. Veillon (1986) has recommended two possible solutions: (1) packaging frozen sample containers in sealed plastic bags containing a few cubes of ice to provide a high relative humidity environment within the freezer; and (2) freeze-drying samples before storage, particularly if subsequent sample processing would benefit from a dry sample. Regardless of the storage procedure used, the investigator should perform a few preliminary storage experiments on the anticipated sample type in order to carefully check possible sample loss or alteration.

Another type of sample alteration that can occur in storage are chemical reactions leading to analyte deterioration and loss. In the reviews of Coutts and Beckett (1977) and Lindeke (1982), extensive discussions were presented on the problems of the alterations of nitrogen-containing analytes during storage and sample work-up. The nitrogen atom in many organic compounds is characteristically unstable, particularly in terms of its susceptibility to chemical alterations via oxidation/reduction reactions. For example, aromatic amine compounds like pyridine or aniline are readily oxidized to N-oxides, N-hydroxyl and other oxidized species on exposure to air. This process is accelerated by the presence of heavy metal ions, such as Cu^{+2}, Fe^{+3} and Mn^{+2}, in the sample matrix (Lindeke, 1982), as is the situation for many environmental samples. Therefore it is important to completely fill sample containers during the sampling process to avoid contact of the sample with air. Alternatively, sampling containers should be flushed with an inert gas such as nitrogen or argon prior to sealing to retard analyte oxidation.

The nitrogen atom in many organic compounds also is a strong nucleophile, thereby increasing the likelihood of these compounds undergoing reactions during storage with many compound types. For example, Beckett and Ali (1979) reported that a variety of alicyclic and aromatic amines react with the dichloromethane extracting solvent during sample storage. A number of chloromethochloride artifacts of the amines were formed within 1 h following extraction of the original sample. The results of this study point to the importance of the need for good quality control in method development as related to the storage of samples. The preliminary experiments should include studies of the effects of sample storage times on sample stability.

Sample stability during storage can be compromised in other ways during the preparation of samples for analysis. Tabor (1989) found that some direct-acting TA100 mutagens, isolated from drinking water samples, were stable for months when stored as the originally isolated complex mixture of organics, but, on separation of the mixture via high performance liquid chromatography (HPLC), the subfraction representing > 80% of the direct-acting TA100 mutagenic activity lost approximately one-half of its activity within six days when stored at -70°C as a solution either in dimethylsulfoxide or in the HPLC elution solvent. Within 11 days of storage, this subfraction had lost all of its mutagenic activity. This was unexpected in that previous studies (Loper and Tabor, 1983) with a different direct-acting TA100 mutagen from another drinking water source had shown that the mutagenic mixtures and HPLC subfractions were stable when stored in this way. Therefore, the results of these two investigations of similar sample types emphasize that preliminary studies of sample stability during storage need to be conducted.

Recommendations for Sample Containers and Storage

These representative problems of analyte losses or artifact formation due to sample storage and sample containers suggest additional general precautions and recommendations to the analyst that will minimize sample alterations (e.g., Gunderson and Anderson, 1979; Watts, 1980; Grob and Kaiser, 1984; Tabor, 1988). As to sample storage, three general precautions are warranted. First, the sample should be protected from light, since exposure to light will accelerate many types of chemical reactions which can lead to destruction of sample analytes. Protection from light can be accomplished by using amber containers for collection and storage, wrapping these containers in coverings such as aluminum foil or storing the containers in enclosures such as boxes. Second, it should be remembered that many reactions causing sample alteration, including microbial growth, will be retarded at lower temperatures. Therefore, quick freezing may be beneficial in preserving sample constituents. Furthermore, the storage of samples at 0° to 4°C in refrigerators or packed in ice during transport to the laboratory will retard sample deterioration. On arrival at the laboratory, storage of the samples in refrigerators or freezing at -20°C or -70°C will retard sample deterioration. However, the suggestions of Veillon (1986), noted above the prevent sample desiccation, should be kept in mind. Third, storage of samples in concentrated form generally will add to analyte stabilization. That is, do not dilute samples until time for analysis. Loss of analytes due to adsorption, e.g., to container surfaces, is generally more pronounced for a dilute solution than from a concentrated solution.

SAMPLE PREPARATION FOR ANALYSIS

Another primary intrinsic issue in the analytical process is associated with the preparation of samples for analysis. As stated previously (Tabor, 1988), the objective of sample preparation techniques is to render the analyte(s) of interest into a form suitable for detection, quantitation or identification. The analyst uses various techniques for sample preparation remembering the following aims: (1) to prevent an alteration in the composition of the analytes of interest; (2) to maximize the yield of all analytes of interest for each step in the sample preparation process; (3) to maximize the separation of the analytes of interest from interfering components, and (4) to minimize the introduction of impurities to the sample from the solvents, etc., used in sample preparation. Two of these specific aims, analyte alteration and impurity introduction, are addressed in the following sections.

Solvents Used in Sample Preparation

The initial step employed in the isolation of sample components of interest from the sample matrix for subsequent clean-up and analyte analysis generally involves one of two approaches: (1) removal of the matrix from the sample as, for example, in evaporation or lyophilization of a liquid matrix and subsequent solubilization of the sample into a solvent for further clean-up; or (2) isolation of the sample from the matrix as, for example, in the isolation of sample components from a liquid matrix by adsorption of the components to a chromatographic stationary phase with subsequent elution of the components using a mobile phase or as in the isolation of sample components from a liquid or solvent matrix by solvent extraction. One feature common to both approaches is the use of solvents either for elution or for extraction. The solvents used for these procedures can be a major source of impurities subsequently interfering with analyte analysis or bioassay of toxicologically significant components.

A typical example of the type and nature of trace impurities found in solvents used for the analysis of environmental samples is presented in a study by Bowers et al. (1981a). Commonly used high quality grade solvents, such as "distilled-in-glass" and "pesticide," were evaluated. Trace organic impurities were found to be present at levels of 1 to 150 ng/ml solvent. Impurities identified via mass spectrometry included phthalate esters, hydrocarbons and chlorohydrocarbons. Furthermore, some pesticide grade solvents were found to contain up to 21 different trace impurity components at maximum concentrations per single component of 30 to 50 ng/ml solvent. Distilled-in-glass grade solvents typically had fewer trace impurities at low concentrations. The significance of these contaminants is that they not only could interfere in chemical analyses of a sample but also could interfere in bioassays. An example of this latter point was emphasized by De Zeeuw et al. (1975), who reported that phthlate acid ester contaminants in their high purity solvents interfered with their anticholinergic assays. Another type of significant solvent contaminant, presenting major problems in bioassays, was described by Shertzer and Tabor (1985). In this study, the analysis of 43 commonly used laboratory solvents showed that a number of solvent classes, including hydrocarbons, esters, alcohols and gavage vehicles commonly used in animal studies, are prone to form reactive peroxides during storage if adequate precautions are not taken. Recommended methods for peroxide removal and for proper solvent storage were discussed. Therefore, contaminants, such as in these examples, can interfere with environmental trace component analyses and bioassays by interfering with the detection of the analyte, such as via electron-capture gas chromatography, by reacting with the analyte of interest to form undetected derivatives, or by compromising a variety of bioassay procedures.

Several methods have been proposed to evaluate solvents for the detection of trace organic impurities that could interfere in analytical or bioassay procedures. Tabor (1985b) described a 10000-fold concentration procedure to evaluate solvents used in sample extraction and preparation for mutagen bioassays. A solvent purity test has been recommended by Bristol (1980) for trace impurities in solvents used in high performance liquid chromatography, whereas Bowers et al. (1981b) and Berezkin and Budantseva (1980) have described methods for evaluation and detection of trace impurities in solvents used in environmental sample preparation procedures for sample analyses via gas chromatography and gas chromatography/mass spectrometry. In addition to these representative procedures, a number of other similar methods for the evaluation of solvent contaminants have been published. It is important that any environmental analytical procedure be based on methods that have undergone vigorous validation and have sound quality control/quality assurance procedures in place.

Sample Preparation-Analyte Alteration and Intrinsic Reactions/Losses

One aim of sample preparation procedures is to maximize the separation of the analytes of interest from interfering components of the environmental sample matrix. In general, procedures for the isolation of components from the sample matrix used in subsequent clean-up steps have a developmental basis in the previous experiences of the analyst or in experiences derived from procedures developed for similar types of studies as reported in the scientific literature. However, these approaches are beset with problems of analyte alteration, analyte losses, unexplained analyte reactions or artifact formation if the analyst does not use vigorous method validation techniques or employ sound quality control/quality assurance procedures. Recommendations for sample clean- up procedures validation during method development have been published by Albro (1981) and Gunderson and Anderson (1979). Likewise, numerous recommendations have been published regarding the development of quality control quality assurance procedures for environmental trace analysis (e.g., Hall, 1984; U.S. EPA, 1987; Taylor, 1987). The following examples illustrate some of the problems of analyte alteration and intrinsic reactions/losses encountered during the development of sample preparation procedures.

One commonly used procedure for the isolation of neutral and polar sample components from solid environmental matrices such as soils, sediments and sludges is a series of acidic, neutral and basic solvent extractions. Lafleur and Pangaro (1981) have investigated the source of anomalous findings of isophorone in a number of soil samples analyzed via EPA priority pollutant analytical methods employing Soxhlet extraction (U.S. EPA, 1979). The acetone solvent used in this extraction procedure produced a number of dimers, trimers, tetramers and other complex homologs in the acidic and basic extracts. Some of the more common substances identified included phorone, mesitylene and isophorone, the latter being listed as an EPA priority pollutant. These data show the role of the catalytic effect of acids and bases on the self-condensation of acetone to yield a varity of products, one of which was the analyte of interest. Studies from our laboratory (Tabor et al., 1985; Tabor and Loper, 1987) to develop methods for the isolation of mutagenic compounds from sludges have shown that similar U.S. EPA procedures for the analysis of priority pollutants in sludges (Billets and Lichtenburg, 1983) not only destroyed labile mutagens in some sludge extracts but also caused the formation of mutagenic artifacts in other extracts during these isolations. The U.S. EPA method involves solvent extractions under extremes of pH, both basic and acidic, conditions that favor many commonly known organic reactions, some of which were discussed (Tabor et al., 1985). These two examples illustrate problems of potential chemical reactions resulting from conditions intrinsic to sample clean-up procedures and emphasize the importance of method validation to ensure that interfering artifacts are not produced or that important labile analytes are not destroyed during sample manipulations prior to analysis.

Another commonly used procedure for the isolation of sample components for analysis is chromatography. Numerous studies have shown that chromatographic fractionation procedures are efficient in terms of separation of compound classes. In an investigation to evaluate the degree of recovery of mutagenic components from complex mixtures of fossil fuel combustion by-products using four types of chromatographic stationary phases, Lafleur et al. (1986) reported that labile direct-acting mutagens are not easily recovered from most stationary phases, including commonly used alumina and silica gel, but that mutagens requiring metabolic activation are more easily recovered. These results are similar to those reported by Tabor et al. (1980) wherein chromatographic stationary phases were shown to be a major source of losses of mutagenic components during the separation of complex mixtures of organics from environmental samples. These two examples emphasize the problems of losses of labile compounds during sample clean-up prior to bioassay and, again, the need for the analyst to validate the clean-up methods used for each sample type to ensure that sample losses or artifact formation are not the results of problems intrinsic to the methods employed.

DISCUSSION AND CONCLUSIONS

The environmental analyst is a key collaborator in the overall assessment and evaluation of environmental contamination by hazardous anthropogenic compounds or in the assessment of the threat of a release of these hazardous materials into the environment. These assessments require analytical and bioassay data that are scientifically sound, both in terms of accuracy and precision and in terms of being based in valid methodology free of artifacts and interferences. During the past 20 years a concerted effort has been made by the scientific community to develop approaches suitable for these required assessments. Many of the methods proposed and developed by environmental analysts need to be reevaluated in terms of the issues addressed in this report, i.e., sample integrity, artifact formation during analysis and contaminant introduction during sample preparation. Due to the public's trust in the scientific community and the legal issues related to these environmental assessments, each individual in a laboratory should take responsibility for the quality of the data. In the acquisition and interpretation of such data, an overall analytical strategy and an understanding of the analytical process are required for the assessment and evaluation of environmental pollution.

ACKNOWLEDGMENTS

Financial support for a portion of this effort was provided by National Institute of Environmental Health Sciences grant ES00159. Thanks are

extended to Dr. H. Kagen, Dr. C.C. Smith and Ms. D. Gorman for valuable
advice and assistance in the preparation of this paper.

REFERENCES

Aitio, A., 1984, Sources of error and quality control in biological monitoring, in: "Biological Monitoring and Surveillance of Workers Exposed
to Chemicals," A. Aitio, V. Riihimaki and H. Vainio, eds., Hemisphere
Publishing Corporation, New York, pp.263-272.

Albro, P.W., 1979, Problems in analytical methodology: Sample handling,
extraction, and clean-up, Ann. N.Y. Acad. Sci., 320:19-27.

Albro, P.W., 1981, Validation of extraction and clean-up procedures for
environmental analysis, in: "Environmental Health Chemistry," J.D.
McKinney, ed., Ann Arbor Science Publishers, Inc., Ann Arbor, MI,
pp.163-175.

Beckett, A.H., and Ali, H.M., 1979, Artifacts produced by using dichloromethane in the extraction and storage of some antihistaminic drugs, J.
Chromatogr., 177:255-262.

Berezkin, V.G, and Budantseva, M.N., 1980, Chromatographic analysis of
major trace components in solvents with concentration prior to sample
injection, J. Chromatogr., 191:309-312.

Billets, S., and Lichtenburg, J.J., 1983, Interim methods for the measure-
ment of organic priority pollutants in Sludges, Physical and Chemical
Methods Branch, Environmental Monitoring and Support Laboratory, U.S.
EPA: Cincinnati, OH, pp.1-70.

Bowers, W.D., Parsons, M.L., Clement, R.E., Eicemen, G.A., and Karasek,
F.W., 1981a, Trace impurities in solvents commonly used for gas
chromatographic analysis of environmental samples, J. Chromatogr.,
207:203-211.

Bowers, W.D., Parsons, M.L., Clement, R.E., and Karasek, F.W., 1981b,
Component loss during evaporation-reconstitution of organic
environmental samples for gas chromatographic analysis, J. Chromatogr.,
207:203-211.

Bristol, D.W., 1980, Detection of trace organic impurities in binary
solvent systems: A solvent purity test, J. Chromatogr., 118:193-204.

Clement, R.E., Karasek, F.W., Bowers, W.D., and Parsons, M.L., 1980,
Correction for artifacts in the analysis of atmospheric aerosols, J.
Chromatogr., 190:136-140.

Coutts, R.T., and Beckett, A.H., 1977, Metabolic N-oxidation of primary
and secondary aliphatic medicinal amines, Drug. Metab. Rev., 6:51-104.

De Zeeuw, R.A., Jonkman, J.H.G., and van Mansvelt, F.J.W., 1975, Plastic-
izers as contaminants in high-purity solvents: A potential source of
interference in biological analysis, Anal. Biochem., 67:339-341.

Fed. Regist., 1984, 40 CFR Part 136 [FRL-2636-6] Guidelines establishing
test procedures for the analysis of pollutants under the Clean Water
Act: Final rule and interim final rule and proposed rule, October 6,
Vol. 49, No. 209, pp.1-210.

Grob, R.L., and Kaiser, M.A., 1982, Environmental problem solving using
gas and liquid chromatography, J. Chromatogr. Libr., 21:1-240.

Gunderson, E.C., and Anderson, C.C., 1979, Development and validation of
methods for sampling and analysis of workplace toxic substances, Report
of contract no., 210-76-0123 to U.S. DHEW Center for Disease Control,
NIOSH, Cincinnati, OH, pp.1, A-1-38, to A-6, B-1 to B-7.

Hall, J.R., 1984, Quality assurance in environmental trace analysis, in:
"Liquid Chromatography in Environmental Health," J.F. Lawrence, ed.,
Humana Press, Inc., Clifton, NJ, pp.1-17.

Lafleur, A.L., and Pangaro, N., 1981, Artifact formation in the soxhlet
extraction of environmental samples with acetone, Anal. Let.,
14:1613-1624.

Lafleur, A.L., Braun, A.G., Monchamp, P.A., and Plummer, E.F., 1986,

Preserving toxicologic activity during chromatographic fractionation of bioactive complex mixtures, Anal. Chem., 58:568-572.

Levine, S.P., Puskar, M.A., Dymerski, P.P., Warner, B.J., and Friedman, C.S., 1983, Cross-contamination of water samples taken for analysis of purgeable organic compounds, Environ. Sci. Technol., 17:125-127.

Lindeke, B., 1982, The non- and postenzymatic chemistry of N-oxygenated molecules, Drug Metab. Rev., 13:71-121.

Loper, J.C., and Tabor, M.W., 1983, Isolation of mutagens from drinking water: Something old, something new, Environ. Sci. Res., 27:165-181.

Meranger, J.C., Hollebone, B.R., and Blanchette, G.A., 1981, The effects of storage times, temperature and container type on the accuracy of atomic absorption determinations of Cd, Cu, Hg, Pb and Zn in whole heparinized blood, J. Anal. Toxicol., 5:33-41.

Peterson, J.C., Freeman, D.H., 1982, Method validation of GC-MS-SIM analysis of phthalate esters in sediment, Int. J. Environ. Anal. Chem., 12:277-291.

QueHee, S.S., Ward, J.A., Tabor, M.W., and Suskind, R.R., 1983, Screening method for Aroclor 1254 in whole blood, Anal. Chem., 55:157-160.

Shertzer, H.G., and Tabor, M.W., 1985, Peroxide removal from organic solvents and vegetable oils, J. Environ. Sci. Health, 20:845-855.

Tabor, M.W., 1985a, Lead: Methods of analysis, in: "Clinical Chemistry - Theory, Analysis and Correlation," L.A., Kaplan and A.J. Pesce, eds., C.V. Mosby, Inc., St. Louis, MO, pp.1369-1376.

Tabor, M.W., 1985b, Drinking water: Mutagenicity sample preparation protocols, in: "Guidelines for Preparing Environmental and Waste Samples for Mutagenicity (Ames) Testing: Interim Procedures," L.R. Williams, Technical Monitor, U.S. EPA, Office of Research and Development, U.S. EPA/600/4-85/058, Las Vegas, NV, pp.67-108.

Tabor, M.W., 1987, Lead (plumbum), in: "Methods in Clinical Chemistry," A.J. Pesce and L.A. Kaplan, eds., C.V. Mosby, Inc., St. Louis, MO, pp.394-404.

Tabor, M.W., 1988, Chemical analysis for assessment and evaluation of human exposure to hazardous anthropogenic compounds, in: "Hazardous Waste: Detection, Control, Treatment," R. Abbou and I. Antoni, eds., Elsevier Science Publishers, Amsterdam, The Netherlands, pp.1065- 1072.

Tabor, M.W., 1989, The role of granular activated carbon in the reduction of biohazards in drinking water, in: "Biohazards of Drinking Water Treatment," R.A. Larson, ed., Lewis Publishers, Inc., Chelsea, MI, pp.213-233.

Tabor, M.W., Loper, J.C., Barone, K., 1980, Analytical procedures for fractionating nonvolatile mutagenic components from drinking water concentrates, in: "Water Chlorination: Environmental Impact and Health Effects, Vol. 3, R.L. Jolley, W.A. Brungs, R.B. Cumming, and V.A. Jacogs, eds., Ann Arbor Science Publishers, Inc., Ann Arbor, MI, pp.899-912.

Tabor, M.W., Loper, J.C., Myers, B.L., Rosenblum, L., and Daniel, F.B., 1985, Isolation of Mutagenic compounds from sludges and wastewaters, Environ. Sci. Res., 32:269-288.

Tabor, M.W., and Loper, J.C., 1987, New methods for the isolation of mutagenic components of organic residuals in sludges, Adv. Chem. Ser., 214:675-692.

Taylor, J.K., 1987, "Quality Assurance of Chemical Measurements," Lewis Publishers, Chelsea, MI, pp.1-328.

U.S. EPA, 1979, Manual of Analytical Quality Control for Pesticides and Related Compounds, U.S. EPA-600/1-79-008, Office of Research and Development, Health Effects Research Laboratory, Research Triangle Park, NC.

U.S. EPA, 1987, Preparation Aids for HWERL Quality Assurance Project Plans, PA QA PP-0007-GFS, Office of Research and Development, Hazardous Waste Engineering Laboratory, Cincinnati, OH.

Veillon, C., 1986, Trace elemental analysis of biological samples, <u>Anal</u>.
<u>Chem</u>., 58:851A-858A.

Watts, R.R., 1980, Analysis of Pesticide Residues in Human and Environmental Samples, U.S. EPA-600/8-80-038, Office of Research and Development,
Health Effects Research Laboratory, Research Triangle Park, NC.

IN SITU TOXICOLOGICAL MONITORING: USE IN QUANTIFYING ECOLOGICAL EFFECTS

OF TOXIC WASTES

W.J. Birge and J.A. Black

Graduate Center for Toxicology and
School of Biological Sciences
University of Kentucky
Lexington, Kentucky

ABSTRACT

A series of investigations has focused on the development and evaluation of short-term tests with fish embryo-larval stages for the purpose of
estimating chronic effects of aquatic contaminants on biota in freshwater
systems. In principle, the procedure involves exposing fish embryos starting at or soon after fertilization and continuing through four days posthatching. A standardized exposure period of eight days has been adopted
for use with the fathead minnow. Usual test endpoints include embryonic
and larval mortality and teratogenicity (i.e., abnormal development). Data
are expressed as LC50s and toxicity threshold (LC10, LC1) or chronic
values. The latter usually have correlated reasonably well with maximum
acceptable toxicant concentrations (MATCs) and chronic values developed for
selected metals and organic compounds in life-cycle studies. Cadmium and
other reference toxicants have indicated good reproducibility of test results using either continuous-flow or static- renewal procedures.

In subsequent investigations, biomonitoring studies were conducted on
two point-source impacted streams. Results of conventional laboratory toxicity tests, in situ toxicity tests, in-stream chemical measurements, and
ecological endpoints were analyzed and compared for sensitivity and reliability for measuring or predicting ecological effects of hazardous
wastes. In each case a series of receiving water stations, ranging from
high to low impact, and reference sites were studied. On-site short-
chronic toxicity tests with fish embryos and larvae produced results that
correlated closely with independent ecological parameters. Toxicity values
obtained in effluent dilution tests were predictive of measured in-stream
effects. Principal reliance was placed on macroinvertebrate species richness, abundance, diversity, and functional group analysis for characterizing ecological effects. Species composition of fish populations was a
useful but less sensitive measure of impact.

In developing test systems for evaluating chronic effects of point-
source discharges regulated under the National Pollutant Discharge Elimination System (NPDES), effluent samples were tested simultaneously in the
field and laboratory. Though optimal sample preservation and minimal storage time (i.e., < 24 hrs) were observed, laboratory-tested samples produced
substantially less biological activity, as measured by embryopathic effects

In Situ Evaluations of Biological Hazards of Environmental Pollutants
Edited by S. S. Sandhu *et al.*
Plenum Press, New York, 1990

on fish and amphibians (e.g., embryonic mortality, teratogenesis). Certain waste samples that were toxic in the field produced no effect in the laboratory, clearly indicating the prospect for "false negative" results in laboratory screening.

Studies to date support the feasibility of a broad-based in situ monitoring program for evaluating ecological and health effects of hazardous substances.

INTRODUCTION

The principal objectives of this study are to examine the concept of in situ toxicological monitoring and provide further documentation of its relevance and usefulness in performing site-specific characterizations of the effects of hazardous wastes on aquatic ecosystems and general environmental health. Emphasis is placed on (1) evaluating short-chronic procedures for estimating chemical and other stresses on aquatic organisms, populations, and communities and (2) comparing the reliability and sensitivity of toxicological and ecological endpoints for impact assessments under NPDES and hazardous waste programs.

In 1978, Peltier presented guidelines for applying acute toxicity testing to complex effluents. Citing high cost and other problems involved in the analysis of the many individual chemicals contained in point-source discharges, he stated "...the most direct and cost-effective approach to the measurement of the toxicity of effluents is to conduct a bioassay with aquatic organisms..." In the same year, Birge (1978) reported on the use of short-chronic fish embryo-larval tests for quantifying the combined effects of a suite of metals contained in coal fly ash leachates, and this biomonitoring concept was extended in 1979 (Birge et al., 1979a).

Toxicological monitoring, particularly when performed on site, has proved to be a valuable means of assessing the biological activity of complex chemical mixtures. Such technology provides direct toxicological readout (e.g., delayed hatch, growth, mortality, terata) for all substances contained in effluents or receiving waters. Net effects of all toxicants, toxic interactions, and other variables are directly reflected in test responses and the resulting toxicity values can be used to estimate the potential for ecological effects (Birge et al., 1989). It appears plausible that the toxicological monitoring approach can be modified and supplemented with biomarker techniques to provide information on potential human health effects of aquatic contaminants. Further consideration of this rationale is given below.

PROCEDURES FOR SHORT-CHRONIC FISH EMBRYO-LARVAL TESTS

Continuous-flow and static-renewal procedures for short-term embryo-larval tests are described for single compounds and complex effluents. The applicability of the short-term test in effluent biomonitoring was investigated in both laboratory and field studies. Most laboratory studies were performed using cadmium chloride as a reference toxicant to evaluate the precision and usefulness of test methods. Further descriptions of procedures have been published earlier (Birge and Cassidy, 1983; Birge et al., 1985), and those methods standardized for NPDES testing are given by Horning and Weber (1985) and Weber et al. (1988).

Static-Renewal Procedures for Single Compounds, Effluents, and Receiving Waters

For static-renewal tests, eggs are deposited in Pyrex deep petri dishes (400-mL capacity) or 600-mL beakers. Studies in progress indicate that, for special needs, tests may be conducted in beakers or petri dishes using 60-100 mL of solution, provided that sample size is restricted to 20 eggs per dish (E.M. Silberhorn, University of Kentucky, 1989, personal communication). Test solutions are replaced at regular 12- or 24-h intervals, depending on persistence of toxicity or economic considerations. During the change, a small volume of solution (e.g., 10%) is retained in the exposure chamber to protect organisms. Effluent tests are conducted with composite or grab samples collected and stored using procedures described by Horning and Weber (1985). It should be noted that toxicity may decrease with effluent storage time (Birge and Black, 1981). In static-renewal tests, dissolved oxygen should be monitored daily. Dilution waters may be aerated prior to use and, if necessary, dry filtered air may be supplied slowly to test chambers at a rate sufficient to maintain minimum dissolved oxygen. A lower limit of 60% saturation is preferred, although a 40% minimum is given in NPDES procedures. Monitoring dissolved oxygen is especially important in tests with high effluent concentrations (i.e., 50% to 100%).

Continuous-Flow Procedures for Single Compounds and Effluents

The short-chronic embyro-larval test may be performed under flow-through conditions using diluters and exposure chambers developed for the 30-day early life stage test, such as described by Benoit et al. (1982). Alternatively, continuous-flow tests with single compounds may be performed using modified deep petri dishes supplied with water and toxicant by peristaltic and syringe pumps (Birge and Cassidy, 1983). Deep petri dishes are modified by adding an inlet tube (10 mm i.d.) near the bottom of the dish. An outlet tube is annealed to the opposite side of the dish just below the neck. The working capacity is approximately 300 mL. When flow rate is set at 200 mL/h, replacement time is 1.5 h. Flow rate may be varied to meet special needs. Each exposure chamber receives water and toxicant from a side-arm flask (e.g., 125 mL), which serves as a mixing chamber and usually is operated with a magnetic stirrer. Graduated flow from a syringe pump is used to administer toxicant to each mixing chamber, and dilution water is supplied by regulated flow from a peristaltic pump. Dilution water is stored in 50-Liter Nalgene containers that receive continuous aeration. The exposure concentration is regulated by adjusting the mixing ratio between toxicant and dilution water or by varying the concentration of toxicant delivered from the syringe pump. Flow rates are determined using timed volumetric measurements taken once or twice daily for syringe and peristaltic pump lines, equipped with rapid disconnects. Special exposure and mixing chambers have been designed for use with compounds of low water solubility or high volatility (Birge et al., 1981).

Effluent tests may be conducted with modified deep petri dishes, as given above. Undiluted effluent and four to five effluent dilutions can be supplied to test chambers using proportional diluters such as those described by Birge and Black (1981) and Peltier and Weber (1985). Effluent may be drawn from the discharge point or from Nalgene reservoirs that are filled daily. Alternatively, peristaltic pumps may be used to supply effluent and dilution waters, as described above. Chambers may be given slow aeration if required to maintain dissolved oxygen within acceptable limits.

Test Conditions and Monitoring Procedures

Laboratory toxicity tests reported below were performed in temperature-regulated environmental rooms. On-site biomonitoring investigations were conducted in a temperature-controlled mobile laboratory in which exposure chambers for warm water species were maintained on open bench tops. Tests with the rainbow trout were performed in a table-top refrigerator.

Reconstituted water (Birge and Cassidy, 1983; Horning and Weber, 1985) is recommended for dilutions, but well or natural surface waters of acceptable quality may be used. Control survival of 90% or more can generally be achieved, but 80% should be set as a minimum. Tests can be conducted in duplicate if probit analysis is to be used, but three to four replicates are recommended, particularly if results are to be statistically analyzed by hypothesis testing. In definitive tests, an exponential series of five or more exposure concentrations is used.

Except in preliminary static-renewal tests, for which nominal exposure conditions can be reported, test solutions should be monitored at regular intervals for temperature, dissolved oxygen, pH, hardness, alkalinity, and conductivity. The first three are the most critical and should be measured daily in at least one replicate chamber for each concentration. Different replicate chambers can be monitored on successive days. Preferably, a minimum of three measurements should be made for each exposure chamber during an eight-day test. Monitoring frequencies should be increased for dissolved oxygen and pH if fluctuations are notable, particularly when static-renewal procedures are used. If necessary, dissolved oxygen should be measured at the beginning and end of each renewal interval in at least one replicate chamber for each exposure concentration.

Water hardness, alkalinity, and conductivity are also useful parameters, particularly when metals are being evaluated, and generally should be monitored at the same frequency as the other water quality characteristics. In static-renewal tests, these analyses may be conducted on solutions that are prepared for routine changes. In continuous-flow studies, measurements can be made on water samples taken from the chamber outlet. Specific requirements for biomonitoring of NPDES effluents, including photoperiod and other test conditions, have been summarized by Horning and Weber (1985). Analytical procedures used for monitoring have been described in earlier studies (Birge and Cassidy, 1983; Birge et al., 1985).

Temperature should be set within the optimum range for any given test species. With the fathead minnow, successful tests can be conducted within the range of 22° to 28°C (Birge et al., 1985). Use of higher temperatures in this range (i.e., 25-28°C) affords the advantage of an earlier hatching time and nearly equal exposure periods for embryonic and larval stages in an eight-day test.

Studies with cadmium designed to evaluate reproducibility of the embryo-larval procedure were performed with continuous-flow and 12-h static-renewal procedures. Cadmium was analyzed using a Perkin-Elmer (model 503) atomic absorption spectrophotometer equipped with a graphite furnace (model HGA 2100). Cadmium was monitored daily for each exposure chamber and mean values with standard errors were obtained by combining data for two replicates. Analytical procedures for other toxicants have been given in previous publications cited below.

In setting up a test, eggs collected for use are placed in clean finger bowls or petri dishes and examined either under a stereoscopic microscope or over a viewbox using a magnifier. Eggs that are clustered should be gently separated from one another as described by Horning and Weber (1985), and unhardened, opaque, or damaged eggs should be discarded. Viable eggs are then transferred to exposure chambers without delay or prolonged exposure to light, using disposable 7.5 mL transfer pipets. Initial samples may include a few extra eggs, and final sample size is determined at the first inspection after initiation of the test. This inspection should be performed within about 12 h after the eggs are placed in the exposure chambers and should start with observations of the control organisms. Any defective eggs that were overlooked in the initial culling operation should be discarded and not included in the sample number. Variation in the number of eggs among dishes generally should not exceed 10%. It is advisable to retain unused viable eggs through this stage of the test in the event that some replacements are necessary. After this step is completed, test chambers may be randomized, if desired. Throughout the test, populations should be observed daily to gauge extent of development, tabulate and remove dead organisms, and inspect cultures for contamination (e.g., fungus).

The principal biological endpoints are frequencies of mortality and terata observed at the end of the test. Determinations of teratic larvae have been limited to gross, debilitating anomalies, as discussed previously (Birge et al., 1979a, 1983). Live terata at the end of the test are counted as mortalities in the final tabulation of test responses. Probit analysis (Finney, 1971) is used to calculate median lethal (LC50) and threshold (LC10, LC1) values. Reliable probit values usually can be achieved when there is an adequate dose-response relationship, involving two to three partial kills within the toxicant exposure range. Responses also may be analyzed by hypothesis testing, as described by Birge et al. (1985) and Horning and Weber (1985). Estimated chronic values can be calculated as the geometric mean of the no observed effect concentration (NOEC) and the lowest observed effect concentration (LOEC). Additional information may be obtained by measuring growth (dry weight), by analyzing test responses at hatching (e.g., hatching time, egg hatchability, terata), or by observing behavior (e.g., balance, locomotion.

Approximately 30 fish and amphibian species have been used with the short-term embryo-larval procedure. Fish most frequently studied include the bluegill sunfish (Lepomis macrochirus), channel catfish (Ictalurus punctatus), fathead minnow (Pimephales promelas), goldfish (Carassius auratus), largemouth bass (Micropterus salmoides), and rainbow trout (Salmo gairdneri). Amphibian species tested most often are the African clawed frog (Xenopus laevis), bullfrog (Rana catesbeiana), and leopard frog (Rana pipiens). Practical factors important in selecting species include sensitivity, seasonal availability of eggs, number of eggs per spawn, temperature requirements, and hatching time. Xenopus, bluegill, largemouth bass, and goldfish have shorter hatching times (e.g., 2 to 3.5 days, 22°-25°C) but generally are less sensitive to toxicant stress. Tests with these species usually can be completed within 6 to 7 days. The more sensitive warm water species generally have longer hatching times, thus requiring 8 to 10 days for the full test depending on temperature selection. Xenopus and fathead minnow eggs are available year-round, although the latter species produces comparatively small numbers of eggs per spawn (Peltier and Weber, 1985). The other species listed spawn larger numbers of eggs, but availability is seasonal. This restriction can be alleviated somewhat by selecting spring and fall spawning strains (e.g., trout) or by hormonal regulation of spawning (e.g., leopard frog, goldfish). In laboratory studies with rainbow trout, the complete test involves a 28-day exposure period (13°C). However, in on-site tests with effluents, good results have been obtained with this species with an exposure period of 8 to 10 days, starting either with freshly fertilized or eyed eggs. The former consistently have given sensitive tests (Birge and Black, 1981).

OBSERVATIONS AND DISCUSSION

In standardizing test conditions and evaluating reproducibility of the eight-day embryo-larval (EL) procedure with the fathead minnow, a series of investigations was performed using cadmium as a reference toxicant (Birge et al., 1985). Initial tests were conducted to determine an optimum temperature range. Toxicity was reduced at 20°C and atypical results occurred at 30°C. However, toxicity values were consistent at temperatures in the range of 22.8 to 27.9°C. A range of 25° to 28°C is workable and generally consistent with recommendations of Horning and Weber (1985). Tests also were performed with and without feeding brine shrimp to larval stages, and there was no measurable effect on cadmium toxicity. Based on these and other studies (Birge et al., 1981), lack of feeding generally does not impair testing (e.g., control survival) if exposure is terminated within the first week after onset of hatching. Withholding of feeding simplifies maintenance of test chambers and minimizes test variables.

Reproducibility of the EL test as determined in continuous-flow (CF) studies with cadmium administered at a water hardness of 100 mg/L $CaCO_3$ is shown in Table 1. The six tests were performed independently over an eight-month period, drawing from different stocks of test organisms, and results were highly consistent. The toxicity threshold (LC1) values and NOECs were reasonable estimates of the chronic toxicity of cadmium to fish, compared to chronic values given in the cadmium criterion document (US EPA, 1980), as discussed below. A series of 11 EL tests was conducted with static-renewal (SR) procedures to develop threshold and no effect concentrations. Results are presented in Table 2. Although values were somewhat lower and more variable than observed in the CF tests, reproducibility was satisfactory, particularly considering the economy of this method. Cadmium toxicity was greater with the SR procedure, and this has been observed with a few other persistent chemicals (e.g., PCBs). For most compounds and effluents, this method usually has underestimated toxicity, as compared with CF results.

The short-term EL test is also adaptable for use with mixed populations of fish or amphibian species, which can be treated simultaneously in the same exposure chambers. However, it is necessary to select species for which early life stages can be distinguished from one another and which can be maintained at the same temperature. To evaluate this application, a multiple species test was conducted with cadmium at a mean temperature of 22.5°C, using the carp, fathead minnow, and largemouth bass. Toxicity threshold values (ug/L) ranged from 4.1 with the carp to 13.9

Table 1. Reproducibility of Eight-Day Embryo-Larval Tests with Cadmium Using the Fathead Minnow in a Continuous-Flow System

Test No.[a]	LC50 (mg/L)	95% Confidence Limits	LC1 (mg/L)	95% Confidence Limits	NOEC (mg/L)
1	0.067	0.056-0.078	0.010	0.006-0.014	0.012
2	0.084	0.073-0.097	0.013	0.008-0.017	0.011
3	0.073	0.062-0.085	0.012	0.007-0.017	0.013
4	0.079	0.068-0.091	0.012	0.008-0.017	0.011
5	0.084	0.072-0.098	0.014	0.009-0.020	0.010
6	0.070	0.060-0.081	0.010	0.006-0.015	0.014

[a] The six tests were performed independently over about eight months, starting November 1982. Reprinted with permission from Environmental Toxicology and Chemistry, vol. 4, 1985, 807-821, Pergamon Press, New York.

Table 2. Reproducibility of Eight-Day Embryo-Larval Tests with Cadmium Using the Fathead Minnow in a Static-Renewal System[a]

Test No.	LC1 (ug/L)	95% Confidence Limits	NOEC (ug/1)
1	2.3	1.3 - 3.3	3.2
2	2.8	1.6 - 4.3	11.0
3	5.4	2.5 - 8.8	13.0
4	5.7	3.2 - 8.7	12.0
5	6.3	3.3 - 9.6	12.0
6	6.8	4.6 - 8.7	3.2
7	7.0	-	10.0
8	7.1	5.1 - 8.7	3.2
9	8.1	3.9 - 11.4	10.0
10	8.2	5.5 - 10.5	10.0
11	13.5	8.8 - 17.8	12.0

[a]All tests performed in moderately hard water (100 mg/L as $CaCO_3$).

with the bass. These data and results obtained in other EL tests with channel catfish and goldfish are presented in Table 3, together with MATCs and chronic values taken from the cadmium criterion document (US EPA, 1980). The range of toxicity values obtained with the short-chronic mixed species test approximated the chronic values for cadmium administered at water hardnesses below 100 mg $CaCO_3$/L. Pickering and Gast (1972) reported a cadmium MATC of 37 to 57 ug/L in a fathead minnow life-cycle study, which was performed at a water hardness of 201 mg/L. To compare results, a short-chronic test (CF) with the fathead minnow was performed with cadmium using reconstituted water at a hardness of 200 mg/L. The LC1 in ug/L (with 95% confidence limits) was 35.8 (20.8 - 50.5). Based on these data, cadmium toxicity decreased at the higher hardness level and the toxicity threshold approximated the NOEC reported in the life-cycle study. Similar comparisons for other toxicants were reported previously (Birge et al., 1981; 1985). Closer correlations result when short-chronic tests are conducted with flow-through procedures, as are conventional life-cycle studies. When static-renewal methods are used, best comparisons are achieved with sensitive animal species, e.g., rainbow trout).

Table 3. Comparison of LC1 and Chronic Values for Cadmium

Species	LC1 (ug/L)	Species	MATC (ug/L)	Chronic value (ug/L)
Largemouth bass	13.9	Smallmouth bass	4.3-12.7	7.4
Channel catfish	8.6	Channel catfish	11-17	13.7
Carp	4.1	Northern pike	4.2-12.9	7.4
Goldfish	3.0	White sucker	4.2-12.0	7.1
Fathead minnow	11.5	Flagfish	4.1-8.1	5.8
		Walleye	9-25	15.0

Range of cadmium LC1 values
Multiple-species test (bass, carp, minnow):
4.1-13.9 ug/L
Five species: 3.0-13.9 ug/L
Range of cadmium chronic values
Six warmwater species: 5.8-15.0 ug/L

[a] LC1 values determined in short-term embryo-larval tests terminated four days after hatching; chronic values taken from the criterion document (US EPA, 1980). Comparisons were restricted to tests in which hardness was below 100 mg/L $CaCO_3$. Reprinted with permission from Environmental Toxicology and Chemistry, vol. 4, 1985, 807-821, Pergamon Press, New York.

The guidelines for establishing water quality criteria for aquatic biota recommend testing with at least eight species (Stephen et al., 1985). This protocol takes into account the variability in response of different aquatic species to the same compound. An example of the hetero-geneity of response among species is given in Table 4, in which embryo-larval stages of 20 different fish and amphibians were exposed to mercury in a static-renewal system (Birge et al., 1979a). The LC50 values varied by more than two orders of magnitude. Furthermore, the order of species sensitivity is not always consistent for different toxicants (Birge, 1981). Therefore, the practice of standardizing on the fathead minnow for biomonitoring studies with fish should be viewed with some caution. An-other available option includes a recently described embryo-larval test with the zebrafish, Brachydanio rerio (Dave et al., 1987). Inclusion of additional species is recommended, when possible, to provide a more ade-quate basis for impact assessments. This is especially pertinent to hazard-ous waste sites for which more comprehensive documentation is desirable.

Table 4. Comparative Toxicity of Mercury to Embryo-Larval States of Fish and Amphibians

Species	LC50 (ug/L)	95% Confidence Limits
Fish		
Salmo gairdneri	4.7	4.2-5.3
Ictalurus punctatus	30.0	26.9-33.2
Lepomis macrochirus	88.7	73.5-106.3
Carassius auratus	121.9	112.3-132.1
Lepomis microlophus	137.2	115.0-162.8
Micropterus salmoides	140.0	128.7-151.9
Amphibians		
Gastrophryne carolinensis	1.3	0.9-1.9
Hyla chrysoscelis	2.4	1.5-3.4
Hyla squirella	2.4	1.5-3.8
Hyla gratiosa	2.5	1.7-3.4
Hyla versicolor	2.6	1.2-4.2
Hyla crucifer	2.8	1.9-3.9
Rana pipiens	7.3	4.8-10.0
Acris crepitans blanchardi	10.4	8.5-12.6
Bufo punctatus	36.8	18.3-51.1
Bufo debilis debilis	40.0	25.6-52.2
Rana heckscheri	59.9	53.8-65.9
Bufo fowleri	65.9	44.0-84.0
Rana grylio	67.2	54.3-79.5
Ambystoma opacum	107.5	72.5-153.5

[a] Exposure was maintained from fertilization through four days posthatching using 12-h static-renewal procedures.

The initial use of the eight-day EL test for evaluating the toxicity of complex effluents involved direct toxicological monitoring studies of coal ash leachates (Birge, 1981). Flow-through leaching of coal fly ash in a model system produced effluent that contained detectable concentrations of more than 20 metals and had physical and chemical characteristics compara-ble to those reported for fly ash settling ponds. One example from these studies is summarized in Table 5. Embryo-larval stages of the largemouth bass were used in 8-day flow-through tests to monitor the toxicity of ash effluent for three different periods during 1000 h of continuous leaching. The final effluent was produced by supplying carbon-filtered tap water at 1.4 L/h to a 37.2 kg sample of precipitator-collected fly ash contained in an 88-Liter flow-through settling chamber. Based on mean values for the three elution intervals, concentrations of most metals decreased with leaching time. Although the chemical and physical data provided a basis for estimating biological effects, embryo-larval monitoring gave a more

precise and direct quantification of the net toxicity of the metal mixtures. The concentration of aluminum and reduced pH likely contributed significantly to the higher toxicity observed during the first leaching period
(Table 5). In situ EL monitoring was also more useful than chemical data
for characterizing toxicity reduction of coal ash effluents that were
either filtered through charcoal or circulated over sediment (Birge et al.,
1978).

Further application of the eight-day EL procedure to effluent monitoring occurred in an investigation designed to evaluate short-chronic testing
for use under the NPDES program. After initial laboratory investigations,
field studies with mobile facilities were conducted on four different industrial effluents (Birge and Black, 1981). These studies were planned to
coincide with acute biomonitoring performed by personnel from EPA Region
IV. A comparison of results of the acute and eight-day EL flow-through
tests is given in Table 6.

Acute tests were performed with fathead minnow larvae (Peltier and
Weber, 1985), and LC50s could not be calculated for the three least toxic
effluents. By comparison, LC50 and threshold (LC1) values with 95% confidence limits were obtained in EL tests for all effluents, providing good

Table 5.   Chemical and Toxicological Characterization of Coal Ash Effluent
            Using Embryo-Larval Stages of the Largemouth Bass

Effluent Characteristics[a]	Elution Interval (hrs)		
	0 - 250	250 - 500	750 - 1000
Total dissolved solids (g/L)	0.62 ± 0.41	0.13 ± 0.02	0.14 ± 0.01
pH	5.58 ± 0.14	6.25 ± 0.11	6.69 ± 0.05
Conductivity (umhos/cm)	343 ± 74	158 ± 3	168 ± 3
Alkalinity (mg/L as $CaCO_3$)	4.0 ± 3.0	12.0 ± 3.3	23.3 ± 5.0
Hardness (mg/L as $CaCO_3$)	206 ± 16	165 ± 8	167 ± 8
Temperature (°C)	18.9 ± 0.2	19.5 ± 0.2	20.0 ± 0.2
Ag (ug/l)	1.2 ± 1.2	<1.0	<1.0
Al (ug/l)	13757 ± 7395	4708 ± 1063	507 ± 78
As (ug/l)	100 ± 100	<10.0	<10.0
Cd (ug/l)	3.5 ± 1.9	1.8 ± 0.5	0.7 ± 0.7
Cr (ug/l)	160 ± 107	13.0 ± 5.0	<1.0
Cu (ug/l)	114 ± 44	51.0 ± 6.4	30.8 ± 3.4
Fe (ug/l)	404 ± 338	3.0 ± 3.0	5.5 ± 5.5
Hg (ug/l)	0.08 ± 0.03	0.03 ± 0.03	0.16 ± 0.07
Mn (ug/l)	120 ± 50	47.0 ± 3.0	52.3 ± 3.3
Ni (ug/l)	59.5 ± 17.7	34.8 ± 6.0	28.0 ± 2.9
Pb (ug/l)	30.0 ± 30.0	<10.0	<10.0
Sr (ug/l)	158 ± 34	160 ± 9	155 ± 10
Zn (ug/l)	160 ± 58	76.8 ± 12.1	68.7 ± 8.5
LC50 (%Effluent)[b] (95% Confidence Limits)	0.9 (0.8-1.2)	12.0 (8.0-18.0)	39.0 (31.0-52.0)
LC1 (%Effluent)[b] (95% Confidence Limits)	0.04 (0.02-0.06)	0.1 (0.04-0.3)	0.2 (0.09-0.4)

[a] Given as mean ± standard error.

[b] Data are presented at four days posthatching for an eight-day exposure
period.

Table 6. Comparative Evaluations of Effluent Toxicity as Determined in Flow-Through Acute and Embryo-Larval Tests with the Fathead Minnow

Effluent Source	Acute LC50 (% effluent)	Embryo-larval values (% effluent)			
		LC50	Confidence Limits	LC1	Confidence Limits
Tannery-sewage treatment plant	8.0	0.3	0.2-0.6	0.001	0.0002-0.006
Synthetic rubber plant	ND[a,b]	8.3	4.6-13.2	0.02	0.002-0.08
Metal plating plant	ND[c]	21.6	13.9-29.5	0.8	0.1-2.0
Chemical manufacturing plant	ND[d]	29.4	22.1-36.4	2.8	1.1-5.1

[a]Not determined (ND).
[b]50% was the highest effluent concentration tested; no mortality observed.
[c]35% mortality in 100% effluent.
[d]No mortality in 100% effluent.

quantification of toxicity. In further studies at the metal plating plant, independent static-renewal tests were conducted on each of the four waste streams which, together with runoff water, comprised the final effluent (Birge et al., 1986). The toxicity values were more definitive than chemical characteristics for comparing impact potential of the different waste streams (Table 7). Michael et al. (1989) described an effluent biomonitoring plan that placed principal reliance on acute testing. Considering data presented in Table 6 and other information, such an approach takes limited advantage of current biomonitoring technology.

One additional aspect of the NPDES study involved a comparison of toxicity for effluents tested in the field with fresh samples and in the laboratory with stored composite samples. To minimize variables, composite samples were collected simultaneously with those used for field testing. The composites were stored in Pyrex containers without airspace and transported to the laboratory at 4°C, allowed to equilibrate to 22°C, and tested within 24 h of collection with the same static-renewal procedures and test organisms as used in the correlated field study. The results, summarized in Table 8, clearly showed a reduction in toxicity for the stored effluents. Concerning stored samples, a characteristic dose-response relationship was obtained only for the highly toxic effluent from the sewage treatment plant that received waste from a nearby tannery. These findings raise questions regarding alterations in the integrity of effluent samples that are subjected to cold preservation and storage. This is of interest as this practice is frequently used in NPDES biomonitoring. These data indicate not only a decrease in effluent toxicity with storage time, but also the prospects for false negative results. This matter also may be of concern for evaluations on mutagenicity and carcinogenicity. There is an obvious need to characterize changes in biological activity that may occur with collection and storage of effluents.

Following the NPDES study, the next step in test validation was to apply the short-term EL procedure to receiving water systems (1) to examine the relationship between toxicity values and ecological endpoints and (2) to evaluate reliability of EL test data for quantifying effluent and receiving water toxicity and for estimating ecological effects. In addition, different ecological endpoints were compared as to their usefulness for characterizing impact of effluent discharges on freshwater biota.

Table 7. Chemicaland Toxicological Characteristics of Effluent Components and Final Effluent from a Metal Plating Plant

Effluent Characteristics[a,b]	Effluent Components[c]				Final Effluent	Dilution Water
	1	2	3	4		
Percent final effluent[d]	3	24	<1	43	-	-
pH	9.2±0.1	9.2±0.2	7.8±0.1	7.2±0.1	8.1±0.1	8.1±0.1
Alkalinity (mg/L as CaCO$_3$)	1844±20.4	157±9	86±5	310±26	273±10	260±7
Hardness (mg/L as CaCO$_3$)	148±13	86±6	163±3	438±7	281±11	272±5
Conductivity (umhos/cm)	12209±1966	866±37	189±3	509±58	716±42	229±5
Cadmium (ug/L)	31±5	9±2	8±1	58±46	6±1	<0.2
Chromium (ug/L)	741±282	484±205	6±3	585±48	296±135	13
Copper (ug/L)	239±44	57±12	14±2	11±1	23±5	<1
Iron (ug/L)	377±66	425±157	44±21	36±5	292±146	31
Zinc (ug/L)	2488±585	893±223	1173±139	74±8	379±115	<2
LC50 (Fathead minnow)	0.01%	0.05%	23.6%	25.4%	44.7%	-

[a] Chemical characteristics expressed as mean ± standard error for a minimum of five determinations.

[b] LC50 values were determined during on-site biomonitoring, using embryo-larval static-renewal procedures.

[c] The final NPDES effluent included sludge-bed filtrate (1), process water (2), brazing process water (3), and cooling water (4).

[d] Remaining component of the final effluent was from surface runoff(~29%).

Reprinted with permission from Environmental Hazard Assessment of Effluents, 1986, 66-80, Pergamon Press, New York.

Receiving systems selected for this study included streams of moderate size with well-established riffle-pool habitats, offering reasonable comparability for sequentially located monitoring stations. Each study site contained a single major effluent of high impact potential. Hydrological measurements and chemical parameters were used to define the pattern of instream effluent dilution. In each case, there was a distinct downstream dilution gradient of effluent constituents.

The first study, conducted in the fall of 1983, involved a secondary sewage treatment plant outfall that discharged into Town Branch (TB). At the time of study, flow was minimal above the outfall and the effluent provided the bulk of the flow for this moderate-size stream. Town Branch entered South Elkhorn (SE), a fourth-order stream, about 14 km below the outfall. The upstream site (TB1) was adequate as a toxicological reference but was not suitable for ecological studies due to low flow volume and habitat alterations resulting from urbanization. Therefore, additional reference sites (SE1, SE2) were situated on the South Elkhorn above the Town Branch confluence. Taking measurements just above the confluence of the two streams, flow volume from the South Elkhorn (SE2) was about one-third of that recorded for Town Branch (TB4). Monitoring stations SE3, SE4, SE5, and SE6 were downstream of the outfall by 14.8, 37.5, 54.1, and 67.6 km, respectively. Toxicological and ecological values obtained at SE6 agreed with those given for the reference sites (Table 9). Percent in-stream effluent at SE6 was calculated to be 33% based on hydrological measurements (Birge et al., 1989).

Downstream of the outfall (TB2 - SE6), there were good correlations between EL toxicity data and the independent ecological endpoints, especially species richness for macroinvertebrates. Inverse correlations were obtained when EL survival was tested against percent effluent and phosphate concentrations. Therefore, EL survival improved downstream as the proportion

Table 8.   Comparative Toxicity of Effluents Determined in On-Site and
            Laboratory Static-Renewal Tests

Effluent Source (Test Species)	Effluent Concentration(%)	Percent Survival[a]	
		On-Site	Laboratory
Chemical Manufacturing	100	46	72
Plant	50	73	72
(Xenopus laevis)	10	81	78
	1	87	84
	0.1	93	87
	Control	97	85
Six-day LC50(%)		~100	ND[b] -
Synthetic Rubber	100	37	93
Plant	50	53	85
(Rainbow trout--eyed	10	75	90
eggs)	1	89	94
	0.1	93	96
	Control	97	95
Nine-day LC50(%)		48.4	ND[b]
(95% Confidence Limits)		(25.5-100)	
Tannery-Sewage	100	0	0
Treatment Plant	50	0	41
(Rainbow trout--freshly	10	4	78
fertilized eggs)	1	47	90
	0.1	67	92
	Control	97	94
Nine-day LC50(%)		0.5	32.1
(95% Confidence Limits)		(0.3-0.8)	(25.5-38.4)

[a] Determined four days after hatching of Xenopus and eyed rainbow trout
stages and after nine days of exposure for freshly fertilized trout eggs.

[b] Not determined.

of effluent constituents decreased. Other chemical parameters measured for
such purposes included polar organics, sodium, and zinc, and the correlation coefficients against EL data were -0.91, -0.94, and -0.59, respectively. Proportions of effluent chemicals persisting at SE6 were calculated to
be 23% to 24% for phosphate and sodium, 21% for zinc, and 15% for polar
organics. For comparative purposes, percent effluent and numbers of macroinvertebrate taxa were tested for correlation with the other categories of
data. Percent effluent gave a good inverse correlation with percent EL survival and with the number of macroinvertebrate taxa (Table 9). There was a
high positive correlation between percent effluent and phosphate concentration. In summary, macro invertebrate species richness (number of taxa)
provided an excellent index to ecological impact. However, invertebrate
density, diversity, dominance, and functional group analyses were also
useful and necessary considerations for assessing extent and pattern of
effects within the system. Under conditions that prevailed at the time of
study, results of the eight-day embryo-larval test with the fathead minnow
provided good quantification of effluent and in-stream toxicity and agreed
well with spatial assemblages of benthic fauna.

In the statistical analyses performed at the time of study with
Dunnett's test, toxicity values reported for TB1, SE1, and SE6 did not differ, but the values for SE6 and SE4 differed significantly (p <0.05). However, in preparation of this paper, toxicity data were reevaluated statistically using arc sine square root transformations with Dunnett's test and
the US EPA computer program recommended by Weber et al. (1988). Differences between SE4 and the other aforementioned values approached but did
not fall within 0.05 probability, as in comparing the results from TB1 and
SE4 (p=0.08). However, the 69% survival at SE3 differed significantly from

Table 9. Comparison of Toxicological, Ecological and Selected Chemical Data for a Sewage Treatment Plant Effluent/Receiving Water System

Field Station	Percent Effluent	% Embryo-larval Survival (mean ± SD)	No. of Fish Species	Invertebrates No. of Taxa	H′	Phosphate (mg/L)
Upstream						
TB1	-	92 ± 3	-	-	-	0.49
Effluent	100	0	-	-	-	3.52
Downstream						
TB2	88	0	0	3	1.05	2.87
TB3	106	0	2	3	1.09	2.85
TB4	73	0	1	6	0.53	2.10
SE3	60	69 ± 6	4	19	2.18	1.88
SE4	53	86 ± 4	10	22	2.50	1.14
SE5	-	-	17	26	3.05	1.04
SE6	33	92 ± 2	11	29	3.84	0.80
Reference						
SE1	-	90 ± 4	11	34	3.70	0.82
SE2	-	-	8	30	3.26	0.69
Correlations[a]						
r vs. % Eff.	-	-0.87	-0.83	-0.94	-0.84	0.96
r vs. % EL	-0.87	-	0.92	0.96	0.93	-0.91
r vs. Inv. Taxa	-0.94	0.96	0.88	-	0.96	-0.96

[a] Correlation coefficients (r) were calculated against other parameters for percent effluent, percent EL survival, and number of invertebrate taxa. Modified from Environmental Toxicology and Chemistry, vol. 8, 1989, 437-450, Pergamon Press, New York.

each other toxicity value. Toxicity tests for SE5 were not performed at the time of study, as reliable values for flow volume and percent effluent could not be determined. A review of field records indicated that the toxicity value reported originally for SE2 was in error (Birge et al., 1989) and, therefore, has been deleted. However, the values obtained for SE1 and TB1 provided adequate toxicological references.

A similar investigation was conducted on a portion of the Red River drainage in Kentucky, which was impacted by an oilfield brine discharge. Chloride was the only predominant toxicant detected. Results of this study are now being prepared for independent publication. However, some of the findings of this investigation are summarized for the seven monitoring stations (Table 10). Stations 1 through 6 were located on the South and Middle Forks of the Red River and were positioned to provide a progressive decrease in chloride concentrations. Station 1 was about 1 km below the outfall and station 6 was approximately 13 km farther downstream. Field station 7 was the reference site and was situated on the Middle Fork of the Red River 3.4 km upstream of its confluence with the South Fork. As with the previous study, good correlations were obtained between EL toxicity data and the ecological findings, particularly the number of macroinvertebrate taxa. The toxicological and ecological endpoints correlated well with chloride concentrations of 0.07 to 3.80 g/L, but less well at higher concentrations. Therefore, correlation coefficients were calculated for all field stations, and for stations 3 through 7. The poorer correlations at higher chloride concentrations were due partly to the fact that major impact was already apparent at station 3, where only three invertebrate taxa were found. The precipitous drop in EL survival between 3.80 and 5.79 g/L total chloride was generally consistent with results in laboratory toxicity tests with chloride.

Table 10. Comparison of Biomonitoring Tests with Macroinvertebrate and Fish Data Obtained in a Chloride Field Study

Field Station	Chloride g/L ($\bar{X}$ ± SD)	Percent EL Survival ($\bar{X}$ ± SD)	No. of Fish Species	Invertebrates No. Taxa	H'
1	17.39 ± 5.28	0	0	2	0.99
2	5.79 ± 3.09	0	8	3	0.25
3	3.80 ± 0.60	67.3 ± 1.2	9	3	0.51
4	3.16 ± 0.68	82.5 ± 1.7	17	10	2.21
5	1.10 ± 0.25	81.5 ± 1.5	11	19	2.96
6	0.10 ± 0.07	90.3 ± 1.8	20	25	2.69
7	0.07 ± 0.02	98.5 ± 1.9	22	31	3.45
Correlations r vs. % EL					
FS 1-7	-0.81	-	0.42	0.77	0.80
FS 3-7	-0.86	-	0.92	0.94	0.91
r vs. Inv. Taxa					
FS 1-7	-0.69	0.77	0.84	-	0.93
FS 3-7	-0.97	0.94	0.78	-	0.91

[a] Correlation coefficients (r) were calculated against other parameters for Field Stations 1-7 and 3-7 using both percent EL survival and number of invertebrate taxa.

The number of invertebrate taxa (species richness) again was the most sensitive and consistent parameter for assessing ecological impact. Nevertheless, it was important to determine both species composition and abundance and to calculate values for such parameters as diversity (H'), evenness, and dominance. The analysis of functional groups of macroinvertebrates also was essential in making overall assessments of biotic conditions.

Results of the EL tests provided reliable quantification of chloride toxicity, as well as a useful basis for estimating the potential for instream effects. As in the previous study, the fathead minnow tests were performed using four replicates and 24-h static-renewal procedures. Toxicity results were statistically analyzed with the above referenced program. All toxicity values differed significantly (p <0.05) from each other, except for comparisons between stations 1 and 2 and between stations 4 and 5.

Indices of community integrity (invertebrates) and biological integrity (fish) were not used in these studies, but they provide additional approaches for characterizing impact on aquatic systems (Karr et al., 1986). Selection of reference stations, habitat comparability, and percent of taxa recovered for study are among the more important variables that affect the precision and validity of these assessments. The designation of tolerant or sensitive species also introduces subjectivity into the integrity index approach, as species sensitivity may vary with different toxicants or stream conditions.

In considering results of the two integrated biomonitoring studies discussed above, it should be noted that they were performed under conditions that optimized the toxicological characterization of receiving systems. In each case, there was a single, highly toxic effluent and a marked downstream dilution gradient. Tests were conducted in the fall under generally stable hydrological conditions, about 2 months after the seasonal low-flow period. Stream flow was still below the annual mean. In addition, it was possible to delineate effluent chemical parameters with which to compare variations in toxicity and biotic conditions within different segments of the receiving streams. Based on results of studies conducted under these conditions, the EL data provided good quantification of effluent and receiving water toxicity and reasonable estimates of the extent of ecological impact. For example, in the chloride investigation, statistically significant decreases in embryo-larval survival correlated closely with observed changes in biotic conditions. However, the usefulness of short-chronic procedures may vary under different environmental conditions (e.g., high flow) or when used with effluents of more moderate impact potential.

Mount et al. (1985) and Norberg-King and Mount (1986) also have evaluated the use of effluent and receiving water toxicity tests for predicting ecological effects, using the fathead minnow seven-day growth test and the seven-day _Ceriodaphnia_ reproduction test. Interpretation of their results is complicated somewhat by the fact that they worked with complex receiving water systems impacted by multiple effluents. Also, in performing the correlations between the toxicity tests and three categories of ecological data, they selected those values that showed the greatest effects at each monitoring station. Using fish, zooplankton, and benthic invertebrates interchangeably in the same correlation does not take into account their unique biological and ecological differences or the fact that these organisms are not necessarily exposed or stressed similarly within the same system. Direct correlations between results of either the _Ceriodaphnia_ or fathead minnow test with each individual category of ecological data might have been useful. Nevertheless, their results support the premise that short-chronic toxicity tests are useful for predicting in-stream biological effects.

Present acute (Peltier and Weber, 1985) and short-chronic procedures (Horning and Weber, 1985) appear adequate to meet biomonitoring requirements for the more toxic effluents. Furthermore, application of US EPA toxicity identification (TIE) and toxicity reduction evaluations (TRE) can be used to facilitate appropriate case-by-case remediation of acutely toxic effluents.

It is less clear, however, whether present biomonitoring technology is adequate for estimating impact in receiving water systems containing substantial proportions of low to moderately toxic effluents. For example, initial results in a current study indicate moderate reductions in macroinvertebrate taxa and substantial decreases in density in a stream system affected by a series of effluents that usually exert low chronic toxicity and no acute effects. These effluents contain low concentrations of metals and other priority pollutants. In addition, they alter and produce temporal fluctuations in stream hydrology, suspended solids, and other conventional water quality parameters (e.g., alkalinity, hardness, conductivity). Such alterations in stream conditions, which often are overlooked in studies with more toxic effluents, might well potentiate toxicant stress or physically affect habitat quality. These situations likely are not uncommon, and it may be necessary to improve chronic testing, develop chronic methods for TRE, and supplement end-of-the-pipe biomonitoring with ambient chemical, ecological, and toxicological studies. Accordingly, appropriate biological criteria should be defined and incorporated into assessment strategies for aquatic systems impacted by hazardous wastes. In addition, biomarker studies show high potential for screening aquatic contaminants for possible human health effects (McCarthy et al., 1989; Sandhu and Lower, 1989). It is also important to emphasize that the biomonitoring procedures used with NPDES effluents are applicable to other point sources, including streams from hazardous waste sites.

ACKNOWLEDGMENTS

We gratefully acknowledge our former project officers, coauthors, and other colleagues as named in those papers from this laboratory that have been included in this review. Their encouragement, guidance, and contributions have been invaluable. We are deeply appreciative for the many contributions made by Albert Westerman to the toxicological investigations and to Terry Short who has the major responsibility for the benthic invertebrate studies.

REFERENCES

Benoit, D.A., Puglisi, F.A., and Olson, D.L., 1982, A fathead minnow Pimephales promelas early life stage toxicity test method evaluation and exposure to four organic chemicals, Environ. Pollut. Ser. A, 28:189.

Birge, W.J., 1978, Aquatic toxicology of trace elements of coal and fly ash, in: "Energy and Environmental Stress in Aquatic Systems," J.H. Thorp and J.W. Gibbons, eds., DOE Symposium Series (CONF-771114) Washington.

Birge, W.J., and Black, J.A., 1981, In situ acute/chronic toxicological monitoring of industrial effluents for the NPDES biomonitoring program using fish and amphibian embryo-larval stages as test organisms. OWEP-82-001, U.S. Environmental Protection Agency, Washington.

Birge, W.J., Black, J.A., and Ramey, B.A., 1981, The reproductive toxicology of aquatic contaminants, in: "Hazard Assessment of Chemicals-Current Developments," J.Saxena and F.Fisher, eds., Academic Press, New York.

Birge, W.J., Black, J.A., and Ramey, B.A., 1986, Evaluation of effluent bio-monitoring systems, in: "Environmental Hazard Assessment of Effluents," H.L. Bergman, R.A. Kimerle, and A.W. Maki, eds., Pergamon Press, New York.

Birge, W.J., Black, J.A., Short, T.M., and Westerman, A.G., 1989, A comparative ecological and toxicological investigation of a secondary wastewater treatment plant effluent and its receiving stream, Environ. Toxicol. Chem., 8:437.

Birge, W.J, Black, J.A., and Westerman, A.G., 1979a, Evaluation of aquatic pollutants using fish and amphibian eggs as bioassay organisms, in: "Animals as Monitors of Environmental Pollutants," National Academy of Sciences, Washington.

Birge, W.J., Black, J.A., and Westerman, A.G., 1985, Short-term fish and amphibian embryo-larval tests for determining the effects of toxicant stress on early life stages and estimating chronic values for single compounds and complex effluents, Environ. Toxicol. Chem., 4:807.

Birge, W.J., Black, J.A., Westerman, A.G., and Hudson, J.E., 1979b, The effects of mercury on reproduction of fish and amphibians, in: "Biogeochemistry of Mercury in the Environment," J.O. Nriagu, ed., Elsevier/North Holland Biomedical Press, Amsterdam.

Birge, W.J., Black, J.A., Westerman, A.G., and Ramey, B.A., 1983, Fish and amphibian embryos--a model system for evaluating teratogenicity, Fundam. Appl. Toxicol., 3:237.

Birge, W.J., and Cassidy, R.A., 1983, Structure-activity relationships in aquatic toxicology, Fundam. Appl. Toxicol., 3:359.

Birge, W.J., Hudson, J.E., Black, J.A., and Westerman, A.G., 1978, Embryo-larval bioassays on inorganic coal elements and in situ biomonitoring of coal-waste effluents, in: "Surface Mining and Fish/Wildlife Needs in the Eastern United States," D.E. Samuel, J.R. Stauffer, C.H. Hocutt, and W.T. Mason, eds., FWS/OBS-78/81, Fish and Wildlife Service, U.S. Department of the Interior, Washington.

Dave, G., Damgaard, B., Grande, M., Martelin, J.E., Rosander, B., and Viktor, T., 1987, Ring test of an embryo-larval toxicity test with zebrafish (Brachydanio rerio) using chromium and zinc as toxicants, Environ. Toxicol. Chem., 6:61.

Finney, D.J., 1971, "Probit Analysis," 3rd ed., Cambridge University Press, New York.

230

Horning, W.B., and Weber, C.I., 1985, Short-term methods for estimating the chronic toxicity of effluents and receiving waters to freshwater organisms. EPA 600/4-85-014, U.S. Environmental Protection Agency, Cincinnati.

Karr, J.R., Fausch, K.D., Angermeier, P.L., Yant, P.R., and Schlosser, I.J., 1986, Assessing biological integrity in running waters--a method and its rationale, Illinois Natural History Survey, Special Publication 5.

McCarthy, J.F., Jimenez, B.D., Shugart, L.R., and Sloop, F.V., Biological markers in animal sentinels. In this volume.

Michael, G.Y., Egan, J.T., and Grimes, M.M., 1989, Colorado's biomonitoring regulation: a blueprint for the future, Journal Water Pollut. Control Fed., 61:304.

Mount, D.I., Steen, A.E., and Norberg-King, T.J., 1985, Validity of effluent and ambient toxicity testing for predicting biological impact on Five Mile Creek, Birmingham, Alabama, EPA 600/8-85/015, U.S. Environmental Protection Agency, Duluth.

Norberg-King, T.J., and Mount, D.I., 1986, Validity of effluent and ambient toxicity tests for predicting biological impact, Skeleton Creek, Enid, Oklahoma, EPA 600/3-86-006, U.S. Environmental Protection Agency, Duluth.

Peltier, W.H., 1978, Methods for measuring the acute toxicity of effluents to aquatic organisms, EPA/600-14-78-012, U.S. Environmental Protection Agency, Cincinnati.

Peltier, W.H., and Weber, C.I., 1985, Methods for measuring the acute toxicity of effluents to freshwater and marine organisms, 3rd ed., EPA/600/4-85/013, U.S. Environmental Protection Agency, Cincinnati.

Pickering, Q.H., and Gast, M.H., 1972, Acute and chronic toxicity of cadmium to the fathead minnow Pimephales promelas, J. Fish. Res. Board Can., 29:1099.

Sandhu, S.S., and Lower, W.R., 1989, In situ assessment of genotoxic hazards of environmental pollution, Toxicol. Ind. Health., 5:73.

Stephan, C.E., Mount, D.I., Hansen, D.J., Gentile, J.H., Chapman, G.A., and Brungs, W.A., 1985, Guidelines for deriving numerical national water quality criteria for the protection of aquatic organisms and their uses, PB85-227049, U.S. Environmental Protection Agency, Washington.

U.S. Environmental Protection Agency, 1980, Ambient water quality criteria for cadmium, EPA 440/5-80-025, U.S. Environmental Protection Agency, Washington.

Weber, C.I., Peltier, W.H., Norberg-King, T.J., Horning, W.B., Kessler, F., Menkedick, J., Neiheisel, T.W., Lewis, P.A., Klemm, D.J., Pickering, Q.H., Robinson, E.L., Lazorchak, J., Wymer, L., and Fryberg, R., 1988, Short-term methods for estimating the chronic toxicity of effluents and receiving waters to freshwater organisms, Draft, EPA-600/4-88-000, U.S. Environmental Protection Agency, Cincinnati.

INTEGRATED CHEMICAL, PATHOLOGICAL AND IMMUNOLOGICAL STUDIES TO ASSESS

ENVIRONMENTAL CONTAMINATION

B.A. Weeks, R. J. Huggett, and W.J. Hargis, Jr.

Division of Chemistry and Toxicology
Virginia Institute of Marine Science
The College of William and Mary
Gloucester Point, Virginia 23062

INTRODUCTION

Pollution problems are common on the North Atlantic coast, stemming from the input of metals, petroleum-derived hydrocarbons, insecticides and herbicides, as well as many other anthropogenic and natural chemicals. Many of these substances are known to be carcinogenic or mutagenic. The occurrence of toxicants in finfish and shellfish habitats is known to produce acute effects, and in addition may predispose organisms to tissue damage, neoplasms and environmentally induced diseases (Sindermann, 1979).

Among the most contaminated estuarine waters found to date is the Elizabeth River (VA), a three-branched tributary of the lower James River. It has been the site of civilian and military shipbuilding, shipping and shoreside commerce, and associated manufacturing and processing industries for almost 400 years. Figure 1 shows the affected areas of the Elizabeth River.

Concentrations of polynuclear aromatic hydrocarbons (PAHs) have been found to exceed thousands of parts per million in the sediments of some areas of the river (Bieri et al., 1986; Hargis et al., 1984). This appears to represent the highest concentration of PAHs in any estuary in the world. Biological data derived from the collection and analysis of organisms collected at these sites reveal that the natural community has been affected and that the impacts are greatest in the areas where the PAHs are in highest concentration (Huggett et al., 1987).

The purpose of this paper is to assess environmental contamination in the Elizabeth River subestuary and to evaluate the effects of pollution on the health of marine and estuarine organisms. Survey data will be examined to assess the distribution of contamination within the river as indicated by sediment loads, and the bioavailability of PAHs in the system. In addition, the results of immunological assay techniques and gross and histopathological studies will be examined to determine the real and potential hazards associated with chemical contamination in the region.

In Situ Evaluations of Biological Hazards of Environmental Pollutants
Edited by S. S. Sandhu *et al.*
Plenum Press, New York, 1990

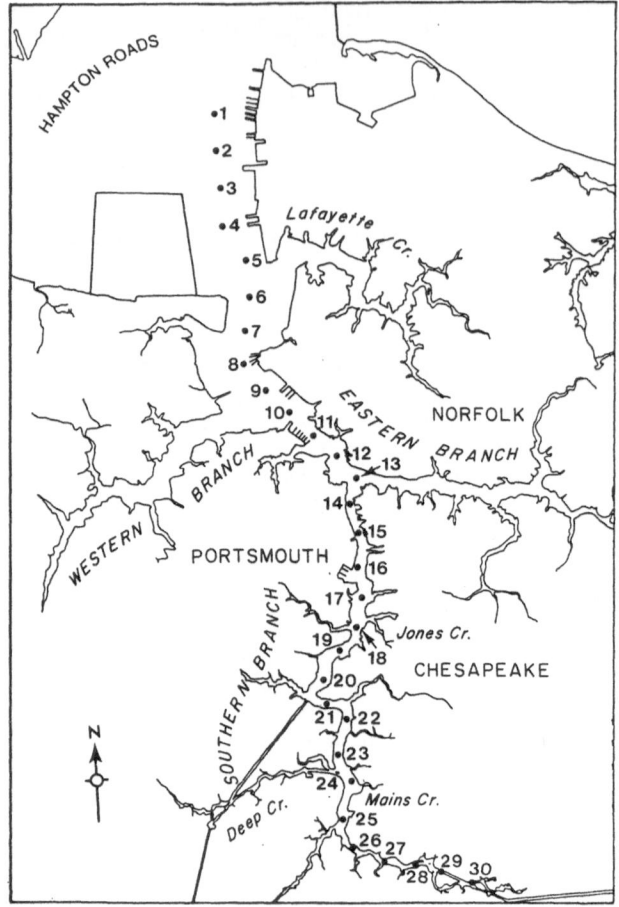

Figure 1. Reference view of Elizabeth River with distance from the
river mouth shown in kilometers.

METHODS AND RESULTS

Analysis of Contamination Load

Surface sediment samples collected at 28 sites along the Elizabeth
River and its branches were freeze-dried and homogenized. The samples were
extracted with dichloromethane to separate organic chemicals from the sediments. These crude extracts were then subjected to gel permeation chromatography to remove organic polymers. The cleaned extracts, containing the
organic compounds of interest, were further separated by high performance
liquid chromatography (HPLC), which yielded fractions containing polar or
polarizable compounds, most of which were aromatic. These fractions were
analyzed by either gas chromatography (GC) or gas chromatography-mass spectrometry (GC-MS). Recovery and precision of the extraction technique were
estimated from analyses of PAH standards added to multiply extracted sediment. The identification of all compounds was based on aromatic retention
indices and mass spectra of reference compounds.

Hundreds of compounds were detected by gas chromatography in the aromatic fraction of some samples. About half of the resolved concentration
sum was accounted for by 14 unsubstituted PAHs, including phenanthrene,
fluoranthene, pyrene, benzo(b)fluorene, benzo(c)phenanthrene, benz(a)-

Table 1. Concentration of Selected PAHs in Surface Sediments from the Elizabeth River (ng/g dry weight)

Sample[a] code	Phe[b]	Fla	Pyr	BbF	BaA	Chr	BFls	BeP	BaP	IPy	BghiP
06	86	290	250	89	93	160	210	90	84	21	28
07	110	300	260	82	79	140	150	57	53	n.d.[c]	n.d.[c]
08	180	460	410	160	190	340	590	280	260	100	76
09	200	840	880	470	320	470	1,000	420	480	360	350
10	130	320	440	160	150	290	870	380	380	260	230
11	410	860	960	290	350	590	1,000	480	520	200	110
12	580	1,500	1,400	530	560	970	1,700	740	740	280	170
13	670	1,900	1,800	720	880	1,400	2,600	1,200	1,200	550	500
14	750	2,200	2,000	900	840	1,400	2,700	1,200	1,200	660	560
15	760	2,200	2,800	1,200	1,000	1,700	4,100	1,300	1,700	940	620
16	2,300	5,500	4,600	2,000	1,900	3,200	4,900	2,000	2,100	730	340
17	950	3,800	3,600	1,500	1,500	3,500	6,300	2,800	2,600	1,400	750
18	710	2,600	2,000	760	940	1,700	2,400	970	1,000	200	84
19	25,000	42,000	28,000	12,000	11,000	19,000	17,000	6,300	8,700	2,100	1,600

[a] Numbers correspond to kilometers upstream from the mouth of the estuary.

[b] Abbreviations: Phe = phenanthrene; Fla = fluoranthene; Pyr = pyrene; BbF = benzo(b)fluorene; BaA = benz(a)anthracene; Chr = chrysene; BFls = benzofluoranthenes (j,b,k); BeP = benzo(e)pyrene; BaP = benzo(a)pyrene; IPy = indenopyrene; BghiP = benzo(ghi)perylene.

[c] n.d. = not detected.

anthracene, chrysene, j, b and k isomers of benzofluoranthene, benzo(e)-pyrene, benzo(a)pyrene, indeno(1,2,3-cd)pyrene and benzo(ghi)perylene. All of these PAHs are combustion products and are components of wood-treatment products such as creosote. Many other compounds, mostly hydrocarbons, a few heterocyclic compounds, and some chlorinated hydrocarbons, were also identified. Low levels of naphthalenes and phenanthrenes indicate a small contribution from petroleum. Levels of selected PAHs are given by station in Table 1 (Huggett et al., 1987). The high PAH levels found in the sediments of the middle reaches of the Southern Branch (stations 16-19) suggest that abandoned and operational wood treatment plants situated nearby are contributory factors.

The concentrations of PAHs in the Elizabeth River sediments are among the highest found anywhere in the nation (Table 2; Bieri et al., 1986). Field data indicate that there is reason to be concerned about the ecological effects of these high concentrations in the Elizabeth River.

Bioavailability of PAHs

The Elizabeth River subestuary, contributing to commercially valuable seed oyster and blue crab spawning sites, is quite sensitive from an ecological standpoint. The biological impact of PAH contamination on the area's aquatic inhabitants must be carefully evaluated.

Oysters (_Crassostrea virginica_) collected in September 1983 from a reference site in the Rappahannock River were transplanted to five stations in the Elizabeth River. At weekly intervals, sub-populations were removed from each site and depurated for 24 h to reduce the PAHs passing through the digestive system. Tissues and fluids were homogenized, freeze dried and extracted with methylene chloride. Oyster extracts were reduced in volume and cleaned by gel permeation chromatography before HPLC separation and GC analysis. Uptake of PAHs into oyster tissues was rapid, increasing linearly for the first 4 weeks of exposure at the 17-km site (station 17). Concentrations decreased at most stations after 6 weeks exposure and then increased again. Analysis of the data for individual compounds shows considerable fluctuation in levels with time, particularly at some locations. The most consistently abundant compounds identified were benzofluoranthene, benzo(a and b)fluorene, benzo(a)pyrene, fluoranthene, pyrene, benz(a)-

Table 2. Concentration (ug/g dry wt) of selected PAHs in sediment from some of the most polluted areas in continental United States

Area BaP	Phe[a]	Fla	Pyr	BaA	Chr	BFls	BeP	
Charles River Massachusetts	3	13	12	21[b]		---	33	---
Elliot Bay (Pier 54) Washington	7	8	11	7	6	5	4	2
Commencement Bay (Hylebo Waterway) Washington	0.4	2	2	2	2	3	1	0.5
Newton Creek New York	14	10	7	6	3	---	1	1
Elizabeth River Virginia	25	42	28	11	19	17	6	9

[a] Phe = phenanthrene; Fla = fluoranthene; Pyr = pyrene; BaA = benz(a)anthracene; Chr = chrysene; BFls = benzofluoranthenes (j, b, K); BeP = benzo(e)pyrene; BaP = benzo(a)pyrene.

[b] Combined values for BaA and Chr.

anthracene and chrysene/triphenylene. Concentrations of benzo(a)pyrene ranged from 2 ug/g at the 17-km site to 0.1 ug/g near the river mouth.

Cellular Immune Function

The cellular immune system was found to be compromised in fish captured in the Elizabeth River and in fish exposed to Elizabeth River sediments in the laboratory (Weeks and Warinner, 1984; Weeks et al., 1986, 1987; Warinner et al., 1988). Spot (Leiostomus xanthurus), hogchoker (Trinectes maculatus), and flounder (Paralichthys dentatus) were captured by bottom trawl from the reach of the Elizabeth River most contaminated by PAHs and from a site in the York River known to be relatively uncontaminated by PAHs. Macrophages were isolated from the anterior portion of the kidney by maceration and centrifugation on Percoll density gradients. Several in vitro functions of the purified macrophages were evaluated.

Phagocytic efficiency in spot and hogchoker was measured by incubating the macrophages with formalin-killed Escherichia coli and determining microscopically the proportion of macrophages which contained intracellular bacteria. The phagocytic activity of macrophages from Elizabeth River fish was found to be significantly lower than the York River controls (Table 3).

A similar decrease was observed in the ability of macrophages from Elizabeth River spot and hogchoker to migrate specifically toward a chemical stimulus (chemotaxis) (Table 3). Chemotactic activity was measured in a double-chambered apparatus in which macrophages were allowed to migrate from the upper chamber through a Nuclepore filter to the lower chamber, which contained E. coli as the chemotactic stimulus. The macrophage chemotactic and phagocytic activities of Elizabeth River fish returned to normal after the fish were held in clean water for 3 weeks.

The pinocytic function of kidney macrophages, a measure of the ability of the cells to take up liquid solutions, was measured using a colorimetric assay of ingested neutral red dye. The pinocytic activity of macrophages

Table 3. Effect of Environment on Macrophage Activity

	Control River	Elizabeth River
Chemotaxis[a] % Cells Migrating (at 90 min)		
Spot	55 ± 5	$33 + 2$
Hogchoker	85 ± 5	56 ± 5
Phagocytosis[a] % Phagocytosis (at 120 min)		
Spot	74 ± 5	19 ± 3
Hogchoker	88 ± 4	32 ± 5
Pinocytosis[a] Neutral Red Uptake (ug/10^6 cells) (at 120 min)		
Spot	1.58 ± 0.12	1.48 ± 0.27
Hogchoker	0.56 ± 0.04	1.25 ± 0.20
Chemiluminescence[b] Peak Amplitude (cpm)		
Spot-exposed to sediments in the wild	350,000	15,000
Spot-exposed to sediments in the laboratory	540,000	220,000

[a] Each value represents the mean $\pm$ SEM of 4 experiments using 6-10 pooled fish specimens per experiment.

[b] Representative experiment.

from Elizabeth River spot did not differ from control values, but there was a two-fold increase in pinocytic activity in Elizabeth River hogchoker (Table 3).

The chemiluminescent (CL) response of fish phagocytes, which is associated with the formation of microbicidal reactive oxygen intermediates, was found to vary with the species tested and the phagocytic stimulus used. Macrophages from spot captured in the contaminated reaches of the Elizabeth River were markedly deficient in generating a zymosan-induced CL response compared with York River reference fish. Spot exposed in the laboratory to contaminated sediments showed a similar decrease in the CL response (Table 3).

Pathobiology

Teleosts were collected by trawl in the Elizabeth River. Stations, arranged along the river to encompass the sites where sediments are most severely contaminated as well as those with less pollution, were sampled monthly for 1 full year in 1983-1984. Other riverwide collections were made in 1982, 1985, 1986, and 1988. The nearby Nansemond River, the lower James River and the York River provided reference collections. Over 95,000 specimens of 49 species were collected from the Elizabeth River in the year-long sampling effort. Over 17,000 specimens of 20 species were taken from the Nansemond River during this same period.

External Pathology. A careful visual examination was made of 71,262 individuals of eight species, taken during 11 of the 1983-1984 monthly

Table 4. External Lesions Observed in Elizabeth River Collections, October-December 1983; April-November 1984

Lens Cataracts

Species	Total Examined	Number with Lesions	% of Total
Spot (_Leiostomus xanthurus_)	42,434	1,254	2.96
Atlantic Croaker (_Micropogonias undulatus_)	8,475	403	4.76
Weakfish (_Cynoscion regalis_)	5,980	184	3.08
Spotted Hake (_Urophycis regia_)	2,925	47	1.61
Gizzard Shad (_Dorosoma cepedianum_)	46	1	2.17

Fin Erosion (Fin Rot)

Spot	42,434	7	0.02
Atlantic Croaker	8,475	25	0.29
Weakfish	5,980	15	0.25
Hogchoker (_Trinectes maculatus_)	10,744	178	1.66
Oyster Toadfish (_Opsanus tau_)	631	40	6.33

Ulcerations

Spotted Hake	2,925	23	0.78
Red Hake (_Urophycis chuss_)	27	1	3.70

cruises. Of these specimens, 3.0% bore one or more of three externally visible lesions including lens cataracts, fin erosion and integumental ulcerations. No lesions were observed in the Nansemond River reference collections. Table 4 shows the prevalence of the three external lesions in these species (Hargis et al., 1988).

Examination of the collections on a station-by-station basis showed that the prevalence of each of the lesion types was positively correlated with contamination by PAHs. When the collections were grouped into five downriver sites and five upriver sites, the difference in the prevalence of all three lesions was even more pronounced. At some stations, the prevalence of cataracts approached 100% in some of the larger size classes of weakfish and Atlantic croaker (Hargis et al., 1988). All three externally visible lesions were found in spot collected at station 20. Control animals found in the York River were completely free of such lesions (Hargis et al., 1984). Fewer individuals from each species were collected upriver, especially at station 20 and nearby stations, than at downriver stations.

Histopathology. Specimens of weakfish, Atlantic croaker, spot, toadfish and hogchoker collected during 1982, 1983, 1984, 1985, and 1986 from the Elizabeth and the Nansemond Rivers were dissected. Gills, livers, kidneys, and eyes were processed histologically and examined by light microscopy. Results of examinations of the gills revealed four primary types of gill lesions: (1) ballooning dilatation of lamellae, (2) hypertrophic responses of lamellar and filament cells, (3) hyperplastic growths of lamellar and filament cells and (4) growth deformities or developmental aberrations of lamellae and filaments. All four types of lesions occurred in all species and at all stations, including those in the reference Nansemond River. The prevalence and mean severity of the lesions were lowest in the Nansemond River collections and highest in those from the Elizabeth River. In the Elizabeth River samples, prevalence and mean severity of ballooning dilatation, hypertrophy and hyperplasia were highest at stations 20 and 23, where sediments are severely contaminated by creosote-related PAHs. The prevalence of growth deformities was highest at station 20. However, the severity of the deformities was highest in the downriver stations, because more toadfish (the species with highest prevalence and mean severity of this lesion) were taken at the cleaner downriver station 10 than at either station 20 or station 23 (Hargis and Zwerner, 1988b).

Internal lesions of the eye, such as enlargement of the choroid rete, and retinal, lens and corneal abnormalities, have also been directly studied (Hargis and Zwerner, 1988a). Results showed these ocular lesions to be directly correlated with the contamination levels in the sediments at station 20 in the Elizabeth. In contrast to results of the gill lesion study, which found some of each type of branchial abnormality in the Nansemond as well as in the Elizabeth River specimens (though in lesser number and severity), Nansemond River fishes exhibited no such eye abnormalities. These findings agree with the results of immune function assays, and closely resemble the pattern of external lesion distribution. Further, preliminary results of histopathological analyses of liver and kidney tissues appear to confirm the distributional pattern of internal and external lesions (Hargis and Zwerner, unpublished data).

CONCLUSION

As a result of these studies, both external and internal lesions as well as immune function levels have been positively related to those stations in the Elizabeth River where sediments are most heavily contaminated with PAHs. There is little question that those sediment-contained contaminants have a marked negative influence on the health of exposed finfishes. Additional long-term exposure studies are needed to determine the sediment contamination levels necessary to cause effects in benthic species, and to refine estimates of chronic toxicity levels.

ACKNOWLEDGMENTS

We thank our colleagues at the Institute, D.E. Zwerner, J.E. Warinner, R.H. Bieri and M.E. Bender for their assistance with these studies. VIMS Contribution No. 1512.

REFERENCES

Bieri, R.H., Hein, C., Huggett, R.J., Shou, P., Slone, H., Smith, C.L., and Su, C.W., 1986, Polycyclic aromatic hydrocarbons in surface sediments from the Elizabeth River subestuary, Int. J. Environ. Anal. Chem. 26: 97-113.

Hargis, W.J., Jr., Colvocoresses, J.A., Zwerner, D.E., Thoney, D.A., and Warinner, J.E., 1988, Pathology of three external lesions of finfishes from the polluted Elizabeth River, VA. Abstract of a paper presented at the 9th Annual Meeting of the Society of Environmental Toxicology and Chemistry (SETAC), November 13-17, 1988, Arlington, VA.

Hargis, W.J., Jr., Roberts, M.H., Jr., and Zwerner, D.E., 1984, Effects of contaminated sediments and sediment-exposed effluent water in an estuarine fish: Acute toxicity, Mar. Environ. Res. 14:337-354.

Hargis, W.J., Jr., and Zwerner, D.E., 1988a, Effects of certain contaminants on eyes of several estuarine fishes, Mar. Environ. Res. 24:265-270.

Hargis, W.J., Jr., and Zwerner, D.E., 1988b, Some histologic gill lesions of several estuarine finfishes related to exposure to contaminated sediments: A preliminary report, in: "Understanding the Estuary," M.P. Lynch and E. C. Krome, eds., Chesapeake Research Consortium, Gloucester Point, VA.

Huggett, R.J., Bender, M.E., and Unger, M.A., 1987, Polynuclear aromatic hydrocarbons in the Elizabeth River, Virginia, in: "Fate and Effects of Sediment-Bound Chemicals in Aquatic Systems," K.L. Dickson, A.W. Maki, and W.A. Brungs, eds., Pergamon Press:New York.

Sindermann, C.J., 1979, Pollution-associated diseases and abnormalities of fish and shell fish: a review, Fish. Bull. 76:717-749.

Warinner, J.E., Mathews, E.S., and Weeks, B.A., 1988, Preliminary investigations of the chemiluminescent response in normal and pollutant exposed fish, Mar. Environ. Res. 24:281-284.

Weeks, B.A., Keisler, A.S., Warinner, J.E., and Mathews, E.S., 1987, Preliminary evaluation of macrophage pinocytosis as a technique to monitor fish health, Mar. Environ. Res. 22:205-213.

Weeks, B.A., and Warinner, J.E., 1984, Effects of toxic chemicals on macrophage phagocytosis in two estuarine fishes, Mar. Environ. Res., 14:327-335.

Weeks, B.A., Warinner, J.E., Mason, P.L., and McGinnis, D.S., 1986, Influence of toxic chemicals on the chemotactic response of fish macrophages, J. Fish Biol. 28:653-658.

USE OF WILDLIFE FOR ON-SITE EVALUATION OF BIOAVAILABILITY AND

ECOTOXICITY OF TOXIC SUBSTANCES FOUND IN HAZARDOUS WASTE SITES

Ronald J. Kendall[1], Jeanne M. Funsch[2], and Catherine M. Bens[1]

[1]Institute of Wildlife and Environmental Toxicology
Clemson University
Clemson, South Carolina 29632

[2]Environmental Toxicology International, Inc.
Plaza 600 Building
6th and Stewart, Suite 700
Seattle, Washington 98101

INTRODUCTION

Many scientists believe that human beings represent the ultimate senti-nel species of a toxic exposure to hazardous waste site contaminants. How-ever, few research methods have been developed that support a causal rela-tionship between exposure to hazardous waste site chemicals in situ and latent or delayed adverse health effects in humans (Grisham, 1986). Wild-life toxicology, being the investigation of the effects of environmental contaminants on the reproduction, health and well being of animals living in a wild, undomesticated state, represents an area of research which holds great promise in evaluating hazardous waste site issues (Kendall, 1982, 1988). The study of wildlife inhabiting or using hazardous waste sites could reduce the uncertainty inherent in risk assessments of contaminated areas. Monitoring wildlife in situ could provide an early warning system to potential human health effects, as well as reflect problems in the nat-ural environment.

The use of sentinel wildlife species represents an area of rapid re-search and development. A sentinel species would be identified based on sensitivity to environmental contaminants and could thus be used to predict impacts on other relevant species. Wild mammal and bird species have been used in situ previously to investigate environmental contamination (Hanson and Kornberg, 1956; Willard, 1960; Henny, 1972; Domby et al., 1977; Sileo et al., 1977; Halford and Markham, 1978; Lower and Plewa, 1978; Ohlendorf et al., 1978, 1988; Cadwell et al., 1979; Craig et al., 1979; Galluzzi, 1981; Halford et al., 1981, 1982; Anthony and Kozlowski, 1982; Lipsky and Galluzzi, 1982; Markham and Halford, 1982; Rowley et al., 1983; Henny et al., 1985a; Leonard et al., 1985; McBee, 1985; Arthur et al., 1987; Clark, 1987; Hoffman et al., 1988). The growing database of the effects of envi-ronmental contaminants on these wildlife species enhances their usefulness as predictors of health threats to humans.

The use of wildlife for in situ evaluation of hazardous waste site con-taminants has several important advantages. First, short-term laboratory

In Situ Evaluations of Biological Hazards of Environmental Pollutants
Edited by S. S. Sandhu *et al.*
Plenum Press, New York, 1990

241

tests are concerned with an acute exposure to high doses of contaminants. These tests do not adequately reflect the real-life situation of chronic exposure to low doses as is evaluated in situ. This is particularly true in sites with mixed contaminants where synergistic or additive effects of chemicals could be occurring. Second, laboratory tests are expensive and time-consuming and the extrapolation of the results to field situations is often unreliable or difficult to verify. Third, for most wildlife species whose lifespans are much shorter than man's, the latent period for cancer development or other effects is much shorter. Consequently, dangerous contaminants in the environment can be identified earlier. Fourth, wildlife species may be exposed to environmental contaminants at hazardous waste sites, but do not indulge in other activities (i.e., smoking, occupational exposure) that have so often confounded results of epidemiological studies.

Further research is needed in the development of bioindicators, sensitive endpoints used to predict detrimental effects of an agent on biological organization (McCarthy et al., 1988). Categories of bioindicators may include ecotoxicological, physiological, biochemical, chemical residues, immunotoxic, genotoxic, radiation, and behavioral endpoints. The use of bioindicators provides a direct measure of detrimental effects of toxic substances on health, reproduction and survival in wildlife inhabiting or using contaminated areas. For instance, biochemical endpoints may include such measurements as inhibition of cholinesterase activity in brain tissue associated with organophosphate insecticide exposure (Robinson et al., 1988), induction of metallothionein with exposure to some toxic metals (Kendall and Scanlon, 1982), and disturbances of glutathione, metallothionein, heme oxygenase, and cytochrome P-450 in relation to metal ion exposure (Eaton, et al., 1980).

An integrated approach has been suggested by Kendall (1982, 1988) to assess environmental contaminant effects on wildlife. This approach combines field data and laboratory observations with the use of appropriate bioindicator endpoints to analyze the bioavailability and ecotoxicity of toxic substances. With an integration of chemical characteristic information, laboratory toxicity data and field toxicity data, a method of evaluating hazardous waste sites in situ using wildlife can be proposed (Figure 1). This evaluation can be applied to hazardous waste site risk assessment.

In this approach, an initial site assessment is used to identify contaminant concentrations and locations, as well as habitat type and wildlife species present on site. With knowledge of the physical and chemical properties of on site contaminants and species present, appropriate bioindicators can be selected for an in situ study. Endpoints can then be used as a direct measure of bioavailability and toxicity necessary for risk assessment and management.

An approach that integrates this information will expand extrapolation of testing procedures, encourage the development of a broader data base on the effects of hazardous wastes, and enhance the prioritization of hazardous waste sites. Long-term wildlife monitoring in comparison to baseline data can be used to evaluate the success of remedial efforts. An integrated approach can allow a more timely and accurate understanding of the hazard associated with a particular site. The sooner the hazard to either human health or the environment is identified, the more rapidly and reliably cleanup operations can be put into place.

METHODS AND BIOINDICATOR ENDPOINTS

In recent years wildlife has been increasingly used to evaluate contaminants in the environment, particularly pesticides (Ohlendorf et al., 1978; Grue et al., 1982, 1983; Robinson et al., 1988; Henny et al., 1985a,

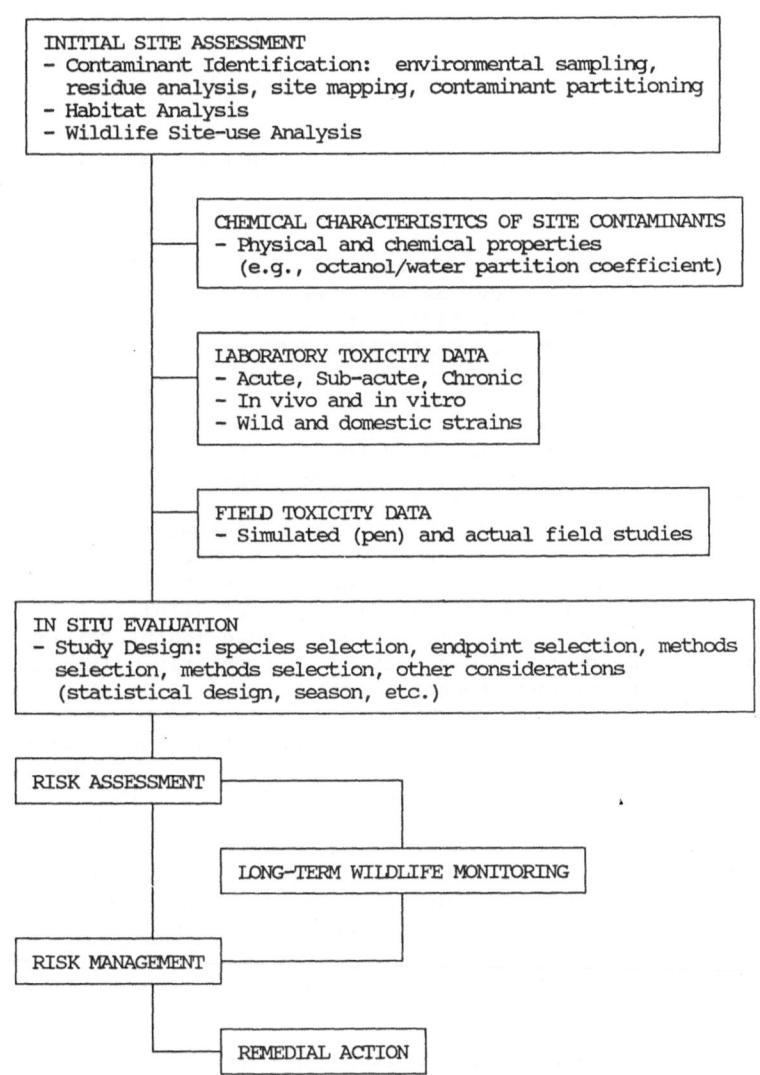

INITIAL SITE ASSESSMENT
- Contaminant Identification: environmental sampling, residue analysis, site mapping, contaminant partitioning
- Habitat Analysis
- Wildlife Site-use Analysis

CHEMICAL CHARACTERISITCS OF SITE CONTAMINANTS
- Physical and chemical properties
  (e.g., octanol/water partition coefficient)

LABORATORY TOXICITY DATA
- Acute, Sub-acute, Chronic
- In vivo and in vitro
- Wild and domestic strains

FIELD TOXICITY DATA
- Simulated (pen) and actual field studies

IN SITU EVALUATION
- Study Design: species selection, endpoint selection, methods selection, methods selection, other considerations (statistical design, season, etc.)

RISK ASSESSMENT

LONG-TERM WILDLIFE MONITORING

RISK MANAGEMENT

REMEDIAL ACTION

Figure 1. An integrated approach to the use of wildlife for assessing hazardous waste sites in situ.

1985b; Brewer et al., 1988a, 1988b; Hooper, 1988). Methods in wildlife capture, handling and measurement have been well developed by the fields of wildlife biology and wildlife management (Schemnitz, 1980; Giron Pendleton et al., 1987). The successful use of wildlife in evaluating site contamination depends on analyzing the contaminants present, selecting appropriate species to be monitored and endpoints to be measured, and choosing the best methods of collection and observation.

Species

The species of choice will depend upon the objectives of the evaluation and the characteristics of the site. The species used may be one or more indigenous to the site, can be one or more brought to the site and released or can be species held on site in field pens. Species can also be attracted to the site by identifying and modifying a limiting factor (e.g., the placement of artificial nesting boxes).

Indigenous species can be identified by an initial site inspection through standard census techniques (Kendeigh, 1944; Emlen, 1971, 1974; Schemnitz, 1980). Marking and recapture techniques can be used to identify individuals and to determine population size (Seber, 1982). Comparison of species richness and density indexes with those of control sites of similar habitat and other resources provides an initial assessment of toxic impacts on wildlife and information on potential species sensitivity. Often, the species chosen for study may be the species most easily monitored or the species offering a sufficient sample size. Extremely sensitive species may no longer be represented on site.

The species found on site will be a function of habitat and other available resources. The species chosen will depend upon the trophic level and food chain of interest. If site contaminants are water soluble or have been measured in water samples and ground water contamination impacts are a major consideration, the species chosen will be directly or indirectly exposed to the aqueous environment (e.g., aquatic invertebrates, shellfish, fish, waterfowl, herons and muskrats). Terrestrial wildlife species would assess soil contamination. Identifying entire food chains for evaluation (e.g., earthworms, shrews and hawks) may be particularly valuable in evaluating food chain transfer and bioaccumulation.

While indigenous species offer the advantage of having a potential lifetime and multimedia exposure, the introduction of a species on site provides the researcher with a greater measure of control. The species chosen can be one in which toxicological information is available from previous field and laboratory research or one in which capture and observation is facilitated. In field-pen studies, animals with known histories can be placed on site and the exposure route controlled. Pens placed off the ground limit exposure to airborne contaminants. Pens placed on the ground can allow for both direct substrate contact and potential ingestion of contaminated food and water sources. The duration of exposure can also be controlled in this approach.

The enhancement of a native population can often be achieved by increasing a limited resource. Nest sites are one of the most common limiting resources for avian species. The placement of nest boxes or platforms can attract such species as European starlings (Sturnus vulgaris) (Robinson et al., 1988), common barn owl (Tyto alba) (Marti et al., 1979), great tit (Parus major) (Greenwood et al., 1979), and eastern bluebird (Sialia sialis) (Mitchell et al., 1953). Enhanced populations can provide a statistically significant number of individuals to study and can facilitate observation and access to both young and adults.

Organisms higher in the food chain are often the best subjects for environmental monitoring because many toxic compounds bioaccumulate. Birds are an excellent choice for biological monitoring. They are easy to observe, sensitive to many toxic compounds, have well-known life histories, and are abundant at many trophic levels in almost all ecosystems (Karr, 1987). Studies of piscivorous birds, such as the great blue heron (Ardea herodias), and raptors, such as the red-tailed hawk (Buteo jamaicensis), have documented the biomagnification and food chain transfer of compounds like organophosphates, organochlorines and PCBs (Henny, 1972; Galluzzi, 1981; Henny et al., 1985a; Henny et al., 1985b; Henny et al., 1987).

Because of their dietary habits and widespread occurrence, great blue herons have a high potential for exposure to various environmental contaminants (Riley et al., 1983; Calambokidis et al., 1985; Henny et al., 1985a). Studies of the great blue heron in western Oregon and British Columbia have provided information on the population and reproductive status, and nesting and feeding requirements of this species in the Northwest (Pratt and Winkler, 1985; Gibbs et al., 1987). Herons nest and feed in urban areas, and colonies are easy to observe. Studies from British Columbia (Simpson, 1984) indicate that this subspecies is resident, with their movements in winter restricted to within 20 to 40 km of the breeding grounds. Research is now underway evaluating the use of salvaged juveniles to monitor the contaminant loads in a breeding colony (Norman, 1988). Salvaged juveniles are those young that fall or are pushed out of the nest. These nestlings usually die from exposure and are often eaten by scavengers. Great blue herons feed mainly on fish and other aquatic animals while feeding young, and therefore juveniles are indicators of the bioavailability of marine contaminants. Each of these factors contributes to the appropriateness of the great blue heron as a biological sentinel.

Raptors such as the red-tailed hawk are also high in the food chain and have been shown to be vulnerable to secondary poisoning (Henny et al., 1985b; Henny et al., 1987). The red-tailed hawk is an opportunistic hunter, feeding on rodents, birds, snakes, and insects (Heintzelman, 1979). It is widely distributed across North America and is found in many ecological biomes (Heintzelman, 1979). Trapping methods for this species and other raptor species are documented (Bloom, 1987) and there is a growing database of its biochemical response to environmental contaminants such as pesticides (Hooper, 1988; Hooper et al., 1989). By monitoring contaminants in well-studied raptors such as the red-tailed hawk, a measure of the bioavailability of terrestrial contaminants is possible.

The European starling is another avian species which may be useful as a biological indicator. The present population of starlings in North America is estimated to be at least 200 million. The starling is common from arctic Canada to the subtropics of Mexico. Starlings are primarily grassland feeders, and usually forage within 200 m of their nest, but tend to forage closer with adequately available food sources. During the breeding season, both the adult and nestling diet consists almost entirely of invertebrates obtained from the surface or from the upper few centimeters of soil (Feare, 1984). Nest boxes are readily accepted as nest sites by starlings (Powell and Gray, 1980; Grue et al., 1982; Robinson et al., 1988). The readiness of starlings to accept nest boxes allows for the concentration of a large sample size in a relatively small area. Starlings are very tolerant of clutch handling and nest box alteration (Grue and Hunter, 1984). For these reasons, the starling is an ideal species for the study of the effects of environmental contaminants on passerines.

Mammals are also useful sentinel species (Scanlon et al., 1983; Ringer, 1988). Species such as the deer mouse (Peromyscus maniculatus) and the vole (Microtus sp.) are found commonly in open grassy areas. Mice of the genus Peromyscus are one of the most widely distributed mammals in North America. Small mammals such as mice have small home ranges (Southern, 1979), making it relatively easy to estimate what contaminants they may have encountered. Most small mammal species burrow in the soil (Hayward and Phillipson, 1979), increasing the potential for exposure to various environmental contaminants. Small mammals have been used extensively in both laboratory (Olson and Hinsdill, 1983, 1984; Olson et al., 1984) and field experiments (Rowley et al., 1983; Watson et al., 1985; Clark, 1987; Block, 1988), providing a large database about their life histories and responses to chemical exposure. Field trapping methods and population estimation techniques are established for many small mammals (Schemnitz, 1980).

Observation/Collection Techniques

Techniques for the collection and observation of wildlife will depend upon the species selected. Extensive advancements in the field of radio-telemetry have made it possible to radio tag many wildlife species to verify site use by individuals. This method allows for the selection of only those individuals with a residence history in the site of interest (Amlaner and Macdonald, 1980; Eberhardt and Cadwell, 1984). Radio-telemetry allows individuals to be located repeatedly for collection and measurement or for observations of activity. The Wildlife Management Techniques Manual (Schemnitz, 1980) provides an extensive coverage of census, trapping and handling methods for many species. Tissue and biological sample collection techniques for residue, enzyme and other analyses are very specific and will not be discussed here.

Endpoints

Biological endpoints or biomarkers in wildlife can be used to assess environmental exposure and to predict ecological and human health effects (McCarthy et al., 1988). Potential endpoints are numerous in an in situ hazardous waste site analysis using wildlife and will depend upon site contaminants and sentinel species selection. Endpoints evaluated will reflect the toxicology of the on-site contaminants. It is ultimately the extent of integration between site contaminant assessment, wildlife exposure assessment and ecotoxicological, biological, chemical and biochemical endpoint analysis that determine the success of the hazard assessment. Some endpoints used to date in hazard assessment are discussed below.

Mortality. The ultimate endpoint of adverse exposure is death. Increased mortality has been found in voles (Microtus pennsylvanicus) in the immediate area of the Love Canal hazardous waste site (Rowley et al., 1983) and in birds inhabiting a DDT contaminated site at Wheeler National Wildlife Refuge (O'Shea et al., 1980). Mortality can be determined by systematic carcass searches on site or more indirectly through an evaluation of species diversity, density and age-class ratios. The on site diversity, density and age-class structure, when compared to a control site, can be used to identify the absence of key species whose presence on site is altered by the site contaminants. These indexes can also be used to assess the health of the population on site in terms of size and reproductive potential.

Due to confounding factors such as immigration, emigration, migration and predation, mortality is seldom the most sensitive measure of effect, although it is often used in pesticide application assessments. An additional factor is the difficulty in locating carcasses (Balcomb, 1986; Mineau and Peakall, 1987; EPA, 1988). Intoxication in wildlife may lead to behavior or physical disabilities which result in death occurring in such a way that finding the carcass is unlikely (e.g., in a burrow or under a bush).

Sudden mortality on site, however, may be an important indicator and guide to further investigation. It is important to note that death can occur off site and with an extended period of time after exposure. Delayed mortality will depend on the toxins wildlife are exposed to, in addition to the physiological state of the animals. The value of mortality as an important endpoint is thus reduced unless telemetry is used to track individuals.

Reproduction. Reproductive success is an extremely important measure of the health of a population. The avian nest box bioassay is an excellent tool for monitoring reproductive success (Kendall, 1982; Robinson et al., 1988). Nest boxes facilitate the measurement of reproductive parameters

with minimal disturbance and maximum ease in observation. Nest boxes can
be built to limit predation from the ground due to human contact with the
nest. Nest boxes can also be placed so that foraging on site is maximized,
reflecting a worst case exposure scenario.

Whether natural or artificial nest sites are examined, eggs and young
can be collected for residue analysis or for embryotoxic and teratogenic
endpoints analyses (Ohlendorf et al., 1986, 1988; Hoffman et al., 1988).
Eggs and eggshells can be collected for eggshell thickness and breaking
strength measurements (Bennett et al., 1988) and for residue analysis.
Adults and juveniles can be weighed and measured, blood drawn or tissues
collected at varying times throughout the breeding season. Food resources
can be verified through photographic record of adults arriving at the nest
with food or analysis of nestling esophageal content. Other reproductive
measurements may include: nesting attempts, clutch or brood size, number
of nestlings hatched, number of birds fledged, juvenile weight, adult atten-
tiveness, number of clutch attempts and gonadal development.

Physiological. Differences in physiological parameters such as body
weight, organ weight (Rowley et al., 1983), body measurements, body temperature, heart rate, fat content, and tissue histopathology (Malins et al.,
1987; Martineau et al., 1988) or the presence of physiological deformities
can be used as a measure of exposure to waste site contaminants. Expected
impairments and measurements will be a function of the basic toxicity of
the compounds present on site.

Biochemical. Biochemical changes can be used to assess xenobiotic exposure in wildlife. With exposure, changes in metabolic enzymes and pathways
may occur. For example, reductions in brain and plasma cholinesterase activity have been extensively used to assess exposure in wildlife to cholinesterase inhibiting compounds such as organophosphate pesticides (Grue et
al., 1983; Grue and Hunter, 1984; Robinson et al., 1988). These analyses
are especially indicative of exposure when integrated with fecal residue
analysis of both parent compound and metabolites (Hooper, 1988; Hooper et
al., in press).

Other biochemical endpoints are increasingly receiving attention for
their ability to indicate exposure to environmental pollutants. The induction of detoxification enzymes can be used to assess exposure to many organic contaminants and many endpoint measurements are available including induction of the mixed function oxidases (Burns, 1976; Lech et al., 1982;
Payne et al., 1984, 1987; Jimenez and Burtis, 1988; Jimenez et al., 1988),
Cytochrome P450 (Parke, 1981), ATPase, monoamine oxidases, NADPH cytochrome
c reductase activities and other cytochromatic enzyme activities.

Changes in immune response may also be indicative of xenobiotic exposure. Recent studies with small mammals (Olson and Hinsdill, 1983, 1984;
Olson et al., 1984; Porter et al., 1984; Fairbrother et al., 1986) and fish
(Spitsbergen et al., 1986) offer insights into the potential use of wildlife and newly developed immunoassays for assessing the impact of contaminated sites on the immune system.

Changes in hormone metabolism may also be used as an indicator of pollutant exposure, including changes in plasma luteinizing hormone concentrations (Rattner and Michael, 1985) and effects on steroidogenesis (Freeman
et al., 1984).

Other biomarkers may include measurement of bile metabolites for PAHs
(Krahn, 1984) and chlorinated phenolics (Oikari and Anas, 1985), glycogen
and sulfhydryl concentrations (Ohlendorf et al., 1988); glutathione per-
oxidase activity (Ohlendorf et al., 1988); perturbations in glutathione

metabolism; alterations in specific heme biosynthetic pathway enzymes and in heme-dependent processes; changes in delta-aminolevulinic acid dehydratase (Kendall and Scanlon, 1982) and porphyrin profile changes.

The uptake, metabolism and elimination of xenobiotics are critical parameters to modeling food chain transfer and risk assessment (Breck and Baes, 1985). Many endpoints are biomedical endpoints and have not traditionally been applied to wildlife in situ but offer exciting new approaches in evaluating impacts to wildlife and in interpreting wildlife results to human risk assessment.

Mutagenic and Carcinogenic Endpoints. In addition to the presence of known carcinogens on site, the oxidative metabolism of many xenobiotics results in metabolic intermediates that are active carcinogens (Ahokas, 1979; Cummings and Prough, 1983). Recent advances in assay development may allow for the use of these endpoints for in situ wildlife monitoring of mutagenesis and carcinogenesis. DNA adducts and cytochrome P450 activation have been found to be indicative of carcinogenic activation using several species and compounds in vivo and in vitro (Shugart and Kao, 1985; Shugart and Matsunami, 1985; Shugart, 1986; Wolf, 1986; Shugart et al., 1987; Martineau et al., 1988). Mutagenic responses have been found in situ in small mammals in association with a lead smelter and a petrochemical waste dump (Lower and Plewa, 1978; McBee, 1985).

In vivo and in vitro mutagenic assays are routinely being conducted on xenobiotics and their application to wildlife could provide a direct evaluation of immunotoxicity, carcinogenicity and mutagenicity of waste site contaminants and a realistic evaluation of the potential impact on man (Stegeman, 1981; Leonard et al., 1985).

Chemical. Chemical analysis of parent compounds and metabolite resdues has been shown to be an extremely effective tool for monitoring chemical uptake and accumulation. Watson et al. (1985) found significant trophic level movement of polychlorinated biphenyls and other organochlorine compounds in a PCB-contaminated waste site. Residues were found in both biota and upper trophic level consumers. Several other researchers have found similar results with voles (Microtus pennsylvanicus) in Love Canal (Rowley et al., 1983), ring-billed gulls (Larus delawarensis) (Sileo et al., 1977), snapping turtles (Chelydra serpentina) (Olafsson et al., 1983) and wild mink (Mustela vison) and otter (Lutra canadensis) (Foley et al., 1988). Heavy metal contamination in small mammals has been associated with waste-water-irrigated habitats (Anthony and Kozlowski, 1982), highways (Blair et al., 1977; Scanlon, 1979) and a mercury contaminated site (Galluzzi, 1981; Lipsky and Galuzzi, 1982). Selenium residues have been found in mammals using California irrigation drainwater (Clark, 1987) and may have implications for other metal- and metalloid-contaminated sites.

Radiation Endpoints. Several researchers have used wildlife to assess the effects of radiation from nuclear waste disposal areas (Hanson and Kornberg, 1956; Halford and Markham, 1978; Halford and Millard, 1978; Craig et al., 1979; Halford et al., 1981). Halford et al. (1982) censused on site wild waterfowl to determine species composition and residence times. Specimens of waterfowl, small mammals and passerine birds were collected and muscle tissue was used to determine radionuclide content. Commercially raised, wing-clipped mallards (Anas platyrhynchos) were released onto radioactive leaching ponds. Attached and implanted dosimeters were used to determine radionuclide content. Species collected enabled researchers to evaluate radionuclide exposure differences between habitat use and ecosystem component (dabbling vs. diving ducks and terrestrial vs. aquatic species) and exposure source (internal vs. external exposure). The hazard to wildlife and to hunters who hunt on site and consume ducks shot on site was evaluated. Conclusions indicated no threat to human health. Other studies evaluated radiation in raptor nestlings and their small mammal food base (Halford and Markham, 1978; Craig et al., 1979), in herons (Domby et al., 1977), in other bird species (Willard, 1960; Cadwell et al., 1979; Markham and Halford, 1982) and in other wildlife (Arthur and Janke, 1986). Wildlife appear to be an excellent monitoring tool for tracking radiation levels on site and for estimating possible offsite movement of contamination.

Behavior. Most avian behavioral research today is concerned with the effects of pesticide exposure, but could be modified to assess waste site contaminant exposure. Behavioral measures include basic motor activity, homing and orientation, breeding behavior, foraging behavior, and predator-prey relationships (Snyder, 1974; Galindo et al., 1985; Buerger et al., 1987; Brewer et al., 1988a; Bussiere et al., 1989). If critical behaviors such as breeding and foraging are altered as a result of contaminant exposure, the survivability of the individual, and possibly the population, is affected. Behavioral measurements taken in situ, in conjunction with other endpoints, could show direct effects of environmental contaminants on wildlife.

EXAMPLES AND DISCUSSION

In situ methods for monitoring wildlife can be valuable for on site hazard assessments and for evaluating the success of site management. The approach suggested here is a series of integrated steps requiring an initial site contaminant assessment followed by a study design centered on carefully chosen wildlife species, food chains and endpoints. Results ultimately lead to a real life assessment of risk that can validate risk management.

Example systems assessing the bioavailability and ecotoxicity of contaminants of a hazardous waste site can be proposed. One approach uses several on site avian and small mammal species. The great blue heron, the red-tailed hawk, and the starling will be discussed in this example. The species selected should be chosen based on preliminary census results. An initial review of the contaminants present will determine the biochemical endpoints used for monitoring each species.

An enhanced starling population could be established on contaminated sites using nest boxes to attract the birds. Uptake (i.e., bioavailability) of toxic substances by adults and their nestlings could be monitored during the reproductive season. Any reproductive or biochemical disturbances would be assessed.

The great blue heron could be monitored and salvaged juvenile tissue residue levels analyzed (for a discussion of the use of salvaged juveniles see Norman, 1988). Evaluation of reproductive success could also be monitored. Possible endpoints for reproductive success include eggshell thickness, clutch size, embryo deformities, number of birds hatched per nest, and nestling weights and body measurements.

Red-tailed hawks could be monitored and evaluated for ingestion of rodents from on-site contaminant locations. Young hawks could be banded in the nest and blood samples routinely taken from both young and adults for assessment of any biochemical responses to toxic substances.

Residue analysis of food items found in avian feeding areas on site can be used to assess the bioavailability of contaminants. By monitoring residue levels in the food base of the starling, the great blue heron and the red-tailed hawk, in addition to residue levels in salvaged juveniles, an assessment of the trophic transfer of both terrestrial and aquatic contaminants is possible.

Residues in soil, water, and vegetation sources can be determined. The combination of studies involving a passerine, a raptor, a wading bird, small mammals, and invertebrates, in correlation with environmental residue analysis, can assess the potential routes of transfer of toxic substances that may occur in wildlife on a contaminated site. Based on these results a comparison between contaminated and uncontaminated sites could be made which would influence remedial action and future land use.

The procedures in this discussion can facilitate delineation of the bio-availability of contaminants on site. If contaminants are bioavailable and endpoint analysis indicates adverse biological impact, remedial action may be called for. By monitoring wildlife long term, quantitative data are available to assess the adequacy of cleanup levels achieved.

REFERENCES

Ahokas, J.T., 1979, Cytochrome P-450 in fish liver microsomes and carcino-gen activation, in: "Pesticides and Xenobiotic Metabolism in Aquatic Organisms," M.A.Q. Khan, J.J. Lech, and J.J. Menn, eds., ACS Symposium Series 99, American Chemical Society, Washington D.C., pp. 279-295.

Amlaner, C.J., and Macdonald, D.W., 1980, A handbook on biotelemetry and radio tracking, Pergamon Press, Oxford, p.804.

Anthony, R.G., and Kozlowski, R., 1982, Heavy metals in tissues of small mammals inhabiting waste-water-irrigated habitats, J. Environ. Qual. 11:20-22.

Arthur, W.J., III, and Janke, D.H., 1986, Radionuclide concentrations in wildlife occurring at a solid radioactive waste disposal area in south-eastern Idaho, J. Environ. Qual. 12:117-122.

Arthur, W.J., Markham, O.D., Groves, C.R., and Keller, B.L., 1987, Radio-nuclide export by deer mice at a solid radioactive disposal area in southeastern Idaho, Health Phys. 52:45-53.

Balcomb, R., 1986, Songbird carcasses disappear rapidly from agricultural fields, Office of Pesticide Programs, U.S. Environmental Protection Agency.

Bennett, J.K., Ringer, R.K., Bennett, R.S., Williams, B.A., Humphrey, P.E., 1988, Comparison of breaking strength and shell thickness as evaluators of eggshell quality, Environ. Toxicol. Chem. 7:351-357.

Blair, W.F., Hiller, A.L., and Scanlon, P.F., 1977, Heavy metal concentra-tions in mammals associated with highways of different traffic densi-ties, Virginia J. Sci. 28:61.

Block, E.K., 1988, The effects of the organophosphate pesticide COUNTER on Peromyscus: An integrated laboratory and field study, M.S. Thesis, Western Washington University, pp.110.

Bloom, P.H., 1987, Capturing and handling raptors, in: "Raptor management techniques manual," Giron Pendleton, B.A. Millsap, K.W. Cline, and D.M. Bird, eds., National Wildlife Federation, Washington, D.C. pp.99-123.

Blus, L.J., Henny, C.J., and Kaiser, T.E., 1980, Pollution ecology of breeding great blue herons in the Columbia Basin, Oregon and Washington, The Murrelet. 61:63-71.

Breck, J.E., and Baes, C.F., 1985, Report on the workshop on food chain modeling for risk analysis. Oak Ridge National Laboratory Report 6051, National Technical Information Service, Springfield, VA.

Brewer, L.W., Driver, C.J., Kendall, R.J., Zenier, C., and Lacher, T.E., Jr., 1988a, Effects of methyl parathion in ducks and duck broods, Environ. Toxicol. Chem. 7:375-379.

Brewer, L.W., Driver, C.J., Kendall, R.J., and Lacher, T.E., Jr., Galindo, J.C., Dickson, G.W., 1988b, Avian response to a turf application of Triumph[R] 4E., Environ. Toxicol. Chem. 7:391-401.

Buerger, T.T., Mueller, B.S., Kendall, R.J., DeVos, T., and Hitchcock, R.R., 1987, in: "The effect of methyl parathion on activity and survivability of bobwhite quail," C. Kyser, J.L. Landers, and B.S. Mueller, eds., Proc. Tall Timbers Game Bird Seminar, Tall Timbers Assoc., Tallahassee, FL, pp.4-9.

Burns, K.A., 1976, Microsomal mixed function oxidases in an estuarine fish, *Fundulus heroclitus*, and their induction as a result in environmental contamination., *Comp. Biochem. Physiol.* 53:443-446.

Bussiere, J.L., Kendall, R.J., Lacher, T.E., Jr., and Bennett, R.S., 1988, Effect of methyl parathion on food discrimination in bobwhite quail (*Colinus virginianus*), Submitted to *Environ. Toxicol. Chem.*

Cadwell, L.L., Schreckhise, R.G., and Fitzner, R.E., 1979, Cesium-137 in coots (*Fulica americana*) on Hanford waste ponds: Contribution to population dose and offsite transport estimates. Pacific Northwest Laboratory Report, PNL-SA-7176. Richland, Washington.

Calambokidis, J., Speich, S., Peard, J., Steiger, G.H., Cubbage, J.C., Frye, D.M., and Lowenstine, L.J., 1985, Biology of Puget Sound marine mammals and marine birds: Population health and evidence of pollution effects, NOAA Technical Memorandum, National Oceanic and Atmospheric Administration, Rockville, MD, p.159.

Clark, D.R., 1987, Selenium accumulation in mammals exposed to contaminated California USA irrigation drainwater, *Sci. Total Environ.* 66:147-168.

Craig, T.H., Halford, D.K., and Markham, O.D., 1979, Radionuclide concentrations in nestling raptors near nuclear facilities, *Wilson Bull.* 91:72-77.

Cummings, S.W., and Prough, R.A., 1983, Metabolic formation of toxic metabolites, *in*: "Biological Basis of Detoxification," J. Caldwell and W.B. Jakoby, eds., Academic Press, New York, pp.1-30.

Domby, A.H., Paine, D., and McFarlane, R.W., 1977, Radiocesium dynamics in herons inhabiting a contaminated reservoir system, *Health Phys.* 33:415-422.

Eaton, D.L., Stacey, N.H., Wong, K.L., and Klaassen, C.D., 1980, Dose-response effects of various metal ions on rat liver metallothionein, glutathione, heme oxygenase, and cytochrome P-450, *Toxicol. Appl. Pharmacol.* 55:393-402.

Eberhardt, L.E., and Cadwell, L.L., 1985, Radio-telemetry as an aid to environmental contaminant evaluation of mobile wildlife species, *Environ. Monit. Assess.* 5:283-290.

Emlen, J.T., 1971, Population densities of birds derived from transect-counts, *The Auk* 88:323-342.

Emlen, J.T., 1974, An urban bird community in Tucson, Arizona: Derivation, structure, regulation, *Condor* 76:184-197.

Environmental Protection Agency, 1988, Guidance document for conducting terrestrial field studies. Ecological Effects Branch, Hazard Evaluation Division, Office of Pesticide Programs, Washington D.C., pp.67.

Fairbrother, A., Yuill, T.M., and Olson, L.J., 1986, Effects of three plant growth regulators on the immune response of young and aged deer mice (*Peromyscus maniculatus*), *Arch. Environ. Contam. Toxicol.* 15:265-275.

Feare, C., 1984, "The Starling," Oxford University Press, New York, p.315.

Foley, R.E., Jackson, S.J., Sloan, R.L., and Brown, M.K., 1988, Organochlorine and mercury residues in wild mink and otter: Comparison with fish, *Environ. Toxicol. and Chem.* 7:363-374.

Freeman, H.C., Sangalang, G.B., and Uthe, J.F., 1984, The effects of pollutants and contaminants on steroidogenesis in fish and marine mammals, *in*: "Contaminants Effects on Fisheries," V.W. Crains, P.V. Hodson and J.O. Nraigu, eds., John Wiley & Sons, New York, pp.197-212.

Galindo, J., Kendall, R.J., Driver, C.J., and Lacher, T.E., Jr., 1985, The effect of methyl parathion on the susceptibility of bobwhite quail (*Colinus virginianus*) to domestic cat predation, *Behav. Neural Biol.* 43:21-36.

Galluzzi, P., 1981, Mercury concentrations in mammals, reptiles, birds, and waterfowl collected in the Hackensack Meadowlands, New Jersey,

Presented at N.J. Acad. of Science, pp.24.

Gibbs, J.P., Woodward, S., Hunter, M.L., and Hutchinson, A.E., 1987, Determinants of great blue heron colony distribution in coastal Maine, The Auk. 104:38-47.

Giron Pendleton, B.A., Millsap, B.A., Cline, K.W., and Bird, D.M., 1987, Raptor Management Techniques Manual, National Wildlife Federation, Washington, D.C., pp.420.

Grisham, J.W., 1986, "Health Aspects of the Disposal of Waste Chemicals," Pergamon Press, New York, p.454.

Greenwood, P.J., Harvey, P., and Perrins, C.M., 1979, Great tit (Parus major) studied in nest boxes, Animal Behav. 27:645-651.

Grue, C.E., and Hunter, B.K., 1984, Brain cholinesterase activity in fledgling starlings: Implications for monitoring exposure of songbirds to ChE inhibitors, Bull. Environ. Contam. Toxicol. 32:282-289.

Grue, C.E., Flemming, W.J., Busby, D.G., and Hill, E.F., 1983, Assessing hazards of organophosphate pesticides to wildlife, Trans. North Am. Wildl. Nat. Resour. Conf. 48:200-220.

Grue, C.E., Powell, G.V.N., and McChesney, M.J., 1982, Care of nestlings by wild female starlings exposed to an organophosphate pesticide, J. Appl. Ecol. 19:327-335.

Halford, D.K., and Markham, O.D., 1978, Radiation dosimetry of small mammals inhabiting a liquid radioactive waste disposal area, Ecology 59:1047-1054.

Halford, D.K., Markham, O.D., and Dickson, R.L., 1982, Radiation doses to waterfowl using a liquid radioactive waste disposal area, J. Wildl. Manage. 46:905-914.

Halford, D.K., and Millard, J.B., 1978, Vertebrate fauna of a radioactive leaching pond complex in southeastern Idaho, Great Basin Nat. 38:64-70.

Halford, D.K., Millard, J.B., and Markham, O.D., 1981, Radionuclide concentrations in waterfowl using a liquid radioactive waste disposal area and the potential radiation dose to man, Health Phys. 38:173-181.

Hanson, W.C., and Kornberg, H.A., 1956, Radioactivity in terrestrial animals near an atomic site, Proc. Int. Conf. Peaceful Uses of Atomic Energy, Geneva, Switzerland, 13:385-388.

Hayward, G.F., and Phillipson, J., 1979, Community structure and functional role of small mammals in ecosystems, in: "Ecology of Small Mammals," D.M. Stoddart, ed., Chapman and Hall, London, pp.135-211.

Heintzelman, D.S., 1979, "Hawks and Owls of North America," Universe Books, New York, pp.33-37.

Henny, C.J., 1972, An analysis of the population dynamics of selected avian species with special reference to changes during the modern pesticide era, Wildlife Research Report No. 1. U.S. Dept. of the Interior, Washington, D.C. pp.99.

Henny, C.J., Blus, L.J., and Hulse, C.S., 1985a, Trends and effects of organochlorine residues on Oregon and Nevada wading birds, 1972-1983, Colonial Waterbirds, 8:117-128.

Henny, C.J., Blus, L.J., Kolbe, E.J., and Fitzner, R.E., 1985b, Organophosphate insecticide (famphur) topically applied to cattle kills magpies and hawks, J. Wildl. Manage. 49:648-658.

Henny, C.J., Kolbe, E.J., Hill, E.F., and Blus, L.J., 1987, Case histories of bald eagles and other raptors killed by organophosphorus insecticides topically applied to livestock, J. Wildl. Dis., 23:292-295.

Hoffman, D.J., Ohlendorf, H.M., and Aldrich, T.W., 1988, Selenium teratogenesis in natural populations of aquatic birds in central California, Arch. Environ. Contam. Toxicol., 17:519-525.

Hooper, M.J., Detrich, P., Weisskopf, C., and Wilson, B.W., 1989, Organophosphate exposure in hawks inhabiting orchards during winter dormantspraying, Bull. Env. Contam. Toxicol., In press.

Hooper, M.J., 1988, Avian cholinesterases: Their characterization and use in evaluating organophosphate insecticide exposure, Ph.D. Thesis. University of California, Davis, CA, p.110.

Jimenez, B.D. , Burtis, L.S., Ezell, G.H., Eagan, B.Z., Lee, N.E., Beauchamp, J.J., and McCarthy, J.F., 1988, The mixed-function oxidase system of bluegill sunfish, Lepomis macrochirus: Correlation of activities in experimental and wild fish, Environ. Toxicol. Chem., 7:623-634.

Jimenez, B.D., and Burtis, L.S., 1988, Response of the mixed function oxidase system to toxicant dose, food and acclimation temperature in the blue gill sunfish, Mar. Environ. Res., 24:45-49.

Jimenez, B.D., and Burtis, L.S., 1989, Influence of environmental variables on the hepatic mixed-function oxidase system in blue gill sunfish, Lepomis macrochirus., Comp. Biochem. Physiol., in press.

Karr, J.R., 1987, Biological monitoring and environmental assessment: A conceptual framework, Environ. Manage., 11:249-256.

Kendall, R.J., 1982, Wildlife Toxicology-Integrated field and lab studies using selected model species might lead to ways of quantifying adverse effects of chemical contaminants, Environ. Sci. Technol., 16:448-453.

Kendall, R.J., and Scanlon, P.F., 1982, Tissue lead concentrations and blood characteristics of rock doves (Columba livia) from an urban setting in Virginia, Arch. Environ. Contam. Toxicol., 11:265-268.

Kendall, R.J., 1988, Wildlife Toxicology: A reflection on the past and the challenge of the future, Environ. Toxicol. Chem., 7:337-338.

Kendeigh, S.C., 1944, Measurement of bird populations, Ecol. Manager., 14:67-106.

Krahn, M.M., Myers, M.S., Burrows, D.G., and Malins, D.C., 1984, Determination of metabolites of xenobiotics in bile of fish from polluted waterways. Xenobiotica, 14:633-646.

Lech, J.J., Vodicnik, M.J., and Elcombe, C.R., 1982, Induction of monooxygenase activity in fish, in: "Aquatic Toxicology," L.J. Weber, ed., Raven Press, New York, pp.107-148.

Leonard, A., Duverger-van Bogaert, M., Bernard, A., Lambotte-vandepaer, M., and Lauwerys, R., 1985, Population monitoring for genetic damage induced by environmental and chemical agents, Environ. Monitor. Assess., 5:369-384.

Lipsky, D., and Galluzzi, P., 1982, The investigation of mercury contamination in the vicinity of Berry's Creek, Manage. of Uncontrolled Hazardous Waste Sites Natl. Sym., Washington D.C., pp.5.

Lower, W., and Plewa, M., 1978, Evidence for an in situ mutagenesis and other toxic effects from a lead smelting operation, Mut. Res., 53:86-87.

Malins, D.C., McCain, B.B., Brown, D.W., Myers, M.S., Krahn, M.M., and Chan, S.L., 1987, Toxic chemical including aromatic and chlorinated hydrocarbons and their derivatives, and liver lesions in white croaker (Genyonemus lineaus) from the vicinity of Los Angeles, Environ. Sci. Technol., 21:765-770.

Markham, O.D., and Halford, D.K., 1982, Radionuclides in mourning doves near a nuclear facility complex in southeastern Idaho, Wilson Bull., 94:185-197.

Marti, C.D., Wagner, P., and Denne, K., 1979, Nest boxes for the management of barn owls, Wildl. Soc. Bull., 7:145-148.

Martineau, D., Lagace, P., Beland, A., Higgins, R., Armstrong, D., and Shugart, L.R., 1988, Pathology of stranded beluga whales (Delphinapterus leucas) from the St. Lawrence Estuary, Quebec, Canada, J. Comp. Pathol., 98:287-311.

McBee, K., 1985, Chromosomal damage in resident small mammals at a petrochemical wastes dump site: A natural model for analysis of environmental mutagenesis, Ph.D. Thesis. Texas A & M Univ., p.118.

McCarthy, J.F., Shugart, L.R., and Jimenez, B.D., 1988, Biological markers in wild animal sentinels as predictors of ecological and human health effects from environmental contamination, in: Eighth Life Science Symposium, on "Biological Indicators: Exposures and Effects," C.W. Gehrs ed., Knoxville, TN, March 21-23, Oak Ridge National Laboratory, Oak Ridge, TN.

Mineau, P., and Peakall, D.B., 1987, An evaluation of avian impact assessment techniques following broad-scale forest insecticide sprays, Environ. Toxicol. Chem., 6:781-791.

Mitchell, R.T., Blagbrough, H., and VanEtten, R., 1953, The effects of DDT upon the survival and growth of nestling songbirds, Jour. Wildl. Manage., 17:45-54.

Norman, D.M., 1988, Organochlorine contaminants in great blue herons collected from the Puget Sound ecosystem. M.S. Thesis, Western Washington University, In progress.

Ohlendorf, H.M., Hothem, R.L., Bunk, C.M., Aldrich, T.W., and Moore, J.F., 1986, Relationship between selenium concentrations and avian reproduction, Trans. N. Am. Wildl. Nat. Resour. Conf., 51:330-342.

Ohlendorf, H.M., Kilness, A.W., Simmons, J.L., Stroud, R.K., Hoffman, D.J. Moore, J.F., 1988, Selenium toxicosis in wild aquatic birds, J. Toxicol. Environ. Health, 24:67-92.

Ohlendorf, H.M., Klaas, E.E., and Kaiser, T.E., 1978, Environmental Pollutants and eggshell thinning in the black-crowned night heron, in: "Wading Birds, Research Report No. 7 of the National Audubon Society," I.V. Sprunt, J.C. Ogden, and S. Winkler, eds., National Audubon Society, New York, pp.63-82.

Oikari, A., and Anas, E., 1985, Chlorinated phenolics and their conjugates in the bile of trout (Salmo gairdneri) exposed to contaminated waters, Bull. Environ. Contam. Toxicol., 35:802-809.

Olafsson, P.G., Bryan, A.M., Bush, B., and Stone, W., 1983, Snapping turtles--a biological screen for PCB's, Chemosphere, 12:1525-1532.

Olson, L.J., and Hinsdill, R.D., 1983, Immunosuppression in deer mice Peromyscus maniculatus fed chlorinated biphenyls arochlor 1254 and chlorocholine chloride, Fed. Proc., 42:Abstract 3811.

Olson, L.J., and Hinsdill, R.D., 1984, Influence of feeding chlorocholine chloride and glyphosine on selected immune parameters in deer mice, Peromyscus maniculatus, Toxicology, 30:103-114.

Olson, L.J., Fairbrother, A., Hinsdill, R.D., and Yuill, T.M., 1984, Immunomodulatory and viral challenge effects of feeding cyclo phosphamide chloro chloro choline chloride and glyphosine to old deer mice Peromyscus maniculatus, Fed. Proc., 43:Abstract 490.

O'Shea, T.J., Fleming, W.J. III, and Chromartie, E., 1980, DDT Contamination at Wheeler Nation Wildlife Refuge, Science, 209:509-510.

Parke, V.D., 1981, Cytochrome P-450 and the detoxification of environmental chemicals, Aquat. Toxicol., 1:367-376.

Payne, J.F., Bauld, C., Dey, A.C., Kiceniuk, J.W., and Williams, U., 1984, Selectivity of mixed-function oxygenase enzyme induction in flounder (Pseudopleuronectes americanus) collected at the site of Baie Verte, Newfoundland, oil spill, Comp. Biochem. Physiol., 79:15-19.

Payne, J.F., Fancey, L.L., Rahimtula, A.D., and Porter, E.L., 1987, Review and perspective on the use of mixed-function oxidase enzymes in biological monitoring, Comp. Biochem. Physiol., 86:233-245.

Porter, W.P., Hinsdill, R., Fairbrother, A., Olson, L.J., Jaeger, J., Yuill, T., Bisgaard, S., Hunter, W.G., and Nolan, K., 1984, Toxicant-disease-environment interactions associated with suppression of immune system, growth and reproduction, Science, 224:1014-1017.

Powell, G.V.N., and Gray, D., 1980, Dosing free living nestling starlings with an organophosphate pesticide, famphur, Jour. Wildl. Manage., 44:918-921.

Pratt, H.M, and Winkler, D.W., 1985, Clutch size, timing of laying, and the reproductive success in a colony of great blue herons and great egrets, The Auk., 102:59-63.

Rattner, B.A., and Michael, S.D., 1985, Organophosphorus insecticide induced decrease in plasma luteinizing hormone concentration in white-footed mice, Toxicol. Lett., 24:65-69.

Riley, R.G., Crecelius, E.A., Fitzners, R.E., Thomas, B.L., Burtisen, J.M., and Bloom, N.S., 1983, Organic and inorganic toxicants in sediment and marine birds from Puget Sound, NOAA Technical Series Memorandum NOS OMS 1. National Ocean Service, NOAA, pp.125.

Ringer, R.K., 1988, The future of a mammalian wildlife toxicology test: one researcher's opinion, Environ. Toxicol. Chem., 7:339-342.

Robinson, S.C., Driver, C.J., Kendall, R.J., and Lacher, T.E. Jr., 1988, Effects of agricultural spraying of methyl parathion on reproduction and cholinesterase activity in starlings (Sturnus vulgaris) in Skagit Valley, Washington, Environ. Toxicol. Chem., 7:343-349.

Rowley, M.H., Christian, J.J., Basu, D.K., Pawlikowski, M.A., and Paigen, B., 1983, Use of small mammals (voles) to assess a hazardous waste site at Love Canal, Niagara Falls, New York, Arch. Environ. Contam. Toxicol., 12:383-397.

Scanlon, P.F., 1979, Lead contamination of mammals and invertebrates near highways of with different traffic volumes, in: "Animals as Monitors of Environmental Pollutants," National Academy of Sciences, Washington, D.C., pp.200-208.

Scanlon, P.F., Kendall, R.J., Lochmiller, R.L., and Kirkpatrick, R.L., 1983, Lead concentrations in pine voles from two Virginia orchards, Environ. Pollut., (Series B), 6:157-160.

Schemnitz, S.D., 1980, "The Wildlife Management Techniques Manual," Wildlife Society, Washington, D.C., p.686.

Seber, G.A.F., 1982, "The estimation of animal abundance and related parameters," Macmillan Publishing Co. Inc., New York, p.654.

Shugart, L., 1986, Quantifying adductive modification of hemoglobin from mice exposed to benzo[a]pyrene, Anal. Chem., 152:365-369.

Shugart, L., and Kao, J., 1985, Examination of adduct formation in vivo in the mouse between benzo[a]pyrene and DNA of skin and hemoglobin of red blood cells, Environ. Health Perspect., 62:223-226.

Shugart, L., and Matsunami, R., 1985, Adduct formation in hemoglobin of the newborn mouse exposed to in utero to benzo[a]pyrene, Toxicology, 37:241-245.

Shugart, L., McCarthy, J., Jimenez, B., and Daniels, J., 1987, Analysis of adduct formation in the bluegill sunfish (Lepomis macrochirus) between benzo[a]pyrene and DNA of the liver and hemoglobin of the erythrocyte, Aquatic Toxicol., 9:319-325.

Sileo L., Karstad, L., Frank, R., Holdrinet, M.V.H., Addison, E., and Braun, H.E., 1977, Organochlorine poisoning of ring-billed gulls in southern Ontario, J. Wildl. Dis., 13:313-322.

Simpson, K., 1984, Factors affecting reproduction in great blue herons (Ardea herodias), M.S. Thesis, Dept. of Zoology, Univ. of British Columbia, p.90.

Snyder, R.L., and Cheney, C.D., 1974, Pesticide effects in wild deer mice. Proc. Utah Acad. Sci. Arts Lett., 49(pt. 2)81-81.

Southern, H.N., 1979, The stability and instability of small mammal populations, in: "Ecology of small mammals," D.M. Stoddart ed., Chapman and Hall, London, pp.103-134.

Spitsbergen, J.M., Schat, K.A., Kleeman, M., and Peterson, R.E., 1986, Interactions of 2,3,7,8-tetrachlorodibenzo-p-dioxin (TCDD) with immune responses of rainbow trout, Vet. Immun. Immunopathol., 12:263-280.

Stegeman, J.J., 1981, Polynuclear aromatic hydrocarbons and their metabolism in the marine environment, in: "Polycyclic Hydrocarbons and Cancer," H.V. Gelboin and P.O. Ts'O, eds., Academic Press, New York, Vol.3, pp.1-58.

Watson, M.R., Stone, W.B., Okoniewski, J.C., and Smith, L.M., 1985, Wildlife as monitors of the movement of polychlorinated biphenyls and other organochlorine compounds from a hazardous waste site, Northeast Fish and Wildlife Conference, Hartford, CT, May 5-8, (14):91-104.

Willard, W.K., 1960, Avian uptake of fission products from an area contaminated by low-level atomic wastes, Science, 32:148-150.

Wolf, C.R., 1986, Cytochrome P-450s: Polymorphic multigene families involved in carcinogen activation, Trends Genet., 2:209-214.

RELEVANCY AND FUTURE

CAPABILITIES AND LIMITATIONS OF APPROACHES TO IN SITU

ECOLOGICAL EVALUATION

Thomas A. Murphy and Lawrence Kapustka

Environmental Research Laboratory--Corvallis
U.S. Environmental Protection Agency
200 SW 35th Street
Corvallis, Oregon 97333

INTRODUCTION

This symposium presents a number of excellent papers on how to gather information on in situ biological conditions, especially at waste sites. Over the years biologists have had little difficulty gathering large amounts of information about our living environment. This is not surprising given the enormous variety and complexity of ecological systems; there are an almost endless number of features to be measured.

The problem comes in drawing useful ecological conclusions from the information we collect. More appropriately, how do we decide what information to collect in order to draw useful ecological conclusions? This point can be illustrated by several examples showing the limitations of field and laboratory data.

Assume we have surveyed the presence and abundance of fish species in a stream above and below the discharge of runoff from a waste site and that we find fewer species and numbers of fish below the discharge point. How do we know these differences are outside the range of natural variation and not due to chance? Ecological systems vary considerably from site to site and from time to time. Barring an almost total absence of fish below the waste site, a single upstream site is normally an insufficient reference for establishing the range of natural variation. While it is ordinarily very easy to measure ecological change, it is not at all easy to determine if the measured change is outside the range of normal variation.

Furthermore, natural populations are influenced by a wide range of natural and anthropogenic stressors. Even if we assume the upstream and downstream differences are outside the range of normal variation, how do we know they are caused by the waste site? Large population differences are often caused by habitat changes, presence or absence of predators, diseases, and other poorly understood factors. Even though we know a change is out of the ordinary, we cannot assume that a physically co-located pollutant or activity is causing the stress.

Limitations also exist for interpreting laboratory data. Suppose we take a soil sample from a waste site, and in the laboratory we measure its toxicity to earthworms, finding that it kills 20% of the animals tested.

In Situ Evaluations of Biological Hazards of Environmental Pollutants
Edited by S. S. Sandhu *et al.*
Plenum Press, New York, 1990

What does this mean in terms of the populations of worms on site? Such a population normally experiences a high mortality rate from predators, drought, winterkill, etc. However, the population can sustain its numbers over time by mechanisms that compensate for mortality. There are circumstances where this additional toxic stress would not be noticed within the natural dynamics of the worm population. On the other hand, there could be circumstances in which this 20% mortality rate measured in the laboratory could be magnified in the field to much greater mortality rate, even to the point of extirpation. This is just one illustration of why it is not possible to make straightforward or simple extrapolations with great confidence from one level of ecological organization to another.

We have a rich history of methods for gathering data and improving our understanding of ecological systems. However, we have very few examples, especially for waste sites, where these data have been used to draw rigorous ecological conclusions. This does not mean that it cannot be done, but that it will require careful design at the inception of an ecological assessment and more blending of different types of biological and ecological information than we have been accustomed to in the past.

The thesis of this paper is that we have the concepts and methods to conduct useful ecological assessments, but that we have not been using our concepts and methods in the most effective manner. Further, the most important improvement we can make is to move away from our tendency first to measure the things we know how to measure and then try to draw conclusions. To draw useful ecological conclusions, we must define our assessment questions at the outset and then gather the data to answer those questions.

LEVELS OF BIOLOGICAL/ECOLOGICAL ORGANIZATION

For clarity we will define the key terms used in this paper. By "ecological assessment" we mean some quantitative evaluation of adverse change in a population, community or ecosystem at a specific location, such as a waste site; and the identification of the probable cause or causes of this change. The term "ecological" refers to the interactions among organisms and their environments. To clarify this, we need to describe the different levels of biological/ecological organization.

Individual

A basic biological unit is the individual organism. It is often very useful to determine how a representative individual plant or animal responds to a stress. This is the usual endpoint of toxicological analysis and essentially the counterpart of human health assessment. Normally, we test the effect of a chemical on a group of fish in a tank, to determine the probable response of a typical individual fish. Drawing from the experience with human health assessment, we are very accomplished in measuring stress at the level of the individual organism. However, from this information alone, one cannot, other than by gross inference, draw ecological conclusions.

Population

From an ecological and evolutionary perspective, the value of an individual is its contribution to a population of which it is a transient member. It is at the population level that much of nature's adaptation to natural environmental stress occurs. This natural stress is considerable. Populations are very dynamic, being influenced by a wide range of stresses such as predation, loss of habitat, climate fluctuations, disease, and starvation. One cannot simply translate an effect of a stress such as a

260

toxicant on an individual organism to a comparable effect at the population level. From experience, we know that the population response can be substantially greater or less. For a number of species, we have the capability for measuring and, to some extent, predicting the level of natural populations. However, this cannot be done with the precision or sophistication with which we can assess effects at the level of the individual organism.

Community

Populations of species interact with other species in the natural environment. Sometimes these interspecies linkages are tight as with the panda, which eats only the bamboo plant. In other cases linkages are weak, as with an opossum, which subsists on many types of foods in a variety of environments. Natural groupings of interacting species we call communities. Socially, we tend to place great value on certain communities, such as certain forest communities. Despite extensive efforts to develop simple indices of community structure, we are not generally capable of effectively characterizing what we mean by a community, especially in terms necessary for assessment. There are, however, some recent advances in this area that offer promise (Karr, 1986).

Ecosystem

A biological community plus its non-living environment is what we call an ecosystem. Examples would be a lake, a wetland, or a forest, including soil, air and water, as well as living components. There are no distinct boundaries for ecosystems. They tend to overlap and vary enormously in scale according to the question being asked. An ecosystem could be a single rotting log in a forest or the entire earth. We normally describe ecosystems in terms of their functions. How much timber will a forest produce? How much pollution will a wetland remove? How much carbon dioxide will an Amazon jungle consume? Socially and scientifically, we express great concern for ecosystems. However, our assessment capability at this level is very limited (Schindler, 1987).

Cearly ecological systems are very complex, operating at many different levels and scales. Also they are highly variable. They are constantly changing, both in response to natural environmental change and to natural ecological succession. Does this complexity and variability mean that useful ecological assessments exceed our grasp? We think not. Certainly there are many questions we will not be able to answer. However, with the right approach and degree of rigor, we believe that we have the capability for answering many useful waste site ecological assessment questions.

PURPOSE OF AN ASSESSMENT

Within the current state of ecological knowledge, there appear to be four useful questions that can be answered for many sites:

1. What adverse ecological change has occurred, on site and off site?

2. What adverse change can be attributed to waste site contaminants or other site activities?

3. What is the extent and severity of adverse change?

4. What adverse change has the potential for abatement as a result of clean-up?

Critical to all four of these questions is the definition of "adverse change." As discussed above, many measured ecological changes are not adverse; they may be within the range of natural variation. Moreover, what may be "adverse" at the level of biological organization we can conveniently measure, that is, the individual may not be adverse at the level of our ecological interest; the population or community. Therefore, the most important single step in an assessment is defining what ecological entity or attribute we wish to assess. This we will call the ecological "assessment endpoint." The assessment endpoint is that feature in the environment for which we consider a change to be undesirable. An assessment may have a single endpoint or more commonly, several endpoints. As assessment endpoint, in addition to being something we value, must be something we are capable of quantitatively assessing. We must be able to either measure the endpoint directly, or to measure something else, from which we can estimate the endpoint. (This definition is similar, but not identical to one developed by Suter, 1989).

It is desirable whenever possible to select an endpoint based on its intrinsic value. At the population level this is often possible: an endangered or threatened species (e.g., the peregrine falcon), a commercially important species (e.g., the razor clam), a recreationally important species (e.g., rainbow trout), an aesthetically important species (e.g., whistling swan), a culturally important species (e.g., the bald eagle). At the community level most intrinsically valued attributes tend to be expressed in terms of a particular species and the set of other species necessary for its sustenance. At the ecosystem level one defines functional attributes such as timber yield for a forested ecosystem, or crop yield for an agricultural system.

Intrinsically valued endpoints are the most useful and effective kind for assessments. They tend to have the highest degree or precision and understandability for assessment. However, we are not able in many cases to define intrinsically valued endpoints. In addition to specific intrinsic ecological attributes, we also socially value the assumption that we don't understand ecological systems well enough to identify all of their important attributes and that we are concerned that ecological systems may have a future value that we don't currently recognize. Therefore, we also use "protective" assessment endpoints. These take the form of questions such as:

° Is there change outside the range of natural variation for that type of ecological system?

° What level of stress (e.g., toxicity) is present that has the potential for causing biological harm?

There, "protective" endpoints have a long and useful history. They are the basis for chemical water quality criteria, as well as regulations for natural resource damage assessment at superfund sites (U.S. Dept. Interior, 1987). Many ecological field surveys are designed around protective endpoints to compare differences between a site of concern and some presumably undisturbed reference site.

On the surface protective endpoints are very attractive. When expressed as "no change" or "no stress" they imply a level of absoluteness that is reassuring. However, these types of endpoints have serious limitations that have become very evident. Natural reference conditions are very difficult to define, because of statistical difficulties and natural ecological systems have been so altered that system unperturbed by modern human influences no longer exist. In practice "no change" or "no stress" are relative rather than absolute concepts. With sufficient effort some change or contamination can always be measured; thus one must always decide how much change or what degree of contamination is acceptable. While we will continue to use protective endpoints, we must be very careful to be explicit about their limitations, and not to use them to draw ecological conclusions beyond these limits.

ASSESSMENT APPROACHES

There are three general sources of data used for ecological assessment of waste sites:

- º <u>field surveys</u> measure the distribution and abundance of taxonomic groups of organisms, and some functional attributes of ecological systems.

- º <u>bioassays</u> measure a direct toxic response in the field or in the laboratory.

- º <u>biomarkers</u> indicate exposure to specific chemicals or classes of chemicals.

FIELD SURVEYS

Generally field surveys are essential for selecting the appropriate ecological assessment endpoints for a site. Also, they are the only way that actual ecological effects can be demonstrated. Sampling designs and protocols for determination of populations of plants, animals, and microbes have been the subject of ecology from its inception. Although no rigid guidelines for sampling are accepted universally, the concepts of adequacy of sample, objectivity, and precision are well entrenched in all field-oriented studies. Protocols can be modified to match the peculiarities of the site and the objectives of the sampling effort.

Field surveys may be restricted at hazardous waste sites because of problems of access and risk to workers. Surveys may not be appropriate at some sites because of their limited size. However, in a large number of sites, field surveys can play a major role.

Conventional on-site surveys of vegetation and animal populations can generate patterns of distribution and abundance of the taxonomic groups. Given the time available for data collection at many hazardous wastes, it is crucial to recognize the large error margins in most survey data. One-time sampling efforts almost always underestimate species richness. Ephemeral populations are easily missed. Quantitative estimates from one-time sampling efforts are static and thus miss the dynamics of the site. There are distinct limitations to the ecological conclusions we can draw from changes in population numbers. Nevertheless, indispensable information can be acquired from field sampling, in some cases even through rather cursory reconnaissance.

In addition to traditional methods of direct censusing of biota, remote sensing can be very useful for field surveys. Vegetation structure and, to some extent, composition can be determined remotely using conventional aerial photography, or more sophisticated radiometric signals that permit analysis of spatial and temporal changes at and surrounding the site. More important, because toxicants can alter spectral reflectance patterns of vegetation, infrared photography and radiometric sensors (Duinker and

Nilsson, 1988) show great promise for use in defining the spatial boundaries of impact.

Table 1 summarizes the capabilities and limitations of field surveys. For more detailed discussion see Kapustka et al. (1989).

BIOASSAYS

Bioassays have traditionally been used to determine the effect of a single chemical on an organism under controlled conditions. Laboratory bioassay conditions cannot replicate the natural environment, in terms of chemical exposure, simultaneous environmental stresses or other ecological conditions. Used in the laboratory, bioassays are essentially an instrument, producing a measured biological response that varies depending upon how the instrument is used. Results have an uncertain relationship with field ecological responses.

Nonetheless, bioassays have been extremely useful for comparing toxicity of chemicals and determining a range of response in different types of organisms. In some cases chronic bioassays have been used to establish what appear to be effective levels of protection for contaminants in aquatic systems (Sloof, 1985). However, experience has shown that bioassays can result in both false negative and false positive findings.

More recently, bioassays have been employed to evaluate toxicity of environmental samples such as wastewater effluents or eluates from waste site soils. In this way, they are able to integrate the response of an organism to simultaneous exposure from multiple differentially toxic agents.

One feature of bioassays is especially important for waste site assessment. Usually they are the only way to experimentally establish a causal relationship between a contaminant and an observed effect. Field surveys

Table 1.   Capabilities and limitations of field surveys (Adapted from Baker 1989).

Capabilities

+   necessary to define endpoints of relevance

+   most direct way to demonstrate adverse change

+   large selection of available sampling techniques permit measurement to specified accuracy

+   reflect the biological integration of all stresses

+   amenable to sophisticated remote sensing technology (at least major vegetation components)

Limitations

-   restricted to correlation analysis, does not establish causation

-   legal and safety concerns restrict access

-   large natural variability may mask significant effects

-   detailed sampling can be expensive

-   "snapshot" view misses time dynamics

---

can only establish correlations between effects and presumed causes. Bioassays provide strong evidence for causation.

However close the laboratory can be made to simulate field conditions,
bioassays remain a surrogate measure of ecological effects. In addition,
laboratory bioassays will always be limited to organisms that can be cultured (see Table 2). Cognizant of such limitations, efforts continue toward developing a broader array of bioassay organisms and toward adapting
existing bioassay procedures so that the test may be performed in situ.
Successful examples of in situ terrestrial bioassays to date include detect
ing and monitoring environmental contamination using honey bees
(Bromenshenck, 1985, 1988) and, earthworm bioassays (C. Callahan, U.S. EPA
ERL-Corvallis, and C. Menzie, Menzie and Associates, Westford, MA, personal
communication; Menzie and Burmaster, 1988). In the near term, it will often be helpful to use a combination of laboratory and in situ assays. This
duplicity is needed in order to provide appropriate calibration of laboratory and in situ measurements. In part this is because of the large data
base already assembled with laboratory bioassays.

In-depth discussions of protocols for specific bioassays and data
analyses/interpretations may be found in Greene et al. (1988) and Weber et
al. (1988).

BIOMARKERS

Biomarkers are measures of molecular and/or physiological features of
organisms that are used to reveal a sublethal (often subtle) response to
some stressor. A given biomarker response may be ephemeral or sustained;
it may be specifically linked to a chemical or it may be associated with

Table 2.   Capabilities and limitations of bioassays (Adapted from Baker,
           1989).

Capabilities

+   can help establish causality

+   extensive laboratory data base and selection of bioassays

+   can integrate multiple, simultaneous chemical stresses

+   can reflect bioavailability

+   test conditions can be manipulated or adapted to meet different
    specification (including adaptation to in situ conditions)

+   chronic assays can more closely mimic field response and actual field
    exposures

Limitations

-   assay conditions (especially in the laboratory) are artificial in terms of
    exposure and other environmental stresses

-   restricted to culturable organisms

-   results cannot be directly translated into population effects in the field

-   results depend upon specific techniques and test conditions

---

a general class of stressors. The biomarker response in most cases is measured in an individual and provides evidence that the individual in question has experienced exposure to the stress. Although this subdiscipline of environmental biology is in its infancy, already excellent tools exist; some have clearly defined relationships between what is measured and the assessment endpoint. DiGiullio (1989) discusses the applicability of biomarkers and presents selected case studies relevant to waste site assessment.

Several key virtues of biomarkers are flexibility for use in the laboratory or in the field as well as on cultured or wild organisms. In this way, biomarkers can aid in defining relationships between laboratory and in situ bioassays as well as relationships between bioassay organisms and the larger array of wild organisms (see Table 3).

Although several limitations to the generalized use of biomarkers exist (e.g., technical uncertainties regarding the sensitivity, interference, general applicability across taxonomic lines), some have been used very effectively to demonstrate adverse effects on organisms due to contaminants. Examples include cholinesterase depression (Hill and Fleming, 1982), mutation frequency in plants (Lower, 1983), karyotype analysis (McBee et al., 1987), flow cytometry to measure cellular DNA content (McBee et al., 1987), DNA unwinding (Shugart, 1988a,b), and analysis of genetic diversity of populations via measurement of allelic distributions of metabolic enzymes (Facemire, et al., 1988; Gillespie and Guttman, 1988; Kopp et al., 1988). In all likelihood as more studies are completed, and as new biomarkers are perfected for field measurements, the theoretical framework to linking biomarker measurements to ecological endpoints will come into sharper focus.

Each of these sources of data has different capabilities and limitations. There is often a tendency to rely too heavily on one source to the exclusion of others--toxicologists arguing for one approach, field biologists for another. In most cases, it is only by combining the strengths of these various data sources into a "weight of evidence" approach that we can effectively bring the power of ecological science to bear on assessment decisions.

Table 3.   Capabilities and limitations of biomarkers. (Adapted from Baker, 1989).

Capabilities

+   evidence of exposure to sublethal concentrations of stressers

+   may be diagnostic

+   amendable to both laboratory and field conditions

+   very active area of research showing great promise

Limitations

-   linkage to ecological effects not inherently clear

-   only a few established biomarker systems available

-   may be operationally complex

CONCLUSIONS

Waste site ecological assessment is still in its formative stages. However, there appear to be no major obstacles preventing it from rapidly developing into a vigorous and useful science. There is no simple or single approach that will be generally applicable. Given the complexity of ecological systems, one must design each assessment carefully using the methods appropriate for the questions being asked. We suggest several guidelines.

1. To determine how organisms are responding in the real world, one must assess at an ecological level of response--the population or above.

2. Careful definition of assessment endpoints at the outset is critical for a useful and scientifically defensible assessment.

    °   intrinsically valued endpoints are best, if they can be defined and assessed.
    °   if protective endpoints are used, their limitations must be recognized.

3. Each of the different types of biological and ecological information used for assessments has limitations. In most cases it is best to use of blend of information from toxicity tests, biomarkers and field surveys.

REFERENCES

Baker, J., 1989, Assessment strategies and approaches, in: "Ecological Assessments of Hazardous Waste Sites: A Field and Laboratory Reference Document, William Warren-Hicks, Benjamin R. Parkhurst, and Samuel S. Baker Jr., eds.,

Bromenshenk, J.J., 1988, Regional monitoring of pollutants with honey bees, in: "Progress in Environmental Specimen Banking, U.S. Department of Commerce, National Bureau of Standards Special Publication 740, pp.156-170.

Bromenshenk, J.J., Carlson, S.R., Simpson, J.C., and Thomas, J.M., 1985, Pollution monitoring in Puget Sound with honey bees, Science, 227:632-634.

DiGiulio, R., 1989, Biomarkers, in: "Ecological Assessments of Hazardous Waste Sites: A Field and Laboratory Reference Document, William Warren-Hicks, Benjamin R. Parkhurst, and Samuel S. Baker, eds.,

Duinker, P., and Nilsson, S., 1988, Proceedings: Seminar on Remote Sensing of Forest Decline Attributed to Air Pollution, International Institute for Applied Systems Analysis, Laxenbrug, Autria (EPRI EA-5715, Project 2661-19).

Facemire, C.F., Osborne, D.R., and Guttman, S.I., 1988, Comparison of the Sensitivity of Starch Gel Electrophoresis and Ecological Indices for the Detection of Stress in Aquatic Ecosystems, Society of Environmental Toxicology and Chemistry, Ninth Annual Meeting, November 13-17, 1988, Arlington, VA.

Gillespie, R.B., and Guttman, S.I., 1988, Effects of environmental contaminants on the frequencies of allozymes in populations of fishes, Reducing Uncertainty in Environmental Risk Assessment, Society of Environmental Toxicology and Chemistry, Ninth Annual Meeting, November 13-17, 1988, Arlington, VA.

Greene, J.C., Warren-Hicks, W.J., Parkhurst, B.R., Linder, G.L., Bartels, C.L., Peterson, S.A., and Miller, W.E., 1988, Protocols for Acute Toxicity Screening of Hazardous Waste Sites, Final Draft, U.S. Environmental Protection Agency, Corvallis, OR, p.145.

Hill, E.F., and Fleming, W.S., 1982, Anti-chloresterase poisoning of birds: Field monitoring and diagnosis of acute poisons, _Environ. Toxicol. Chem._, 1:27-38.

U.S. Department of Interior, 1987, CERCLA 301 Project, Type B Technical Information Document, Injury to fish and wildlife species.

Kapustka, L., LaPoint, T., Fairchild, J., McBee, K., Bromenshenk, J., 1989, Field Assessments, _in_: "Ecological Assessments of Hazardous Waste Sites: A Field and Laboratory Reference Document, William Warren-Hicks, Benjamin R. Parkhurst, and Samuel S. Baker, Jr., eds.

Karr, J.R., Fausch, K.D., Angermeier, P.L., Yant, P.R., and Schlosser, I.J., 1986, Assessing biological integrity in running waters: A method and its rationale, Illinois Natural History Survey Special Publication No. 5, Illinois Natural History Survey, Champaign, IL., pp.28.

Kopp, R.L., Guttman, S.I., and Wissing, T.E., 1988, Trends in Allelic Frequencies as Indicators of Acid Stress and Tolerance in Fish Populations, Reducing Uncertainty in Environmental Risk Assessment, Society of Environmental Risk Assessment, Society of Environmental Toxicology and Chemistry, Ninth Annual Meeting, Nov. 13-17, 1988, Arlington, VA.

Lower, W.R., Drobney, V.K., Aholt, B.J., and Politte, R., 1983, Mutagenicity of the environments in the vicinity of an oil refinery and a petrochemical complex, _Terat._, _Carcinog._, _Mutagen._, 3:65-73.

McBee, K., Bickham, J.W., Brown, K.W., and Donnelly, K.C., 1987, Chromosomal aberrations in native small mammals (_Peromyscus leucopus_ and _Sigmondon hispidus_) at a petrochemical waste disposal site: I. Standard Karyology, _Arch. Environ. Contam. Toxicol._, 16:681-688.

Menzie, C.A., Burmaster, D.E., 1988, Evaluation of Environmental Risk Assessment Methods, Society of Environmental Toxicology and Chemistry, Ninth Annual Meeting, Nov. 13-17, 1988, Arlington, VA.

Schindler, D.W., 1987, Detecting ecosystem responses to anthropogenic stress, _Can. J. Fish. Aquat. Sci._, 44 (Suppl. 1):6-25.

Shugart, L., 1988a, An alkaline unwinding assay for the detection of DNA damage in aquatic organisms, _Mar. Environ. Res._, 24:321-325.

Shugart, L., 1988b, Quantitation of chemically induced damage to DNA of aquatic organisms by alkaline unwinding assay, _Aquat. Toxicol._, (in press).

Sloof, W., 1985, The role of multispecies testing in aquatic toxicology, _in_: "Multispecies Toxicity Testing," J. Cairns, Jr., ed., Pergamon Press, Elmsford, NY, pp.45-60.

Suter, G., 1989, Ecological Endpoints, _in_: "Ecological Assessments of Hazardous Waste Sites: A Field and Laboratory Reference Document," William Warren-Hicks, Benjamin R. Parkhurst, and Samuel S. Baker, Jr., eds., U.S. Department of Interior.

Weber, C.I., Horning, W.I., Klemm, D.J., Neiheisel, T.W., Lewis, P.A., Robinson, E.L., Menkedick, J., and Kessler, F., 1988, Short-term methods for estimating the chronic toxicity of effluents and receiving waters to marine and estuarine organisms, (DRAFT) EPA/600/4-87/028, Environmental Monitoring and Support Laboratory, Office of Research and Development, U.S. EPA, Cincinnati, OH.